aspasia

aspasia

aspasia

The International Yearbook of Central, Eastern and Southeastern European Women's and Gender History

◆

VOLUME 5, 2011

EDITORS

Francisca de Haan, Krassimira Daskalova, and Marianna Muravyeva

Berghahn Books

NEW YORK • OXFORD

Aspasia is an international peer-reviewed yearbook dedicated to publishing the best new scholarship in women's and gender history of Central, Eastern, and Southeastern Europe (CESEE). It aims to transform European women's and gender history by expanding comparative research on women and gender to all parts of Europe, creating a European history of women and gender that encompasses more than the traditional Western European perspective. *Aspasia* particularly emphasizes research that examines the ways in which gender intersects with other categories of social organization and advances work that explores transnational aspects of women's and gender histories within, to, and from CESEE. The journal also provides an important outlet for the publication of articles by scholars working in CESEE itself. Accordingly, contributions cover a rich variety of topics and historical eras, as well as a wide range of methodologies and approaches to the history of women and gender.

COPYRIGHT © 2011 Berghahn Books

ISSN 1933-2882 (PRINT) • ISSN 1933-2890 (ONLINE)
ISBN 978-0-85745-378-5 (PAPERBACK)

ONLINE
Aspasia is available online at www.journals.berghahnbooks.com/asp, where you can browse online tables of contents and abstracts, purchase individual articles, or recommend the journal to your library. Also visit the *Aspasia* website for more details on the journal, including full contact information for the editors and instructions for contributors.

ADVERTISING
All inquiries concerning advertisements should be addressed to the Berghahn Journals editorial office: advertising@berghahnbooks.com

INDEXED/ABSTRACTED
America: History and Life, British Humanities Index, Current Abstracts, Feminist Periodicals, Genderwatch, Historical Abstracts (excludes N. America) Index Islamicus, International Bibliography of Book Reviews of Scholarly Literature on the Humanities and Social Sciences (IBR), International Bibliography of Periodical Literature (IBZ), MLA Directory of Periodicals, MLA International Bibliography, MLA Master List of Periodicals, Sociological Abstracts, TOC Premier

aspasia

The International Yearbook of Central, Eastern and Southeastern European Women's and Gender History

◆

VOLUME 5, 2011

Forum

The Birth of a Field: Women's and Gender Studies in Central, Eastern and Southeastern Europe, Part II (Part I in *Aspasia* 4)
Edited by Krassimira Daskalova

Review Essays

BOOK REVIEWS

NEWS AND MISCELLANEA

Editorial

It is a pleasure to present volume 5 of *Aspasia*, which consists of six articles, nine Forum contributions about regional women's and gender studies, two review essays, nine book reviews, and an In Memoriam of Richard Stites, the influential historian of women in Russia.

The volume begins with a thematic section on "Gendering the History of Spiritualities and Secularisms in Southeastern Europe," guest edited by Pamela Ballinger and Kristen Ghodsee, and with articles by Ballinger and Ghodsee, Maria Bucur, Melissa Bokovoy, and Mary Neuburger. As Ballinger and Ghodsee argue in their Introduction, the articles in this special section contribute to larger debates about the multiethnic and historically pluralistic region of Southeastern Europe, and demonstrate how secularization projects are intricately interwoven with gender relations. Taken together, the articles in this section "urge readers to draw connections between the shifting spiritual cartographies, state formations, and definitions of appropriate masculinity and femininity of particular Southeastern European societies."[1]

The next two articles deal with Ottoman women's political history and contemporary feminist art. Historian Serpil Atamaz-Hazar focuses on Ottoman women's periodicals, which she argues can be crucial sources for uncovering women's history during the Ottoman Constitutional era. Her careful reading of some of the main journals of the early twentieth century leads her to conclude that these journals became "a powerful vehicle for women to make their voices heard, to become visible in public, and to initiate social and cultural change."[2] Atamaz-Hazar emphasizes that using these journals will allow historians, who have hitherto hardly included women's experiences in their narratives about the 1908 Revolution, to integrate the history of women into the history of the Constitutional era and thereby to much better understand "how revolutionary the 1908 Revolution actually was."[3]

The last article focuses on contemporary artist Tanja Ostojić, whose performances are analyzed by Jehanne-Marie Gavarini. Gavarini is herself a visual artist whose work has been exhibited internationally and who also writes about art, cinema, and visual culture. Gavarini argues that Ostojić's work exposes the gendered power dynamics that rule the Western-dominated art world and the European Union.

In the previous volume of *Aspasia* we published the first of a two-part Forum series on women's and gender studies in Central, Eastern and Southeastern Europe (CESEE). This volume contains the second part of that Forum. Together, these contributions present and discuss the different paths of women's and gender studies in fourteen countries of CESEE: Hungary, Poland, Romania, Serbia, and Ukraine (published in

aspasia Volume 5, 2011: ix-x
doi:10.3167/asp.2011.050101

Aspasia 4) and Bosnia and Herzegovina, Bulgaria, the Czech Republic, Estonia, Greece, Latvia, Lithuania, Macedonia, and Turkey (see this volume). Krassimira Daskalova, who served as the main organizer and editor of the two-part Forum, ends the series with a concluding article about "'The City of Gender Studies' in Central, Eastern and Southeastern Europe."

The Forum contributions contain a wealth of information about the development of the field and its institutionalization since 1989 (or, in the case of Turkey, much earlier). In addition to demonstrating profound differences between the countries—in some cases, for example, gender studies is still not recognized as an academic discipline at all, while in others it has been successfully institutionalized—the articles also point to a number of interesting similarities. These include the importance of transnational collaboration in establishing women's and gender studies, the ongoing lack of sufficient material resources, ranging from books to salaries[4]—and a number of important attempts to develop context-related concepts and theories. These concepts and theories, it seems, have not yet had a significant impact on hegemonic Western women's and gender studies. In the words of Katerina Kolozova, a professor of gender studies and philosophy in Macedonia, "the wider region of Eastern Europe is yet another context with respect to which gender studies should establish a culturally and politically sensitive approach that will inform the field in new and innovative ways, both in terms of epistemology and of contents."[5] Finally, it is important to note that many of the texts included here most likely and in the not-so-far future will become important historical documents themselves, because they will allow us to identify and analyze shared discourses on women's and gender studies in former socialist countries in Europe.

The *Aspasia* editorial team has been expanded significantly over the last year, which has much increased our possibilities and networks. The editors as always thank those who have reviewed manuscripts for *Aspasia* for their crucial contribution to maintaining high-level scholarship, and express our thanks for the ongoing support from the Editorial Board. Committed as we are to the goal of transforming European women's and gender history, we continue to welcome contributions in the field of interdisciplinary women's and gender history of the region. These contributions can deal with all periods, as long as there is a clear historical question or approach. Notes for contributors can be found on the inside back cover of this volume. For more and updated information about *Aspasia*, please visit http://journals.berghahnbooks.com/asp/.

Francisca de Haan, on behalf of the *Aspasia* editors

◊ Notes

1. Kristen Ghodsee and Pamela Ballinger, this volume, 1.
2. Serpil Atamaz-Hazar, this volume, 105.
3. Ibid.
4. The low salaries of academics in the region, including feminist scholars, are among the most important and influential structural inequalities between these academics and their colleagues working in "the West."
5. Katerina Kolozova, this volume, 187.

articles

Gendering the History of Spiritualities and Secularisms in Southeastern Europe

◆

Introduction

Pamela Ballinger and Kristen Ghodsee

◆

Scholars of religion have increasingly brought secularism within the framework of critical studies of spirituality, analyzing the dialogic relationship between religions and secularisms past and present. This emerging field of "postsecularist" studies examines the multiple meanings and practices that different cultures and societies attach to the concepts of "religion," "faith," and "piety."[1] The articles presented in this special section of *Aspasia* contribute to these larger academic debates by focusing on the multiethnic and historically pluralistic region of Southeastern Europe, an area too often ignored in larger scholarly discussions that have focused primarily on Western Europe and the so-called Third World. More important, the articles in this volume demonstrate how secularization projects are intricately interwoven with gender relations in any given society. Collectively, the articles urge readers to draw connections between the shifting spiritual cartographies, state formations, and definitions of appropriate masculinity and femininity of particular Southeastern European societies.

This collection of articles is the intellectual product of two workshops funded with a grant from the American Council of Learned Societies: the first held at Bowdoin College in Brunswick, Maine in October 2009 and the second at the Faculty of Political Science at the National School of Political Sciences and Public Administration in Bucharest, Romania in June 2010. The first workshop brought together social scientists and historians from across the United States, scholars specifically pursuing research projects in Southeastern Europe examining the intersections of gender, modernity, and secularist discourses in Bulgaria, Serbia, Romania, and former Yugoslavia, as well as the historical legacies of the Ottoman and Habsburg empires. Our scholarly deliberation focused on the ways in which the long history of Balkan religious pluralism challenges many of the currently fashionable accepted histories of secularism and their lack of attention to the fundamentally gendered nature of secularizing projects. Many of these current studies unwittingly reproduce Occidentalist biases by grounding the history of secularism in a Protestant/Roman Catholic Christian tradition, ignoring the importance of Eastern Orthodox Christianity,[2] while simultaneously neglecting the ways that secularist projects differentially affected men and women in Balkan societies. The second workshop allowed two representatives of the first workshop to solicit feed-

aspasia Volume 5, 2011: 1-5
doi:10.3167/asp.2011.050102

back from scholars of Orthodoxy in the region and to lay the groundwork for future collaborations between scholars based in the United States and in Eastern Europe.[3]

Much of the emergent discourse in the interpretive social sciences has sought to produce genealogies of secularism that demonstrate its complicity with the ambition to political hegemony of Western European powers. This selective genealogy rests primarily on assumptions of a Western philosophical tradition derived from ancient Greece and Rome and refracted through Renaissance and Enlightenment thought. Although acknowledging this history, we contend that alternative histories of secularity would be generated if one took seriously the local debates—which are intrinsic to all religions and societies—about how to engage others; this category encompasses both foreigners and those who, though familiar, do not share the same conceptions of the world or notions of social justice. The history of doctrinal debates and socio-cultural change in Southeastern Europe demonstrates that most religions represented here carry within themselves the seeds of their own auto-critique. The articles included here examine the ways in which various populations in Southeastern Europe have attempted to actualize these possibilities, thus facilitating peaceful coexistence with others for extended periods of time, contrary to the stereotypes that plague the Balkans. Studying this historically recurrent possibility has great relevance to the present moment, when the apparently divergent claims of religion and secularism seem resistant to any mediation, and when the need for strategies of co-being is so great.

At the same time, the collection of articles presented here go beyond the romantic and idealized visions of coexistence that have characterized some recent reassessments of pluralist regions, such as the Balkans or the pre-Reconquista Iberian Peninsula.[4] In multiconfessional empires like those of the Habsburgs and Ottomans, as well as pluralist states such as socialist Yugoslavia, religious coexistence could at times prove precarious, particularly given the ways in which religious identities increasingly intersected with ethnic and national identities from the nineteenth century on. Anthropologist Robert Hayden has described such coexistence in terms of "antagonistic tolerance,"[5] a formulation that underscores the potential for religious difference to become a flashpoint, as has occurred in different moments in the recent history of Southeastern Europe.

Key to rethinking this delicate balance between religious and other identities, as well as between religiously defined communities, is a critical eye on various state formations that have sought to manage, co-opt, or contain religious identities and movements in the region. The ways in which both states and their subjects found religious identity useful—or not—helps explain many seeming paradoxes or contradictions in Southeastern Europe. The Slavic pirates or *uskoks* sponsored by the Habsburgs in the sixteenth and seventeenth centuries along the Military Frontier between the Ottoman and Habsburg empires, for example, could act out of sincere motivation as Christian warriors against Islam. This occurred even as they attacked the ships of (Christian) Venetians—Habsburg rivals—and engaged in practices of blood brotherhood with Muslim neighbors who shared their warrior code.[6] Likewise, the Habsburg Emperor Joseph II (1741–1790) maintained a Catholic identity even as he introduced wide-ranging reforms, such as the Patent of Toleration and sought to subordinate church to an enlightened state. Joseph's particular version of church/state separation preceded the

French Revolution, thereby unsettling one enshrined narrative of the history of the secular state. A century later, the *Kulturkampf* would continue to pit liberals against conservatives in the empire over the issue of the appropriate relationship between church and state.[7] Within the context of these formal, state-directed policies there thus unfolded a wide range of contests and accommodations on the ground, which articulated with other understandings of "cultural" and political identity, such as language, nation, and particularly gender.

Scholars have recently described the Habsburg Monarchy as characterized by the "parallel realities" of ethnic-national identification and loyalty to the dynastic, multinational state.[8] This phrasing does not assume from the outset an oppositional relationship between nationalism and supranational/imperial loyalty but rather seeks to determine the complex interplay and tensions between such ideologies. Other scholars have similarly questioned dichotomies such as cosmopolitanism and traditionalism (or nationalism) in the region and beyond.[9] The articles in this special section do not treat religion and secularism as inevitable opposites, thus complicating the simple historical narratives that have often defined the past and present realities of religion and secularism in Southeastern Europe. In doing this, the articles implicitly build on an earlier generation of critical scholarship that questioned common myths about the millet system in the Ottoman Empire and the antipathy of Balkan religions to Enlightenment ideologies.[10]

Southeastern Europe offers a particularly apposite site for uncovering alternative genealogies and understandings of secularism and religious diversity, as it reveals in dramatic fashion both the perils and promises of religious pluralism. The region has witnessed a wide range of state types within which such pluralism has been negotiated, including empires, nation-states, socialist states, and now European Union (EU) member states. All of the states in the region host significant religious minorities, which can mobilize both past narratives of coexistence and contemporary international norms about religious protections and tolerance. In contemporary Southeastern Europe, articulations and experiences of secularism and religion take place in a landscape unevenly transformed by capital, warfare, and recent violence, and access to EU resources as well as an ever-shifting terrain of local gender systems.

In the first article of the thematic section, Ballinger and Ghodsee use the examples of socialist Bulgaria and Yugoslavia to propose new directions for rethinking scholarly understandings of secularism and the ways in which socialist secularizing projects were intricately intertwined with questions of gender equality. This article takes two cases of Balkan states to explore the theoretical contours of what the authors refer to as "socialist secularism." Although Bulgaria and Yugoslavia's experiences of socialist secularism differed in the degree of their coerciveness, this article demonstrates that there are important similarities in the conceptualization of the secularizing imperative and the rhetoric used to justify it, specifically the rhetoric of communist modernization and women's liberation from "traditional" religious backwardness.

In the case of communist Romania, Maria Bucur explores the feminization of Eastern Orthodox piety against the socialist secularizing impulses of the government. Bucur examines in depth the seeming paradox that despite the antireligious campaigns of the Ceaușescu regime, the number of women entering the Orthodoxy nunnery increased substantially during the communist era. Historically, Romanian women had always provided a spiritual foundation for the Romanian Orthodox Church and Bucur investigates the fascinating continuities between the precommunist and communist periods. Her article is a reflective one; using her own experiences as a child in Romania and drawing on an emerging body of oral history scholarship, Bucur aims to outline the contours of a research agenda that will inspire new scholars from the region.

Historian Melissa Bokovoy directs our attention to underutilized materials, such as photographs, in the study of spiritualities and secularisms in Southeastern Europe. Bokovoy examines the tensions that arise when nation-states (or, more precisely *nationalizing* states) seek to appropriate ritual practices of mourning grounded in both "universal" religious doctrine and deep (and gendered) local histories. In an empirically rich analysis, Bokovoy explores how the emerging field of photography simultaneously drew on and competed with women's practices of remembering the dead in Serbia during and after World War I, in particular women's laments derived from epic poetic traditions. The efforts of the Serbian and successive Yugoslav state to harness local Orthodox traditions of grief in service to a project of national identity proved a key aspect of a secular (if not necessarily a secularizing) project. The gendered celebration of sacrifice in service to the nation proved common to states throughout Europe in the aftermath of World War I, pointing to further directions for comparative research.

Focusing on questions of consumption and identity in Ottoman Bulgaria and the early independence era, historian Mary Neuburger explores the local construction of "successful" masculinity in the public spaces of the tavern and the coffee house. Neuburger skillfully uses Western archival sources to examine the discursive construction of Christian Bulgarians as lazy drunkards compared to the supposedly sober, industrious Turks. Early Bulgarian communists found unlikely allies in American and British Protestant missionaries hoping to spread the message of temperance to the Balkans. Despite the intransigence of local masculine ideals of sociability, which required Bulgarian Christian men to demonstrate their spiritual and ethnic allegiances through inebriation, the Turkish coffee shop did eventually gain popularity among Bulgarians eager to engage in international trade and commerce. Neuburger's article aptly demonstrates that both religious and ethnic allegiances, and the practices associated with their maintenance, can easily melt away in the face of emerging commercial opportunities.

The articles presented here are meant to be read as a collection, with each contribution building on and expanding on the others. The authors come from different disciplinary perspectives, and the editors hope that together these articles will stimulate further research and discussion on the interwoven relationships of gender, religion, and secularism in Southeastern Europe. The articles encompass a wide range of cases, methods, and intellectual traditions, pointing to the productiveness of intellectual ecumenism and pluralism on these topics. We hope that scholars of Southeastern Europe will bring their expertise in the particular histories of religion and secularism in the region to bear on broader debates from which the Balkans have been largely absent. Finally, the examples from Southeastern Europe explored here point to the centrality of gender to secularist projects, an insight that surprisingly few scholars of secularism have explored. Far from being a marginal or peripheral case, the Balkans emerge here as an exciting new site of innovation for the larger study of spiritualities and secularisms.

◊ Notes

1. For key texts, see Talal Asad, *Genealogies of Religion* (Baltimore: Johns Hopkins University Press, 1993); Philip Blond, *Post-Secular Philosophy: Between Philosophy and Theology* (London: Routledge, 1998); Talal Asad, *Formations of the Secular* (Stanford: Stanford University Press, 2003); Saba Mahmood, *Politics of Piety: The Islamic Revival and the Feminist Subject* (Princeton: Princeton University Press, 2005); and Lara Deeb, *An Enchanted Modern: Gender and Public Piety in Shi'i Lebanon* (Princeton: Princeton University Press, 2006).

2. Kristen Ghodsee, "Symphonic Secularism: Eastern Orthodoxy, Ethnic Identity and Religious Freedoms in Contemporary Bulgaria," *Anthropology of East Europe Review* 27, no. 2 (2009): 227–252.

3. We would like to thank Adrian and Mihaela Miroiu for generously agreeing to host the second part of the workshop in Romania.

4. On Bosnia, see Michael Sells, *The Bridge Betrayed: Religion and Genocide in Bosnia* (Berkeley: University of California Press, 1998). On Iberia, see Maria Rosa Menocal and Harold Bloom, *The Ornament of the World: How Muslims, Jews, and Christians Created a Culture of Tolerance* (New York: Little, Brown, 2002). For a critique of such romanticization, refer to Pamela Ballinger, "Imperial Nostalgia: Mythologizing Habsburg Trieste," *Journal of Modern Italian Studies* 8, no. 1 (2003): 84–101.

5. Robert Hayden, "Antagonistic Tolerance: Competitive Sharing of Religious Sites in South Asia and the Balkans," *Current Anthropology* 43, no. 2 (2002): 205–231.

6. Catherine Wendy Bracewell, *The Uskoks of Senj: Piracy, Banditry, and Holy War in the Sixteenth-Century Adriatic* (Ithaca: Cornell University Press, 1992).

7. Nancy Wingfield, "Emperor Joseph II in the Austrian Imagination up to 1914," in *The Limits of Loyalty: Imperial Symbolism, Popular Allegiances, and State Patriotism in the Late Habsburg Monarchy*, ed. Laurence Cole and Daniel Unowsky (Oxford: Berghahn Books, 2007), 62–85.

8. Laurence Cole and Daniel L. Unowsky, "Introduction," in *The Limits of Loyalty*, ed. Cole and Unowsky, 1–10.

9. Pheng Cheah and Bruce Robbin, eds., *Cosmopolitics: Thinking and Feeling beyond the Nation* (Minneapolis: University of Minnesota Press, 1998); Pamela Ballinger, *History in Exile: Memory and Identity at the Borders of the Balkans* (Princeton: Princeton University Press, 2003).

10. Traian Stoianovich, "The Conquering Balkan Orthodox Merchant," *Journal of Economic History* 13, no. 4 (1953): 398–411; Benjamin Braude and Bernard Lewis, eds., *Christians and Jews in the Ottoman Empire: The Functioning of a Plural Society* (New York: Holmes and Meier, 1982); Paschalis Kitromilides, *Enlightenment, Nationalism, and Orthodoxy: Studies in the Culture and Political Thought of South-Eastern Europe* (Aldershot: Variorum, 1994).

Socialist Secularism

Religion, Modernity, and Muslim Women's Emancipation in Bulgaria and Yugoslavia, 1945–1991

Pamela Ballinger and Kristen Ghodsee

ABSTRACT

This article uses the examples of socialist Bulgaria and Yugoslavia to propose some new directions for rethinking scholarly understandings of "secularism" and the ways in which socialist secularizing projects were intricately intertwined with questions of gender equality. Current scholarly debates on the genealogy of secularism root its origins in the Catholic/Protestant West, and systematically ignore cases from the former communist world. This article takes two cases of Balkan states to explore the theoretical contours of what we call "socialist secularism." Although Bulgaria and Yugoslavia's experiences of socialist secularism differed in the degree of their coerciveness, this article examines the similarities in the conceptualization of the secularizing imperative and the rhetoric used to justify it, specifically the rhetoric of communist modernism and women's liberation from religious backwardness.

KEYWORDS: Bulgaria, secularism, state socialism, women's emancipation, Yugoslavia

In his influential work on secularism, Talal Asad contends, "An anthropology of secularism should thus start with a curiosity about the doctrine and practice of secularism regardless of where they have originated."[1] Despite this, however, the growing body of critical work on secularism has devoted relatively little attention to secularism's genealogies in the Second World of socialist states, in particular Southeastern Europe. Furthermore, scholars have considered the centrality of claims about female emancipation to secularist ideologies and projects in fairly limited and restricted ways. Contemporary European defenses of secularism that posit Islam as inherently premodern, nondemocratic, and oppressive to women and debates over veiling/headscarves in European Union states, for example, dominate the discussion of gender in relation to secularism. Less remarked upon are the deeper entanglements of Orientalism with secularism that have played an important role in European states with significant his-

aspasia Volume 5, 2011: 6–27
doi:10.3167/asp.2011.050103

torical Muslim minorities, as well as Muslim majorities (such as Turkey).[2] This article uses the examples of two European states with Muslim minorities—socialist Bulgaria and Yugoslavia—which represent two ideological poles on the spectrum of state socialism. The article draws on these two cases to propose new directions for rethinking scholarly understandings of secularism and the ways in which socialist secularizing projects became intricately intertwined with questions of gender equality. Through a specific focus on Muslim women in Bulgaria and Yugoslavia, it argues that socialist regimes deployed the secularist discourse in unique and fascinating ways that deserve inclusion in the larger scholarly debates.

In many cases, secularism becomes conflated with secularization, a waning in religiosity that is a result of modernization.[3] Such secularisms then become viewed as derivative of a Western genealogy that locates secularism's origins in the Catholic/Protestant split (thereby neglecting Orthodox Christianity)[4] and "closely connected with the rise of a system of capitalist nation-states"[5] as well as modernity. This tendency in secularism studies explains, in part, the scholarly neglect of the particular histories and experience of secularism in socialist states, including those of Southeastern Europe. Just as Southeastern Europe or the Balkans occupy an ambiguous space in the discursive configuration that marks out the Orient from the Occident,[6] so too do the histories of secularism in the region complicate scholarship that all too frequently rests on a dichotomized view of the world as divided between a "secular West" and a "religious rest."[7] In this view, some non-Western societies import, emulate, or have imposed on them secularism. This approach characterizes even the excellent studies of secularist projects in Soviet Central Asia of scholars like Gregory Massell and Douglas Northrop, work that proves exceptional in its attention to the targeting of Muslim women by Bolshevik reformers.[8] Following their lead, we also focus on Muslim women but we treat secularism in the Bulgarian and Yugoslav case as much more organic, drawing on each country's long history of religious pluralism.

The paucity of critical analyses of what we deem "socialist secularism" reflects problematic assumptions and blind spots that ignore the long history of pluralism. Furthermore, state socialism's promotion of industrialization along with secularism likely appeared as an all too obvious confirmation of the secularization thesis.[9] Alternatively, perhaps more fundamental omissions in secularism/secularization scholarship meant that observers of state-religion relations in socialist regimes took not just atheism but also the nature of the state for granted.[10] Whatever the reasons for the (relative) neglect of socialist experiences of secularism, the consequences have been serious not just for the scholarship on secularism but also for understanding of the nature of both socialism and postsocialist change.

The end of state socialism in Europe prompted many scholars to study the transformations in religious practice and expression in former socialist societies, but they often did so through an uncritical lens of "religious revival."[11] Scholars attempting to classify states and societies along a gradient of religiosity and secularism, for example, often mistook the rhetoric of Marxist atheism for the practice of state socialism. Barrett's worldwide studies of state-religion relationships in the 1970s and 1980s, for example, categorized all Eastern European states as "atheist" as the result of communism's "animus against organized religion."[12] Analysts of "religious revitalization" after socialism who oversimplify the complexities of socialist secularism risk mischaracterizing the degree of change, particularly at the level of institutionalized religion.

More important for the purposes of this article, a study of socialist secularism highlights the ways in which women's emancipation was a core justification for the control and, in some instances, outright oppression of religious communities. Attending to the importance of women's issues in the socialist context offers a new lens through which to examine different permutations of secularism. In both Bulgaria and Yugoslavia, as well as in Central Asia, religious belief was blamed for the lack of social progress for women, particularly for Muslim women, although all religions were recognized as promoting "backward" gender relations. In both Bulgaria and Yugoslavia, Muslim women became the particular targets of the state's secularizing efforts but for different reasons. In Bulgaria, the history of Ottoman imperial domination rendered Islam a (perceived) threat to Bulgarian national identity whereas in Yugoslavia Bosnia's Muslims symbolized the Yugoslav promise of a pluralist socialist society.

Scholarly debates on Western secularisms assume a bifurcated public/private sphere, where religion is relegated to the latter in order to maintain pluralism and tolerance both within and between different religious communities. The socialist secularist case allows us to examine a state that controls religion in both the public and private spheres not only to ensure what John Locke called "toleration," but also to promote women's greater independence within the family for the greater good of society. This takes place even as the behavior of all citizens becomes highly regulated by and subordinated to what anthropologist Katherine Verdery has called the socialist "parent-state."[13] Here we find a critical difference between secularism in Turkey or Western Europe and socialist examples. For instance, in France or in Turkey symbols of religious affiliation such as the headscarf are officially banned in public institutions but not in private settings. In socialist Bulgaria, Muslim women's attire was prohibited in all public and private spaces.

A comparison between Bulgaria and Yugoslavia proves productive, given that the two states lay on opposite ends of the socialist spectrum. After the Tito-Stalin split in 1948, Yugoslavia modeled itself as an exceptional case, offering a "third way" that combined limited market practices and relative freedoms (such as freedom to travel abroad) with "unique" socialist principles (such as workers' self-management). The Yugoslav federation became increasingly decentralized from the 1960s and 1970s on. In contrast, Bulgaria was the Eastern European communist country with the closest ties to the Soviet Union and had one of the most centralized, totalitarian regimes. Despite these differences, both states implemented very similar policies with regard to religious communities. The socialist state in both places tried to capture and control religious institutions rather than eradicate them. Furthermore, gender equality and the emancipation of women from cultural "backwardness" were key goals that motivated many state secularist policies, particularly with regard to Muslim minority populations. Of course, we recognize that an article of this length cannot provide a comprehensive discussion of the many different examples of socialist secularism and how it operated in both countries. Rather, our goal here is merely to paint the contours of a broader conceptual framework that will provide the foundation for future research on this topic in the region.

Secularism and State Socialism

It is important to address the issue of secularism within state socialism within the framework of theoretical/philosophical statements about the role of religion in socialism—whether those ideas are grounded in Marxist-Leninism as in Bulgaria, or in Marxist-Leninism together with other ideologies such as Titoism or Yugoslavism in Yugoslavia—[14]versus the practice and lived experience of secularism. Such an analysis must be precise in its use of the term "secularism." Given that there are many definitional debates over the terms "secularism" and "secularization," a clear definition of the parameters of socialist secularism is required (1) as entailing the legal separation of church and state and (2) as the embrace of a "worldly" or non-religious ideology (in this case, that of Marxist materialism), and its relationship to the communist modernization paradigm and women's emancipation.[15] This article employs both definitions, examining how socialist secularism was a core ideological imperative of both the Bulgarian and Yugoslav communist regimes and their efforts to forge a new socialist man and woman.

Religion-state relationships varied considerably throughout the Eastern Bloc depending on the specific local histories, degree of religious difference, and specific configurations of gender relations. Different regimes treated different religions differently. Such treatment depended, for instance, on whether churches were "national" (as in the case of the Orthodox Churches of Serbia or Bulgaria) or "international" and thus potentially created foreign allegiances (as with Catholicism in Croatia or Poland). Other crucial factors included the size and influence of the religions and the behavior of the churches during World War II.[16] Ramet notes that throughout the communist states Eastern Orthodox churches tended to enjoy general tolerance but they were also among the most co-opted of the churches, in part because of the relative scarcity of powerful patrons and supporters in the West.[17] The Bulgarian Orthodox Church (BOC) offers a perfect example of a church that was largely controlled by the communists for their own goals. In the 1950s, the BOC newspapers ran articles that openly supported Stalin and the goals of socialism,[18] and the communist government could appoint and dismiss patriarchs and archbishops at will.[19] Ramet goes as far to argue that this represented an inversion of the tradition of Caesaro-papism, a partnership between church and state whose origins may be traced variously to the Ottoman millet system, to the Spiritual Reglement of 1721 in Tsarist Russia, or even back to the Great Schism of Eastern and Western Christianity of 1054.[20]

By contrast, the Catholic Church in Eastern Europe inherited an oppositional stance toward the state that grew out of the Church's championing of "suppressed nationalities" in the nineteenth century. In the years immediately following World War II and the establishment of socialism in Eastern Europe, a number of states urged (with little success) the Catholic Church to create "national" churches independent of the Vatican.[21] In Yugoslavia (and particularly the Yugoslav Republic of Croatia), the Croatian Church bore the additional burden of its association with the brutal Ustaša regime of Ante Pavelić and the (not unfair) perception that the Vatican favored Italian claims in the Italo-Yugoslav dispute over Venezia Giulia (1945–1954). In Bulgaria, Muslims were likewise seen as belonging to an international religion that forged problematic links between the local Muslim population and NATO-allied Turkey. Furthermore, because the Bulgarian communists feared Turkish irredentism in the Balkans, they were particularly keen to assimilate the Turks of Bulgaria as well as the Muslim Slavs and Roma into a secular, nationalist worldview.[22]

In general, state-religion relationships relaxed considerably over time in all of the communist regimes. In the Soviet bloc, Stalin's death ushered in a new emphasis on co-optation rather than outright repression of religious groups. In many socialist states forms of religious instruction remained possible but were subject to strict state controls (or were prohibited from the state educational sector at various moments), more an assertive attempt by the state to control religion than to eliminate it altogether (or to eliminate it from the public realm). In Bulgaria, the state created a Directorate of Religious Denominations, which oversaw all religious affairs in the country. Similarly, in Yugoslavia, the regime sought to tightly circumscribe the space in which religion held sway. Commissions for Religious Affairs were established at both the federal and republican levels. Until Aleksandar Ranković's removal as minister of the interior in 1966, the Commissions were closely connected to the secret police. As post-Ranković liberalization coincided with increasing devolution to the republican level, the Federal Commission for Religious Affairs became more of a coordinating body than one involved in direct monitoring and repression of religious institutions and believers.[23]

Observers critical of Yugoslav policies toward religion maintained that the "form [of religion] is protected," even as its essence and its institutional bases were depleted.[24] At the same time, as Michael Petrovich has observed, "religion [in Yugoslavia] was not so much a matter of private conscience as one's public identity."[25] We might consider this a "paradox" of Yugoslav socialist secularism, although it only appears paradoxical when viewed through normative understandings of secularism (in which religion belongs to a clearly delineated "private" sphere). Part of the problem stems from the conceptual confusions over how to define the "public" sphere and what Asad has identified as the Western assumption of a secular public sphere as a neutral arena.[26] Intermingled as it was with other aspects of identity in both Bulgaria and Yugoslavia, religion could not be disregarded or discarded by the socialist regime in either country. Rather it required harnessing or "taming," to use anthropologist Kim Shively's term for Islam in Kemalist Turkey.[27] In the Bulgarian case, there was a strong nationalist myth that the Orthodox Church had preserved Bulgarian language and culture during the 500 years of the Ottoman domination and many important national heroes were members of the clergy (i.e., Ivan Rilski, Vasil Levski, Patriarch Eftimii). Under the Ottoman millet system, the Bulgarians were largely defined by their Christianity and a kind of secular Eastern Orthodoxy remained a core facet of Bulgarian national identity in the communist period.

In the case of the Muslim identity in Bosnia and Yugoslavia, this identity became more public or visible over time as the result of the creation in 1971 of the census category "Muslim, in the sense of a nation" (i.e., Muslim as *narod*).[28] Here an ostensibly religious identity became more prominent even as it increasingly acquired cultural (or secular) connotations; devout Muslims often distinguished between communist (or secular) Muslims and believers, however, pointing to a tension in Bosnian Muslim identity that remained unresolved throughout the Yugoslav period.[29] In actively promoting a distinct (secular) Bosnian Muslim identity, the Communist Party or LCY (League of Communists of Yugoslavia) and the state sought to balance competing religious identities and claims.[30] Religious leaders in Yugoslavia often perceived efforts by the regime to orchestrate "national symmetry" between various faiths in Yugoslavia as an attempt to divide and weaken religious communities, who saw themselves as competing with one another as well as with a secularist program.[31] Religious leaders suspected that a supposed policy of national symmetry actually made for asymmetry in the treatment of specific religious communities.[32] At the same time, taming religion and either neutralizing its political power or putting it in service to a socialist state became entangled with the project of liberating women from religion's patriarchal structures. In reality, this project meant that the revolutionary and oppositional potential of women was also "tamed," a critique made vocally by Yugoslav feminists in the 1970s and 1980s.

Taming Religion in Yugoslavia

From its very beginnings, the socialist regime in the Federal People's Republic of Yugoslavia made quite explicit in both theory and practice its aim to contain religion's sectarian tendencies, thereby taming it and containing its potential for political opposition. On the one hand, the Constitution laid out the principles of separation of church and state and religious freedom (Article 25).[33] On the other hand, the document warned, "The abuse of the Church and of religion for political purposes and the existence of political organization on a religious basis are forbidden."[34] In an important qualification, however, the law stipulated, "scientific criticism of religion in general and criticism of improper actions of religious representatives ... cannot be regarded as incitement to religious hatred,"[35] thereby opening an authorized space for proponents of a particular form of secularism (Marxism/Titoism) to critique religion and its followers. The socialist authorities also provided themselves with the legal weapons for cracking down on religious officials deemed guilty of incitement (whether those charges were genuine or trumped up), declaring, "It shall be regarded as an aggravating circumstance of [sic] the offence of incitement to religious hatred is committed by ecclesiastical representatives."[36]

Proponents of socialism in Yugoslavia thus rejected religious sectarianism in favor of what they saw as their own version of religious pluralism. Pro-regime authors like Ivan Lazić stressed the ways in which all religions received equal protection within socialist Yugoslavia, in contrast to the prewar Yugoslav kingdom in which only those religions legally recognized in the state at the moment of its creation (in December 1918) were protected while other religions were persecuted.[37] Critics would instead rebut that under socialism *all* religions were discriminated against, though not in equal measure, just as individual believers enjoyed greater or less degrees of religious freedom depending on their positions in society.[38]

Regime-sponsored authors like Rastko Vidić further contended that those churches that had previously suffered persecution (or, at the very least, did not enjoy a privileged position in the Kingdom of Yugoslavia) welcomed the institution of genuine freedom of religion under socialism. In contrast, some churches "could not easily acquiesce in the loss of the privileged position which they had enjoyed up to that time."[39] These were the same churches that had, in Vidić's estimation, not only colluded with the royal regime but also the quisling states during World War II. As Vidić's comments suggest, from the point of view of the Yugoslav regime, the task of taming religion proved more urgent for some faiths than others.

At different moments in Yugoslav history, the task of taming religion also acquired greater urgency. Many observers have noted that more direct forms of repression and control later gave way to a greater emphasis on ideological pressure, both in the form of providing an alternative (Titoism and Yugoslavism as civil or secular religions) and in valorizing modernization and the abandonment of backwardness, particularly gender inequality. The early socialist state in Yugoslavia was modeled on Stalin's Soviet Union,[40] sharing with the USSR a ruthless attitude toward enemies, real and imagined. Religious functionaries figured prominently among such enemies. In June 1945, for instance, trials of Catholic, Orthodox, and Muslim clergy took place in military courts in Croatia.[41]

Religious institutions came under assault along with religious personnel, who became seen as sources of both potential political disorder and agents of traditional forms of class and gender oppression. In rural areas with low development of class consciousness, the regime could point to "oppressed" women and peasants—what Massell, writing about the Soviet context in Central Asia, deemed a "surrogate proletariat."[42] In parts of Central Asia, notes Northrop, "women's emancipation ultimately came in many ways to exemplify the entire Bolshevik Revolution."[43] In Yugoslavia, too, the simultaneous harnessing of religion and liberation of women became a potent symbol of progress and modernity.

Providing its own forms of education through literacy classes and state schools, the regime closed down religious schools throughout the federation, as well as Muslim cultural associations and Islamic courts, and made schools available equally to men and women.[44] The 1946 land nationalization law hit religious institutions particularly hard, depriving them of key sources of autonomous income. In Sarajevo, for example, the properties of the Baščaršija (the historic Ottoman urban center) controlled by the Muslim institution of the *vakuf* now came under state ownership.[45] The message sent by such expropriations, as well as by the trials of priests, was that no resistance to or even autonomy from the regime would be brooked, particularly opposition that sought to reinforce the link between religion and nationalism. In part, the communist authorities justified this repression in light of the controversial role played by the Catholic Church and clergy (as well as some Muslims) in the independent Ustaša state established under Ante Pavelić during World War II. Similarly, the support of royalist Orthodox Serbs for the Chetniks meant that Orthodoxy had to be disentangled from a specifically Serbian nationalist project.

Yugoslavia's break with the Soviet Union did not bring about an immediate relaxation of church-state relations. The regime continued to struggle to bring priests and imams under its direct supervision. Although the authorities enjoyed some success in Slovenia and Bosnia and Herzegovina in persuading priests to join government-controlled priests' associations (and imams and *hodzas* to participate in the state approved Islamic Association), many Catholic priests in Croatia refused to join.[46] Through such associations, the Yugoslav socialist state hoped to contain religion by making religious leaders not only submit but actively collaborate in the work of the state, underscoring the need to consider socialist secularism in terms of what scholars like political scientist Ahmet Kuru deem an "assertive" or active secularism.[47] In the 1950s, for example, it was the Islamska Zajednica (Islamic Association) that banned dervish orders and oversaw their dismantling.[48] With regard to faiths like Islam and Catholicism, the state aimed to privilege believers' loyalty to specifically Yugoslav religious institutions (government-controlled institutions) and traditions rather than permit the dual loyalty of nation and universal church.

Several significant moments of liberalization in church-state relations modified the early picture of repression and direct control of religious institutions by the state. The 1958 Program of the League of Communists of Yugoslavia, for example, reiterated a view of religion as antithetical to a Marxist worldview and a sign of lingering backwardness. Such backwardness, argued Minister of the Interior Ranković, was to be combated primarily through ideological work from this point on. Ranković's subsequent fall from power in 1966 meant further relaxation of state monitoring over the Commissions for Religious Affairs and the activities of believers. As a result of these shifts, Yugoslavia has been perceived as relatively liberal in comparison to Bulgaria and other socialist states in its policies toward organized religion. This is particularly true when compared to Bulgaria, where religion became the primary vehicle for the promotion of a socialist national identity.[49]

Taming Religion in Bulgaria

When Bulgaria achieved its independence from the Ottoman Empire, the Turnovo Constitution established Bulgarian Orthodoxy as the traditional religion of the new state in 1878.[50] This gave the Bulgarian Orthodox Church a privileged position in Bulgarian society, but this also meant that it was fused with the Bulgarian state during the years of Bulgaria's alliance with Nazi Germany in World War II. The communists who would subsequently come to power inevitably held some of the BOC clergy accountable for their collaboration with the fascist regime. Immediately after the communist victory on 9 September 1944, state officials began the process of removing all religious education from public schools.[51] The church strongly opposed the secularization of education, but the communist government, and the country's first communist Premier, Georgi Dimitrov, closely followed Soviet antireligious policies and felt that religious education would undermine communist efforts to modernize Bulgarian society. In 1946, the government also instituted new laws that only recognized marriages performed by civil authorities and no longer recognized those performed by the church. Furthermore, the various charitable functions of the church were taken over by the state with the nationalization of many church properties. Direct repression of the Bulgarian Orthodox clergy began in 1944 and continued until the mid-1950s; many priests were targeted in a series of purges.[52]

A serious blow to the Bulgarian Orthodox Church came with the passage of the 1947 Dimitrov constitution, which firmly established the separation of church and state by removing the language that had once recognized Orthodoxy as the traditional religion of Bulgaria. Exarch Stefan (the spiritual leader of the BOC) opposed these changes, and was forced into retirement by the communist government in 1948. State officials replaced him with a church leader more favorably disposed to their agenda. At the end of that same year, the Fifth Congress of the Bulgarian Communist Party decided to take a firm stand against the lingering "bourgeois" and religious influences on society by nationalizing monasteries and seminaries. During this time, many priests who opposed the regime were also imprisoned or sent to work camps.[53]

Despite these changes, it must be remembered that the Bulgarian government did not want to destroy the church, but rather to control it and use it as a tool of the state. The passage of the 1949 Denominations Act unequivocally placed the Bulgarian Orthodox Church under the authority of the communist government. Although Article 1 of the 1949 Denominations Act states that "[a]ll people in Bulgaria are given the opportunity of freedom of religion and conscience,"[54] the law heavily proscribed the activities of all religious groups in Bulgaria, including that of the BOC.

Article 12 of the Act allowed the state (through the Directorate of Religious Denominations) to dismiss any clergy member who was suspected of working against the "democratic order of the country" or of jeopardizing the public safety or good morals of the population. As in Yugoslavia, this in effect gave the Bulgarian state complete control over the religious establishment of all denominations. Article 18 stated that: "The mentioning of the Supreme Power and its organs by religious denominations during sermons, rituals and celebrations may be done only by expressions previously approved by the Directorate for Religious Affairs." This article allowed the government to directly interfere in the writing of church sermons and texts; phrases used to speak about God and his powers during religious ceremonies had to be vetted by communist officials beforehand or priests risked arrest and imprisonment. Article 20 banned religious education in schools and Article 22 banned contacts with religious communities in other countries without the express written consent of the Directorate of Religious Denominations, effectively isolating Bulgaria's religious communities from sister spiritual communities abroad. This was particularly detrimental to Bulgaria's Catholic and Muslim populations.[55]

During the 1940s, antireligious propaganda was ubiquitous and the communist newspapers often included reports of drunk or immoral priests. Church newspapers also became an outlet for communist propaganda, publishing articles such as "Socialism: The Eternal Ideal of Humanity,"[56] "The Communality of Religion and Socialism,"[57] and "Long Live the Task of Great Stalin."[58] Thus, rather than the separation of the church and state called for in the Dimitrov constitution, the Bulgarians placed religious institutions firmly in control of the state as it pursued its scientific socialist goals.

After Stalin's death, there was a relaxation of religious persecution in Bulgaria and many of the priests who had been sent to labor camps were granted amnesty. Still, religion was frowned upon in public life and social mobility was tied to atheism. All life cycle rituals previously performed by the church (baptisms, marriages, funerals) were replaced by socialist rituals created by the state, which were devoid of spiritual meaning.[59] In Muslim communities, certain rituals and ceremonies were officially banned and few young people had access to religious education. As a result, one scholar claims that by 1962 the Bulgarian "government was able to boast that the dramatic decline of religious belief in the country was unmatched in the Soviet Bloc," with two million Bulgarians claiming to be atheists and less than one-third of the population claiming to be religious.[60] Thus, socialist secularism in the Bulgarian context was more about the church supporting the state's goals, than a true secular separation of these institutions. The aim was to limit the influence of the "backwardness" of religious mysticism and faith in favor of science and rationality in order to achieve a more modern society, as well as the emancipation of Bulgarian women.

Socialist Secularism: Women and Religious "Backwardness" in Yugoslavia

In 1947, a regime-sponsored publication, *Women of Yugoslavia*, proudly announced, "One of the greatest achievements won by the peoples of Yugoslavia in their fight for liberation was undoubtedly the emancipation of women from age-long injustice and oppression."[61] Published in Serbo-Croatian, English, Italian, and French, this booklet posited socialist Yugoslavia's many achievements in women's equality, ranging from the significant role played by women in the partisan struggle, to the establishment of suffrage and constitutionally guaranteed equality in socialist Yugoslavia's first constitution, to policies to protect mothers and children, to literacy campaigns targeting women. As this pamphlet illustrates, female emancipation served as a key symbol of socialist modernization. In Yugoslavia, as in many other socialist (and capitalist) societies, women's status—concretized in the visual marker of women's dress—thus became metonym of both tradition/backwardness and modernity/progressivism. Within this discursive configuration, women's backwardness figured as a result of traditional structures such as the patriarchal family and religion, thereby legitimating various projects aimed at reform.

In *Women in Yugoslavia*, Olga Kovačić depicted Yugoslav women as both agents and subjects of their and their country's emancipation. On the one hand, the text detailed "what new Yugoslavia has given to women," implicitly portraying the regime as a "parent-state" doling out benefits and rights to its grateful children.[62] Literacy courses, for example, were said to free women "from the chains and prejudices which had fettered them till now." The most abject of these women, in the socialist imaginary, were rural women who had "lived in ignorance, enslaved, just like the sheep which they tended."[63] On the other hand, Kovačić stressed women's agency in winning their rights, beginning with the heroism of women during the partisan war. This was seen to mark a new consciousness for women. "Millions of new citizens, socially aware, have arisen from mediaeval darkness, proud of the confidence shown them and full of the desire to justify that confidence."[64] A later generation of feminist critics in Yugoslavia, however, would argue that women's emancipation was a top-down process that sought to tame women's energies and channel them into sanctioned political activities.[65]

Interestingly, for much of her text, Kovačić does not emphasize the link between religion and women's oppression, in contrast to some of the writings of well-known female Party activists like Vida Tomšič.[66] Nonetheless, regions of the country described as steeped in backwardness were those associated with Islam and Ottoman rule. In singling out areas where the large majority of women did not know how to read before the socialist revolution, for example, Kovačić highlights Bosnia (with 85 percent of women illiterate), Macedonia (88 percent female illiteracy), and Kosovo and Metohija (90 percent female illiteracy). In such descriptions, Kovačić draws an implicit connection among backwardness, religion (especially, but not only, Islam), and female oppression. Only in the final pages of the book does Kovačić make this explicit, discussing the practice of women casting off the veil.

Published three years before wearing of the veil became illegal in Yugoslavia, Kovačić stresses the abandonment of the veil as a voluntary act, one consonant with Yugoslav women's newfound empowerment and sense of agency as (putatively) socialist subjects. She describes "women of the national minorities" (i.e., Muslims) as casting off the symbols of "the wretched past."

> The movement for the removal of the veil has widely affected the Moslem women in the republic of Bosnia, the republic of Macedonia, and the region of Kosovo and Metohija. Villages and towns compete among themselves to "remove darkness from their eyes forever" as one of thirty Bosnian women who went to work on the building of the railway, expressed it.[67]

Photos accompanying the text depicted "Moslem women who removed their veils at a course for the illiterate" and "Moslem girls from the People's Republic of Bosnia and Herzegovina working on the Brcko-Banovici Youth railway." Though unveiled, the women in the latter image wear headscarves and long dresses, suggesting one of the limits to the regime's "emancipation" campaign as a result of its efforts to contain but not completely eradicate religion.

In her account of youth and adulthood in Tito's Yugoslavia, Munevera Hadžišehović describes the same events as Kovačić but from the perspective of a woman raised as a practicing Muslim in the Sandžak who later became a member of the Party. She notes the divisions created within Muslim communities by the regime's secularist programs, as some Muslims themselves became advocates of abandoning the veil and other symbols of the "backward" past. Even before the passing of the 1950 law forbidding the veil, a number of Muslim women—particularly wives of officials and members of partisan families—began to go uncovered. As Hadžišehović puts it, "Muslim communities, at the least the first ones and those who joined right after the war, 'liberated' themselves from religion as they worked their way through the communist literature … They justified their changed outlook on the grounds that they were tired of the backwardness of our people, which they attributed to the long Turkish rule and to Islam."[68] She adds, "Their longing for Muslim equality with the other nationalities in Yugoslavia, as well as for a better life, led them to identify with communist ideas about society."

In order to set an example for other Muslims, prior to the 1950 de-veiling law women from respected families who had not "liberated" themselves from Islam were strongly encouraged to abandon the veil voluntarily. With this in mind, the authorities approached Hadžišehović's mother or Nanna. When she refused to de-veil, she was called before the local secret police, but nonetheless remained adamant. Only with the passing of the law did Nanna finally remove her veil. This particular example reveals the uneasy balance between coercion and consent in Yugoslav efforts to contain religion in the name of female emancipation.[69]

Over time, explicit attention to the so-called woman question faded in socialist Yugoslavia, eventually sparking the rise of feminist groups in the 1970s who critiqued the Party's unwillingness to render the educational system an effective tool for transforming gender relations. This new generation of feminists increasingly went outside the structures sanctioned by the state and the Party (the structures within which female activists like Kovačić and Tomšič worked), instead adopting tactics of self-help and public protest. They also recuperated older traditions of feminism from presocialist Yugoslavia, genealogies that the socialist regime had sought to efface or delegitimize.[70] The second and third wave feminists loudly protested what they saw as the failures of the socialist project with regard to gender relations. From a somewhat different angle than that of Yugoslav feminists, ethnographic work revealed how structural transformations in Yugoslav society such as industrialization and migration to urban areas did much to modify family structures, even as women remained saddled with "traditional" burdens together with new demands.[71]

As Yugoslav society modernized, women's dress remained a key marker of progress. In contrast to Bulgaria, however, the regime trod much more carefully in singling out Muslims as emblems of backwardness, given the need to balance competing ethnoreligious claims and the increasing importance accorded to Bosnian Muslims as a constituent people. At the level of everyday life, non-Muslims pointed to gendered and religious dress as evidence of backwardness. In her study of a mixed Catholic and Muslim village in late socialist Bosnia, anthropologist Tone Bringa reported the reactions of local Catholic women who took the traditional dress of Muslim women in the village as a sign of otherness and a rejection of modernity. Yet many older Catholic women in the village dressed in a style almost indistinguishable from that of their Muslim neighbors (with long skirts and headscarves). In the minds of many Yugoslav citizens, then, tradition, women, and religion remained linked until the last days of socialist Yugoslavia's existence.

Socialist Secularism: Women and Religious "Backwardness" in Bulgaria

As in Yugoslavia, it was the Muslim women who were the most associated with religious "backwardness," and from the earliest years of communist rule Muslim communities were subject to persecution.[72] Bulgaria's Muslim community was quite heterogeneous, made up of ethnic Turkish, Slavic Muslim (Pomak), and Romani communities. Although 95 percent of the Muslims in Bulgaria were technically Hanafi Sunni, there was a very wide spectrum of beliefs represented during the communist era. There was a small heterodox Shi'a population, called the Alevis, and a wide variety of Sufi brotherhoods such as the Bektashis, which had a long history in the country. Most important to the Bulgarian communists, however, was that the Muslims in Bulgaria were remnants of the Ottoman Empire, the vast majority of them ethnic Turks who spoke Turkish as their mother tongue.

Efforts to secularize Bulgaria's Muslims populations and to emancipate Muslim women were inevitably bound up with the communist desire to distance itself from what it considered a feudal Ottoman past and to protect itself from perceived Turkish irredentism, in particular after Turkey invaded Cyprus in 1974.[73] In addition to infiltrating the official religious hierarchy of the Muslim community, the communists also invested heavily in education programs. Communist Party officials targeted individual Muslim men and women to create a cadre of loyal Party members who would spread the communist message among their faith mates. In a retrospective look at the communists' educational policies toward the Muslims, the Bulgarian historian Vera Mutafchieva writes:

> Directed by the Turkish department of the BCP's [Bulgarian Communist Party] Central Committee, a Turkish *nomenklatura* came into existence, enjoying the status and privileges due it. It was isolated from the mass of the Turkish minority who remained untempted by either free laic education or the perspective for its young people to run against their parents and tradition ... [This] education influenced the traditional mentality—secular thinking undermined the religious outlook.[74]

These educational campaigns were even more intensely targeted at the Pomaks, who spoke Bulgarian as their mother tongue and were believed to be the descendants of Bulgarians who had been forcibly converted to Islam during the Ottoman era. One specific example of an education campaign directed at Pomak women is a lesson plan from 1961.[75] The outline for "forty-five day courses for Bulgarian Muslim women activists of the Fatherland Front"[76] includes a variety of lectures, including four hours on Bulgarian and world geography and four hours dedicated to personal hygiene for women and how proper socialist women should dress (without headscarves, of course). There were also two hours each to be dedicated to the "unscientific and reactionary essence of religion," "religion and the woman," "the harm of observing religious holidays and traditions," and "religion and nationality." These lectures were followed by eight hours of lectures on the "morals of the socialist family," the work of the Fatherland Front among women, and the responsibilities of women activists within the Front as well as the role of women in building socialism more generally.[77] In addition to making them more educated and productive workers, these courses were aimed at convincing Muslim women to renounce their allegiance to Islam to embrace scientific socialism in their private as well as public lives.

As in neighboring Yugoslavia, efforts to combat Islam were also intertwined with modernization projects to improve the material conditions of Muslim lives in hopes that economic prosperity would weaken their spiritual commitments.[78] The Bulgarian government published a variety of propaganda materials focusing on the demonstrably rising living standards of the Bulgarian Muslims. One such book published by the Bulgarian Communist Party Press in 1960 targeted the Pomaks and was filled with propaganda lectures describing the ever-improving material conditions of life in the Rhodope Mountains, complete with before and after pictures of Muslim villages and photos of Muslim women smiling for the camera in modern Bulgarian clothing.[79]

The book also included a series of before and after cartoons that visually depicted socialism's superiority over Islam. One such cartoon showed a man smoking a cigarette and riding a donkey while his wife, a barefoot figure wrapped up like a mummy and carrying several large parcels, a baby and two hoes, walks behind him. The bottom panel shows a modern woman holding the hand of a young boy with a helium balloon. She is followed by her husband who is carrying a large household appliance out of a department store,[80] emphasizing not only the material progress made by socialism but the improvement in women's position in society. Another cartoon shows an angry looking man in a turban holding an open book and prayer beads standing in the balcony of a minaret. The caption reads: "Builder of Paradise in the Sky." The bottom panel shows people working together to build a modern socialist city in the mountains and the caption reads, "Builders of Paradise on Earth."[81]

Although the communist propaganda was somewhat successful in winning Pomak converts to scientific socialism, Islam proved to be more tenacious than anticipated. A 1971 book, *Novite gradove v N. R. Bulgaria* (New cities in the People's Republic of Bulgaria), proclaimed that economic development had been so successful in Madan that the Pomaks had voluntarily changed their names back to Bulgarian ones, that the men no longer wore fezzes and that the women had cast off their headscarves.[82] But the situation was not as positive as the communists portrayed it. The early attempts at educating Pomaks to help sway their neighbors to the benefits of modernization were not enough to eradicate persistent local Muslim traditions. The intransigence of these local forms of Islam caused the government to intensify its assimilation campaign, and as the 1970s wore on, they began to ban certain Muslim rituals and the question of Muslim names became paramount.

In comparison to Yugoslavia, the Bulgarians took a much more heavy-handed approach to secularizing Muslim communities, although the justifications for doing so proved remarkably similar. Rather than starting with a complete ban, however, the Bulgarian state first put forward their version of rational arguments against Islamic practices.[83] For instance, fasting during the month of Ramadan was said to inhibit the immune system's ability to fight off disease, making Muslims more susceptible to sickness. This self-inflicted predisposition to illness and the general weakness that accompanied fasting was also said to make workers less productive on the job. The ritual sacrifice of lambs during the *Kurban Bairyam* (Feast of the Sacrifice) was said to be unhealthy for Muslims who could suffer from severe stomach upsets from eating too much fatty meat at one time; it was also economically wasteful because the lambs could be sold for hard currency abroad. Circumcision was considered a barbaric process only practiced by the most backward of peoples, and the communists argued that it was unhealthy and dangerous for boys. Most significant for our article, headscarves and *shalvari* (Turkish trousers) for women were considered signs of their subservience to men, and therefore had to be removed.[84] The communist's goal of equality between men and women justified the ultimate ban on all Islamic dress.

The Bulgarian government would eventually institute prohibitions on the celebration of the two big Muslim holidays *Kurban Bairyam* and *Ramazan Bairyam*.[85] Women were forcibly de-veiled. The minarets on mosques around the region were torn down and destroyed; the mosque buildings were either closed or converted into cultural houses or museums. Islamic burial practices were made illegal, and in 1978 the government created and disseminated new guidelines for "socialist rituals" to be performed at funerals and other important life events.[86] Local imams and *hodzhas* who defied these new rules could be arrested and imprisoned.

Unlike in Yugoslavia where oppression of religious groups and of the Muslim community in particular gradually lessened through the 1970s and 1980s, the Bulgarian government took an increasingly firm stand against its Muslim minorities. Forcible name changing campaigns against the Pomaks and the Turks as well as the so-called Great Excursion (i.e., expulsion) of Turks in May of 1989 are only a few examples of the Zhivkov regime's growing intolerance toward Bulgaria's Muslim minorities. Although nationalism and fear of Turkish irredentism informed these policies, they were also driven by communist attempts to modernize what it considered "backward" religious communities for their own good, particularly for the good of Muslim women.

Conclusion

Given the renewed scholarly interest in the history of secularism and its various impacts on both the Western and non-Western world, it is striking how little has been written about the experience of secularization during the state socialist era in Eastern Europe. Likewise, even careful analyses of the role of religion in post-socialist Eastern Europe often rely on cartoonish or dismissive accounts of religious repression and Marxist promotion of atheism during the socialist period. Just as scholars have debunked the assumption that communism placed a lid on ethnonational divisions by showing the ways in which various socialist regimes actually institutionalized ethno-national identities, so too must researchers take socialist secularism seriously and consider the ways in which it reaffirmed (even as it sought to contain) religious practices and institutions. Doing so enriches our understandings of how socialist regimes approached both gender and religion in ways that illuminate the persistence and transformation (rather than "revival") of supposedly traditional practices after socialism's end.

In writing this article, we were led by several key questions. What does it do to our understandings and genealogies of secularism if we consider socialist experiences? What happens to our understandings of socialism, in turn, if we consider its histories through the lens of secularism? Does this shift the ways in which we think about religion and gender in Southeastern Europe and, if so, how? Although it is clear that Bulgaria and Yugoslavia's experiences of socialist secularism differ in the degree of its coerciveness, it is instructive to examine the similarities in the conceptualization of the secularizing imperative and the rhetorics used to justify it.

The centrality of the figure of Muslim women in socialist Bulgaria and Yugoslavia and the particular attention paid to gendered religious dress (veils, headscarves, baggy pants) highlight parallels with current debates over Western European secularisms (France, Germany, and beyond), as well as Kemalist versions of secularism in Turkey. In *The Politics of the Veil*, historian Joan Scott argues that the ban of the Islamic headscarf for French Muslims attending public schools took shape in a climate of growing French angst about the loss of cultural autonomy in the face of growing European supranationalism.[87] The French—compelled to redefine their national identity within the confines of a newly enlarged "Europe"—mobilized discourses on race, gender, and sexuality in relation to a visible, internal "Other" (i.e., North African Muslim women) as a way to redefine the bounds of appropriate "Frenchness." As Scott reminds us, it is no coincidence that the status and position of women in a society is often seen as a measure of that society's modernity. The examples of Bulgaria and Yugoslavia show that this was also true during the communist era, where a feminized and backward East as represented by Islam was in need of a masculine, modern West as embodied in the discourses of socialist secularism.

Although there are certainly important similarities to the French and Turkish cases, we argue that a study of secularism in the Balkans can provide a new lens through which to view these debates. The socialist example suggests that scholarly genealogies of secularism, grounded in the Catholic/Protestant split, need to engage more directly with the literature on Orientalism. This seems particularly urgent given the centrality of the "oppressed" Eastern female to both classic Orientalist thought as described by Edward Said and Western secularism, which often rests on a view of tolerance that celebrates the protection of such Eastern women *from* religion as well as *to* religion. At the same time, scholars of secularism would do well to take account of what Maria Todorova calls "Balkanism," seen not as mere variant on Orientalism but its own discursive configuration founded on a sense of the region as a space of transition between Europe and the Orient. Scholars can begin to do this by including the history of state-church relations in nations with Eastern Orthodox Christian traditions, nations that are systematically excluded from the existing genealogies of secularism in Europe.

Indeed, the Balkans in general are a place with a centuries old tradition of religious pluralism, under the Byzantine and subsequent Ottoman and Habsburg empires. Socialist reformers in both Bulgaria and Yugoslavia justified their often heavy-handed approach in the belief that their version of secularism further decreased rather than increased sectarian divisions in society. Furthermore, unlike in France or Turkey, where state efforts to impose secularism were focused exclusively on the public sphere, the Bulgarian and Yugoslav cases suggest that the socialist secular model was far more comprehensive in trying to radically reshape the material conditions of everyday life. This difference reflects, in part, the nature, role, and perception of the so-called private and public spheres under state socialism.[88] The cases of Bulgarian and Yugoslav secularist socialism stand diametrically opposed to models of secularism associated with the United States, in which separation of church and state is premised on the need to protect religion from the incursions of the state.

These findings point to the critical purchase that the concept of "socialist secularism" has even between as diametrically opposed cases as Bulgaria and Yugoslavia. By considering these two cases together, we question the frequent exclusion of Yugoslavia from broader discussions of Eastern European socialism and post-socialism. Furthermore, identifying commonalities for Bulgaria and Yugoslavia in both the content and practice of socialist secularism—in spite of the Yugoslav "difference"—highlights the productiveness of the concept of "socialist secularism" for scholars of gender, religion, and beyond.

◊ Acknowledgments

Pamela Ballinger thanks the Center for Advanced Study in Behavioral Sciences (CASBS) at Stanford University for a residential fellowship that offered time and space in which to develop a new research interest quite different from my previous work on the former Yugoslavia. My colleagues at the weekly CASBS gender table proved lively interlocutors. In particular, Saba Mahmood offered useful references on the topic of secularism and I thank her and my other CASBS colleagues. Kristen Ghodsee thanks the National Council for Eurasian and East European Research (NCEEER) and the Institute for Advanced Study in Princeton for their generous support of the research in Bulgaria. Special thanks to Joan Scott and the members of the School of Social Science "Third World Now" seminar for inspiring my initial interests in the study of secularisms. This article is derivative of a conference that was funded by an American Council for Learned Societies Southeastern European Studies workshop grant and generously hosted at Bowdoin College in the fall of 2010. Both authors would also like to acknowledge Scott Sehon and Anne Clifford who read and commented upon several early drafts of this paper.

◊ About the Authors

Pamela Ballinger is an associate professor of anthropology at Bowdoin College. She is the author of *History in Exile: Memory and Identity at the Borders of the Balkans* (Princeton University Press, 2003). She has published in journals such as *Balkanologie, Comparative Studies in Society and History, Current Anthropology, Ethnologia Balkanica, History and Memory, Journal of Genocide Research, Journal of Modern Italian Studies,* and *Journal of Southern Europe and the Balkans.* Ballinger has received fellowships from the American Academy in Rome (the Rome Prize), American Council of Learned Societies, Center for Advanced Study in the Behavioral Sciences at Stanford, Fulbright, Italian Academy at Columbia University, MacArthur Peace and Security, National Council on Eurasia and East European Research, National Endowment for the Humanities, National Science Foundation, and Wenner-Gren. Email: pballing@bowdoin.edu

Kristen Ghodsee is the John S. Osterweis Associate Professor in gender and women's studies at Bowdoin College. She is the author of three books: *The Red Riviera: Gender, Tourism and Postsocialism on the Black Sea* (Duke University Press, 2005), *Muslim Lives in Eastern Europe: Gender, Ethnicity and the Transformation of Islam in Postsocialist Bulgaria* (Princeton University Press, 2009), and *Lost In Transition: Ethnographies of Everyday Life after Communism* (Duke University Press, 2011). Ghodsee has received fellowships from the National Science Foundation, Fulbright, National Council on Eurasian and East European Research, International Research and Exchanges Board, and the American Council of Learned Societies as well as residential research fellowships at the Woodrow Wilson International Center for Scholars, the Max Planck Institute for Demographic Research in Rostock, Germany, the Institute for Advanced Study in Princeton, and the Radcliffe Institute for Advanced Study at Harvard University. Email: kghodsee@bowdoin.edu

◊ Notes

1. Talal Asad, *Formations of the Secular* (Stanford: Stanford University Press, 2003), 17.

2. Ibid., 10, 168–170, 199. That said, important studies on Turkey and Egypt that deal with secularism and gender have appeared in recent years. For example, see Saba Mahmood, *The Politics of Piety: The Islamic Revival and the Feminist Subject* (Princeton: Princeton University Press, 2005); Nilüfer Göle, *The Forbidden Modern: Civilization and Veiling* (Ann Arbor: University of Michigan Press, 1997); Joan Wallach Scott, *The Politics of the Veil* (Princeton: Princeton University Press, 2007).

3. Many scholars trace the origins of this secularization thesis to Enlightenment belief in the progressive triumph of reason and Weber's seminal work *The Protestant Ethic and the Spirit of Capitalism*. See Janet R. Jakobsen and Ann Pellegrini, "Introduction: Times like These," in *Secularisms*, ed. Janet Jakobsen and Ann Pellegrini (Durham: Duke University Press, 2008), 2–3; John T.S. Madeley, "America's Secular State and the Unsecular State of Europe," in *Religion, State, and Society: Jefferson's Wall of Separation in Comparative Perspective*, ed. Robert Fatton and R.K. Ramazani (New York: Palgrave, 2009), 109–111. Neo-Weberian sociologists like Peter Berger championed the secularization thesis into the 1970s. Peter Berger, Grace Davie, and Effie Fokas, *Religious America, Secular Europe? A Theme and Variations* (Aldershot, UK: Ashgate, 2008), 10. For a spirited and thoughtful defense of the secularization framework, consult Jose Casanova, "Secularization Revisited: A Reply to Talal Asad," in *Powers of the Secular Modern. Talal Asad and his Interlocutors*, ed. David Scott and Charles Hirschkind (Stanford: Stanford University Press, 2006), 12–30.

4. Kristen Ghodsee, "Symphonic Secularism: Eastern Orthodoxy, Ethnic Identity and Religious Freedoms in Contemporary Bulgaria," *Anthropology of East Europe Review* 27, no. 2 (2009): 227–252.

5. Asad, *Formations of the Secular*, 7.

6. Maria Todorova, *Imagining the Balkans*, 2nd ed. (Oxford and New York: Oxford University Press, 2009); Pamela Ballinger, "Definitional Dilemmas: Southeastern Europe as 'Culture Area'?" *Balkanologie* 3, no. 2 (1999): 73–91.

7. One notable exception is Jeff Haynes, "Religion, Secularisation and Politics: A Postmodern Conspectus," *Third World Quarterly* 18, no. 4 (1997): 726.

8. Gregory Massell, *The Surrogate Proletariat: Moslem Women and Revolutionary Strategies in Soviet Central Asia, 1919–1929* (Princeton: Princeton University Press, 1974); Douglas Northrop, *Veiled Empire: Gender and Power in Stalinist Central Asia* (Ithaca: Cornell University Press, 2004).

9. Scholars of Yugoslavia noted declining religiosity, a phenomenon that was sometimes related to variables such as urbanization and politicization. Dean E. Frease, "A Politicization Paradigm: The Case of Yugoslavia," *Sociological Quarterly* 16, no. 1 (1975): 33–47.

10. Madeley, "America's Secular State," 110.

11. For one exception see, Lavinia Stan, "Church-State Relations and Secularism in Eastern Europe," in *Quo Vadis Eastern Europe? Religion, State and Society after Communism*, ed. Ines Angeli Murzaku (Ravenna: Longo Editore, 2009), 89–106.

12. Madeley, "America's Secular State," 125.

13. Katherine Verdery, *What Was Socialism and What Comes Next?* (Princeton: Princeton University Press, 1996), 64.

14. Ramet discusses key differences in Marx's writings on religion and Marxist-Leninist interpretations of Marx. Pedro Ramet, "Conclusion," in *Religion and Nationalism in Soviet and East European Politics* (Durham: Duke University Press, 1989), 424. The distinct brand of Yugoslav socialism developed after Yugoslavia's split with the Soviet Union and centered on principles such as self-management and non-alignment is often deemed Titoism. This differs from Yu-

goslavism, an ideology of South Slav unity, which has enjoyed various articulations since the nineteenth century and which shifted considerably in form, content, and prominence during the lifespan of socialist Yugoslavia. For a detailed discussion of Yugoslavism in the socialist context, see Andrew Wachtel, *Making a Nation, Breaking a Nation: Literature and Cultural Politics in Yugoslavia* (Stanford: Stanford University Press, 1998).

15. For just a small sampling of efforts to define secularism and the secular, see Graeme Smith, *A Short History of Secularism* (London: I.B. Tauris, 2008), 2–8; C. John Sommerville, *Religion in the National Agenda: What We Mean by Religious, Spiritual, Secular* (Waco, TX: Baylor University Press, 2009), 195–204. In Bulgarian, see Dimitur Kirov's *Bogoslovie na obshtestvenia zhivot* (Theology of the social life) (Sofia: St. Kliment and Ohridski University Press, 2002).

16. Pedro Ramet, *Cross and Commissar: The Politics of Religion in Eastern Europe and the USSR* (Bloomington: Indiana University Press, 1987), 16; Stan, "Church-State Relations and Secularism in Eastern Europe," 192.

17. Ramet, "Conclusion," 414–415.

18. Evgenija Garbolevsky, *A Church Ossified? Repression and Resurgence of Bulgarian Orthodoxy 1944–1956* (Sofia: Aconi-Izdat, 2005), 16–18.

19. James Lindsay Hopkins, *The Bulgarian Orthodox Church: A Socio-Historical Analysis of the Evolving Relationship between Church, Nation, and State in Bulgaria* (Boulder, CO: East European Monographs, 2009), 202–205.

20. Ramet, "Conclusion," 415. Ghodsee, "Symphonic Secularism," 227–252.

21. Ramet, *Cross and Commissar,* 29.

22. Mary Neuburger, *The Orient Within: Muslim Minorities and the Negotiation of Nationhood in Modern Bulgaria* (Ithaca: Cornell University Press, 2004).

23. Klaus Buchenau, "What Went Wrong? Church-State Relations in Socialist Yugoslavia," *Nationalities Papers* 33, no. 4 (2005): 548.

24. See the letter from the Italian Legation in Belgrade to the Minister of Foreign Affairs in Rome; Ministry of Foreign Affairs Archive, Rome, Italy; Serie Affari Politici 1946–1950 Jugoslavia, b. 71, Telespresso "La Lotta Contro la Chiesa Cattolica in Jugoslavia" (26 April 1950). For a similar judgment, refer also to the Telespresso of 23 October 1948 from the Italian Legation in Belgrade to the Ministry of the Foreign Affairs in Rome; Ministry of Foreign Affairs Archive, Rome, Italy; Serie Affari Politici 1946–1950 Jugoslavia, b. 42.

25. Cited in Vjekoslav Perica, *Balkan Idols: Religion and Nationalism in Yugoslav States* (Oxford: Oxford University Press, 2002), 5.

26. See Kim Shively, "Taming Islam: Studying Religion in Secular Turkey," *Anthropological Quarterly* 81, no. 3 (2008): 687.

27. Shively, "Taming Islam: Studying Religion in Secular Turkey."

28. Noel Malcolm, *Bosnia: A Short History* (New York: New York University Press, 1996), 199.

29. Tone Bringa, *Being Muslim the Bosnian Way* (Princeton: Princeton University Press, 1995), 8–11; Munevera Hadžišehović, *A Muslim Woman in Tito's Yugoslavia*, trans. Thomas J. Butler and Saba Risaluddin (College Station: Texas A&M University Press, 2003), 111, 123.

30. The regime also responded to Bosnian Muslims' insistence on their distinctiveness. As Dijana Alić and Maryam Gushesh put it, "the construction of Bosnia's national identity was informed by a dialogue between competing narratives: the more dominant Yugoslav notion of Bosnia as symbol of a united Yugoslavia, and the persistent claim of a unique cultural heritage distinct from the neighboring states of Serbia and Croatia." Dijana Alić and Maryem Gusheh, "Reconciling National Narratives in Socialist Bosnia and Herzegovina: The Baščaršija Project, 1948–1953," *Journal of the Society of Architectural Historians* 58, no. 1 (1999): 12.

31. Buchenau, "What Went Wrong?," 553–554. The regime's support in 1959 for a Macedonian Orthodox Church autonomous from the Serbian Orthodox Church, for instance, rankled

many within the Serbian church hierarchy. Pedro Ramet, "The Interplay of Religious Policy and Nationalities Policy in the Soviet Union and Eastern Europe," in *Religion and Nationalism*, 313.

32. Buchenau, "What Went Wrong?," 555.

33. Jugoslovenska Knjiga, *Freedom of Religion in the Federative People's Republic of Yugoslavia. Documents* (Belgrade: Stampa Jugoslovensko stamparsko preduzece, 1947), 5–6.

34. Ibid., 6.

35. Ibid., 8.

36. Ibid. For a brief prehistory of these constitutional clauses providing for freedom of conscience and separation of church and state in socialist Yugoslavia, see the pro-regime account in Ivan Lazić, "The Legal and Actual Status of Religious Communities in Yugoslavia," in *Religions in Yugoslavia*, ed. Zlatko Frid (Zagreb: Binoza, 1971), 50–57. Lazić locates the origin of these policies in the "Foca Documents" promulgated by the Supreme Staff of the People's Liberation Army in December 1942.

37. Lazić, "The Legal and Actual Status," 45.

38. Individuals holding certain occupations, such as teachers or soldiers, often could not publically practice religion, though this varied greatly depending on where one lived in Yugoslavia and became more relaxed by the mid-1960s. Stella Alexander, *Church and State in Yugoslavia since 1945* (Cambridge: Cambridge University Press, 1979), 225.

39. Rastko Vidić, *The Position of the Church in Yugoslavia* (Belgrade: "Jugoslavija," 1962), 10.

40. Ivo Banac, *With Stalin against Tito: Cominformist Splits in Yugoslav Communism* (Ithaca: Cornell University Press, 1989), 17.

41. Alexander, *Church and State in Yugoslavia since 1945*, 62.

42. Massell, *The Surrogate Proletariat*.

43. Northrop, *Veiled Empire*, 9.

44. Alić and Gusheh, "Reconciling National Narratives," 12.

45. Ibid., 17.

46. The regime used both carrots and sticks to compel priests to join these groups, which provided the clergy with benefits such as pensions and social security. Miroslav Akmadza, "The Position of the Catholic Church in Croatia, 1945–1970," *Review of Croatian History* 1 (2006): 104. On recently available archival documentation of the frequent fear or opportunism that produced greater participation among Orthodox priests, see Buchenau, "What Went Wrong?," 556.

47. Ahmet Kuru, *Secularism and State Policies toward Religion: The United States, France, and Turkey* (Cambridge: Cambridge University Press, 2009), 11.

48. Bringa, *Being Muslim*, 221.

49. Even as direct intervention and control gave way to subtler forms of ideological pressure, religious faithful often found it difficult to reconcile the supposedly "private" world of religion with participation in socialist secularism. Various scholars have identified as central to Yugoslav secularism a civil religion centered on Tito. This civil religion had its own calendar (New Year, May Day, Tito's birthday, Day of the Republic); rituals (most famously, the annual countrywide relay race to celebrate Tito's birthday); ideology (brotherhood and unity, self management); and sacred history (the partisan struggle, Yugoslavia's "daring" break from the Soviet Union, nonalignment). This civil religion did not supplant traditional religion but competed as an alternative. See Sergej Flere, "The Broken Covenant of Tito's People: The Problem of Civil Religion in Communist Yugoslavia," *East European Politics and Societies* 21 (2007): 681–703; Perica, *Balkan Idols*, 223–226.

50. Ghodsee, "Symphonic Secularism," 227–252; Daniela Kalkandjieva, *Bulgarskata pravoslavna tsurkva i durzhavata, 1944–1953* (The Bulgarian Orthodox church and the state, 1944–1953) (Sofia: Albatros, 1997); Stefan Chureshki, *Pravoslavieto i komunistite v Bulgaria (1944–1960)* (Or-

thodoxy and communists in Bulgaria [1944–1960]) (Sofia: Prosveta, 2004); Momchil Methodiev, *Mezhdu viarata i kompromisa. Bulgarskata pravoslavna tsurkva I komunisticheskata durzhava (1944–1989)* (Between faith and compromise. The Bulgarian Orthodox Church and the communist state [1944–1989]) (Sofia: Institut za blizkoto minalo i izdatelstvo "Siela," 2010).

51. Central State Archive, Sofia, Bulgaria, f. 28, op. 1, a. e. 27 (Memorandum no. 36, 6 October 1944).

52. Hopkins, *The Bulgarian Orthodox Church,* 196–197.

53. Tzvetan Todorov, *Voices from the Gulag: Life and Death in Communist Bulgaria* (University Park: Pennsylvania State University Press, 1999).

54. People's Republic of Bulgaria, *State Gazette* no. 48, 1 March 1949. http://www.minelres.lv/NationalLegislation/Bulgaria/Bulgaria_Denominat_Bulgarian.htm (accessed 21 January 2010).

55. For the specific effects of the policy on the Muslim population see Kristen Ghodsee, *Muslim Lives in Eastern Europe: Gender, Ethnicity and the Transformation of Islam in Postsocialist Bulgaria* (Princeton: Princeton University Press, 2009), 109–115.

56. *Pravoslaven pastir* (Orthodox shepherd), no. 6 (June 1949): 224, cited in Garbolevsky, *A Church Ossified,* 28.

57. *Tsurkovan vestnik,* 12 May 1947, no. 20-21, 91, cited in Garbolevsky, *A Church Ossified,* 28.

58. *Tsurkovan vestnik,* 16 March 1954, no. 11, 2, cited in Garbolevsky, *A Church Ossified,* 28.

59. Klaus Roth, "Socialist Life-Style Rituals in Bulgaria," *Anthropology Today* 6, no. 5 (1990): 8–10.

60. Janice Broun and Grazyna Sikorska, *Conscience and Captivity: Religion in Eastern Europe* (Washington, DC: Ethics and Public Policy Center, 1988), 45–66. Of course, what they said to communist officials and what they believed are two different things. Despite this, post-1989 surveys confirm high levels of atheism/agnosticism as well as low church attendance well after the communist rule was over. See Petar Kanev, "Religion in Bulgaria after 1989: Historical and Socio-Cultural Effects," *South-East Europe Review* 1 (2002): 75–96; and Phil Zuckerman, *The Cambridge Companion to Atheism* (Cambridge: Cambridge University Press, 2006).

61. Olga Kovačić, *Women of Yugoslavia* (Belgrade: Jugoslovenska Knjiga, 1947), 5.

62. Verdery, *What Was Socialism,* 61–82.

63. Kovačić, *Women of Yugoslavia,* 49.

64. Ibid., 6.

65. Jill Benderly, "Feminist Movements in Yugoslavia, 1978–1992," in *State-Society Relations in Yugoslavia, 1945–1992,* ed. Melissa K. Bokovoy, Jill A. Irvine, and Carol S. Lilly (New York: St. Martin's Press, 1997), 184–185.

66. See, for example, the 1980 UN-sponsored report in which Tomšič blames the failure to achieve women's equality in Yugoslavia on "economic underdevelopment, primitivism, religious beliefs and other conservative prejudices, and private-ownership relations which still affect life and family." Vida Tomšič, *Woman in the Development of Socialist Self-Managing Yugoslavia* (Belgrade: Jugoslovenski stvarnost, 1980), 67, 17. Lawyer, writer, politician, partisan, and Yugoslav Party activist, Tomšič championed women's equality, in particular reproductive freedom and the right to abortion, even as she remained within the framework of a Yugoslav socialist project that sought to dictate the ways in which women's emancipation would take place. For a useful discussion of the tensions in Tomšič's thought and politics, see Mateja Jeraj, "Vida Tomšič," in *A Biographical Dictionary of Women's Movements and Feminisms: Central, Eastern, and South Eastern Europe, 19th and 20th Centuries,* ed. Francisca de Haan, Krassimira Daskalova, and Anna Loutfi (Budapest: Central European University Press, 2006), 575–579.

67. Kovačić, *Women of Yugoslavia,* 70.

68. Hadžišehović, *A Muslim Woman in Tito's Yugoslavia,* 110–111.

69. On the regime's efforts to "bribe" women to abandon the veil by offering low cost textiles and tailor services, see Alexander Dragnich, *Tito's Promised Land, Yugoslavia* (New Brunswick, NJ: Rutgers University Press, 1954), 154.

70. Benderly, "Feminist Movements in Yugoslavia, 1978–1992," 184–186. Benderly details the various waves of feminist thought and activity in Yugoslavia.

71. William G. Lockwood, "Converts and Consanguinity: The Social Organization of Moslem Slavs in Western Bosnia," *Ethnology* 11, no.1 (1972): 55–79; Rose M. Somerville, "The Family in Yugoslavia," *Journal of Marriage and the Family* 27, no. 3 (August 1965): 350–363.

72. Ali Eminov, *Turks and Other Muslim Minorities in Bulgaria* (New York: Routledge, 1997), 26.

73. Neuburger, *The Orient Within*, 192.

74. Vera Mutafchieva, "The Turk, the Jew and the Gypsy," in *Relations of Compatibility and Incompatibility between Christians and Muslims in Bulgaria,* ed. Antonina Zhelyaskova (Sofia: International Centre for Minority Studies and Intercultural Relations, 1994), 31.

75. "Uchebna programa na 45-dnevite kursove za Bulgarki Mohamedanki aktivistki na Otechestveniya Front" (Educational program of the forty-five day courses for Bulgarian Muslim women activists of the Fatherland's Front). State Archive, Blagoevgrad, f. 628, o. 3, a.e. 81, l. 44–47.

76. Originally a coalition of various parties and organization opposed to the Germans in Bulgaria during World War II, the Fatherland Front eventually became a communist mass organization. See John D. Bell, *The Bulgarian Communist Party from Blagoev to Zhivkov* (Washington, DC: Hoover Institution Press, 1985), 70.

77. "Informatsiya za rabota na shkolata za Bulgarki Mohamedanki v Gotse Delchev za vremeto ot 16 Noemvri do 29 Dekemvri 1961 godina" (Information about the course for Bulgarian Muslim women in Gotse Delchev for the period from 16 November to 29 December 1961). State Archive Blagoevgrad, f. 628, O. 3, a.e. 81, l. 6, 7.

78. Lenka Nahodilova, "Communist Modernisation and Gender: The Experience of Bulgarian Muslims, 1970–1990," *Contemporary European History* 19, no. 1 (2010): 37–53.

79. Petar Petrov, Asen Spasov, and Donio Donev, eds., *Rodopa: Bulgarski tvardina (Sbornik besedi) Uchebno pomagalo* (Rodopa: Bulgarian fortress [Collection of talks]. Teaching materials) (Sofia: BKP Izdatelstvo, 1960).

80. Ibid., 227.

81. Ibid., 237.

82. Ignat Penkov, *Novite Gradove v N.R. Bulgaria* (New Towns in the People's Republic of Bulgaria) (Sofia: Izdatelstvo Narodna Prosveta, 1971), 124.

83. Eminov, *Turkish and Other Muslim Minorities*, 58.

84. Ibid., 59.

85. Neuberger, *The Orient Within*, 142.

86. Roth, "Socialist Life-Style Rituals," 8–10.

87. Scott, *Politics of the Veil*, 18.

88. Susan Gal and Gail Kligman, *The Politics of Gender after Socialism: A Comparative-Historical Essay* (Princeton: Princeton University Press, 2000).

Gender and Religiosity among the Orthodox Christians in Romania

Continuity and Change, 1945–1989

Maria Bucur

ABSTRACT

This article questions the claim that in Romania, the post-1990 period was one of radically greater freedom in religious matters, as well as greater religiosity on the part of the population. Instead, it suggests that continuity better encapsulates the development of religiosity—religious beliefs and their embodiment in specific practices—among Orthodox Christians in Romania in the twentieth century. It also makes visible important imbalances, gaps, and faulty assumptions about the importance of institutions in the daily religious practices and beliefs of most Orthodox populations in the historiography on Orthodoxy in Romania. Scholars have failed to see continuities and have embraced analytical frameworks that stress change, especially around the communist takeover period (1945–1949) and the fall of communism (1989–1990). Central to re-evaluating this trajectory are two aspects of Orthodoxy in Romania: (1) most believers live in the countryside; and (2) women have remained central to the development and maintenance of religious practices in ways that cannot be accounted for through any institutional analysis of the Orthodox Church, because of its both implicit and explicit misogyny.

KEYWORDS: communism, gender, Orthodox Church, religiosity, Romania, women

Of Continuity and Change

Some of my oldest memories from childhood are linked to the dark tall cupolas, strong incense, and cold, drafty air of the churches where my grandmother, who had been raised as a deeply religious person in her small village near Oradea, would take me on Saturday mornings to have *colivă* blessed by the priest.[1] I knew close to nothing about Christianity, but I loved eating *colivă*, knew it was special and that I should say a prayer[2] and the name of our relative for whom it had been prepared when accepting

aspasia Volume 5, 2011: 28–45
doi:10.3167/asp.2011.050104

it. I also knew to be quiet and lower my head before the priest. Starting at age seven I often went into churches on my own, especially on the way home from school. My school was five minutes from the Patriarchy and from another old church in downtown Bucharest, Sf. Vineri [St Friday], located across the street from the Theological Institute.[3] Most of the time, these churches were practically empty. At most a few older women dressed in black would sit on the hard seats along the walls of the *pronaos*[4] (seldom anywhere close to the altar) or would be in the area where candles for the living and the dead would burn—praying, lighting a candle, or standing around crying. Had I not been introduced to religion by my grandmother and thus become comfortable with and curious about Orthodoxy, I probably would have never known or cared much about this aspect of life in Romania. My city-dwelling parents were both non-religious, and at most engaged in the once-a-year ritual of going to Easter midnight mass, an event more social and culinary (the mass would always be followed by a great feast) than spiritual in how they approached it.

The experience of religious life and institutions of the urban generations growing up under full-fledged Romanian communism (those born in the 1960s) was similarly mediated by either the presence of an older relative—most likely a grandmother who, being retired,[5] had the time to both see to religious rites and also take care of small children—or by the absence of such an important force, and thus more vaguely aware of religious customs and beliefs.[6] Religious rites and self-identification existed somewhere within the normal range of referents in one's life, an institution one could easily comprehend and accept in terms of important rites of passage and ritual. Yet for many children growing up in urban environments during this period, Orthodoxy was not powerfully present, nor clearly situated in relationship with other aspects of our life—playing, friendship, school, dreams for the future, vacations. Religiosity, however, was a different business. One learned (or not) to pray at home, and to identify principles of belief—faith in God, fear, respect, love—rather informally and through one's family circle. Sunday school, weekly church attendance, a programmatic relationship to the Church as an institution and to learning its theology and internalizing its principles was not something I experienced or saw evidence of among most of my urban cogenerationists who were officially Orthodox, meaning that they had actually been baptized in an Orthodox church (most often before the age of two).[7]

Therefore, the growing presence of the Orthodox Church since 1990 in the daily lives of all citizens in Romania, from ritual and personal involvement in activities of the Church to compulsory religious education and a vast presence of the Church in mass media[8] seems like an important change in religion and religiosity in postcommunist Romania.[9] Many scholars, from theologians to political scientists and anthropologists, have viewed the end of communism as a break in the development of Orthodoxy in the former Soviet bloc. They identified the post-1990 period as one of radically greater freedom in religious matters, as well as greater religiosity on the part of the population.[10] This article questions this claim, made both explicitly and implicitly in some of the existent scholarship,[11] and instead suggests that continuity better encapsulates much of the development of religiosity—religious beliefs and their embodiment in specific practices—among Orthodox Christians in Romania in the twentieth century.[12] Implicitly, my argument also suggests that secularization in this country has not been tied as much to the relationship between the communist regime and the Orthodox Church as some have claimed.[13] I also make visible important imbalances, gaps, and faulty assumptions about the importance of institutions in the daily religious practices and beliefs of most Orthodox populations in the historiography on Orthodoxy in Romania. Because of inadequate focus on practices on the ground, scholars have failed to see continuities and have embraced analytical frameworks that stress change, especially around the communist takeover period (1945–1949) and the fall of communism (1989–1990). Central to re-evaluating this trajectory are two aspects of Orthodoxy in Romania: (1) most believers live in the countryside; and (2) women have remained central to the development and maintenance of religious practices in ways that cannot be accounted through any institutional analysis of the Orthodox Church, because of its both implicit and explicit misogyny.[14] My article suggests an agenda for further research on this topic.

The History and Historiography of Orthodoxy before 1945

Continuity as a trope of understanding religiosity during the communist period can only be fully articulated by looking at a longer period of time preceding 1945. The history of the Romanian Orthodox Church identifies it as one of the oldest markings of "Romanianness" north of the Danube, starting in the seventh century A.D.[15] Much of the history of the Church during the middle ages and early modern period stresses the importance of continuity under duress of the institution and also of the practices linked to it. Evidence of such continuity is often linked to the presence of religious funerary symbols, to oral history and folklore, as well as rituals described in historical sources starting in the early modern period, focusing especially on birth, marriages, and death.[16] Such claims and evidentiary basis served the nationalists of the nineteenth and early twentieth century well. These intellectuals, many of whom were clergy or educated through religious institutions, wished to craft a sense of Romanian identity closely linked to the Orthodox Church, through a Janus-faced process of separating ethnic Romanians from other religious-linguistic groups living in the Romanian Principalities and Transylvania (such as Bulgarians, Jews, Russians, Roma, Germans, Protestants, Catholics, Ukrainians, and Hungarians), as well as unifying them around a common set of beliefs and practices.[17]

The claim of equating Romanianness with Orthodoxy could not be sustained through institutional-based links, as many Romanian populations, especially in Transylvania and the Banat, had lived for centuries outside of direct connection with the Orthodox Patriarchates of Ohrid, Constantinople, and Moscow.[18] Many of these populations had not been overseen by priests, did not have the privilege of worshiping in churches—unless they made pilgrimages to places where such churches existed, and were not in the direct care of a church bureaucracy as part of parishes and bishoprics.[19] Many Orthodox priests were regularly harassed in the Habsburg Empire and a majority had the barest education, while some remained illiterate and thus unable to function as active participants in an institution that relied almost exclusively on written documents for its official existence.[20]

In short, for a majority of Orthodox Christians, themselves overwhelmingly illiterate until the twentieth century, religious identity was linked primarily to traditional rituals learned locally from the older generation and passed down primarily by lay believers through food, clothing, and feasting, which women generally coordinated among rural communities.[21] The gendered aspect of religious acculturation and transmission of rituals from one generation to the next has not been the focus of much interest in Romania, though scholars of Orthodoxy elsewhere (especially anthropologists and folklorists) have drawn attention to the central role women played in everyday religiosity.[22] For Romania, a handful of studies have begun to draw attention to the ways in which women were central in both enacting and educating younger generations about rituals centered on religious holidays and rites of passage (birth/baptism, marriage, death), and blending pagan and dogmatic elements.[23] Food in particular, as a central component of both proper engagement with the liturgy and the cult of the dead, has continued to be women's domain. The priest played an important role in blessing food in order to endow it with its symbolic Christian meaning, but in most rural areas, until the twentieth century, the priest was not always there to perform this role. Yet ritualistic food, as suggested by ethnographic evidence from the turn of the twentieth century, was an integral part of how Orthodox Christians in Romania understood religious rituals and norms.[24]

Traditionally, other elements of the cult of the dead, such as wailing, preparing the body for funeral, and proper memorial rituals that continue for seven years after a person's death, were also overwhelmingly the responsibility of women, both customarily and normatively. In the most impressive analysis of rites of passages among populations living in Romania and Transylvania (and focusing almost exclusively on Orthodox Christians) published at the turn of the twentieth century, the author identified women's participation in death rituals as an art—something to be learned and perfected from one generation to the next—as well as something exclusively female.[25] For instance, wailers were to be only women, and if the deceased did not happen to have any immediate female kin, female neighbors and friends were to perform this role. Considering the fact that most of the people involved in the very complex rituals and incantations to be performed were illiterate, the continuity of such precise and normatively inscribed gender roles is striking. It points to the centrality of such issues in the everyday life of a rural community, and in particular in women's lives, who, as described by Romanian scholar Simion Florea Marian, seemed to have spent a great deal of their lives learning and performing such religious rituals, taught primarily by the older generation of women that surrounded them. If, in addition, one also remembers the importance of patrilocality among these women—they tended to follow their husbands after getting married, so would have to leave behind some learned rituals and traditions and learn new ones to pass them down to their own daughters and daughters-in-law—the life-long learning process of engaging with religious rituals and local traditions becomes an even more impressive feat.[26]

Scholars examining religiosity among Orthodox peasants, from current studies to those produced a century ago, go so far as to claim that there is such an entity as a Peasant Church.[27] In a recent study, Costin Nicolescu states that "the Ancient Law [capitalized in the original] of the Romanians has grown from the new law of love and of grace professed by Christ, without actually identifying directly with it [the new law]. In addition, the Ancient Law comprises the whole ensemble of spiritual gestures that refer to the practice of living Christianity, with the proper specific details (national, regional, local, etc.), which enrich it in multiple ways, through force and authenticity, its forms of expression."[28] In this narrative "authenticity" and the "national" speak to the author's specific nationalist assumptions about religious practice and identity. Yet what emerges from this and other similar studies of theology is an awe and recognition of the importance of popular traditions and rural practices in giving specific shape to what Orthodoxy came to mean in the culture of much of the population that inhabited the Romanian Principalities and Transylvania.[29]

The fate of the Orthodox Church as an institution central to the politics of modern Romania was closely related to its own long-standing principle of *symphonia*—the harmonious coexistence of the Church as an institution of spiritual authority alongside the institution of the Byzantine emperor as representing state authority in the pre-Ottoman period (before 1453).[30] Under the Ottoman millet system the Orthodox Church continued to retain important elements of autonomy also by working with the sultan. In the Romanian Principalities (Wallachia and Moldova), which were never directly under Ottoman control, by the 1860s the Orthodox Church was the largest single landowner and greatly benefited from tax exemptions and other legal privileges. Rulers adamant about curbing the great power of this institution, especially Alexandru Ioan Cuza (r. 1859–1864), succeeded only in part in reducing the Orthodox Church's privileges. Cuza forced the Church to close down many monasteries and nationalized the lands they controlled, but he was soon ousted.[31] Subsequent rulers, including the first king of Romania, Charles I (r. 1864–1916), adopted a more respectful attitude of non-involvement in internal Church affairs together with paying homage to the Church as an important national institution.

It is unclear to what extent such institutional struggles over land and economic power/privilege resonated with Orthodox believers. Social elites (the boyar class) were interested in understanding the issue from their position of privileged landowners and with an eye toward possible land reforms that affected them.[32] Many of the populations who had been working on Orthodox-owned lands were relieved to have the harsh regime of the Church administrators removed, but it is not clear that the alternatives were significantly better.[33] By and large, the lessening of the Church's economic power meant that in the countryside priests and the parishes they oversaw had fewer goods flowing from the Church to the community and had to depend more heavily on the generosity of an overwhelmingly poor and uneducated population. Religiosity remained thus closely (and possibly even more so) connected to traditional practices of the community rather than to rituals and practices overseen directly by an educated clergy.[34] Away from the center of political and economic institutional struggles, religious life continued relatively unchanged. However, the image of popular religiosity in relationship to the institutional changes affecting the Church in the nineteenth century remains blurry and demands more sustained focus by scholars.

Romania's victory at the end of World War I enabled King Ferdinand (r. 1914–1927) and especially Queen Marie (r. 1914–1938), both converts to Orthodoxy, to position themselves as overseers and embodiments of Orthodoxy as national religion in Romania.[35] The Liberal and Peasant parties worked to include nods toward religious freedom/ toleration in the postwar Constitution (1923), as both parties had sizeable non-Orthodox populations (especially Jews and respectively Greek Catholics) among their ranks.[36] The dynastic and war official commemorations of the interwar period marked the close connection between the Romanian state and Orthodoxy and reinforced it, cleverly incorporating many elements of popular religious practice, as a means to gain support among the population and sustain legitimacy through such cultural-religious alignments.[37]

Yet, even as the religious institutions of Greater Romania afforded the Orthodox Church a central privileged position, there is evidence that, especially in rural areas, religious customs of Orthodox believers continued unabated and at times in tension or even conflict with centrally mandated practices. For instance, in dictating the use of the modern (Julian) calendar, the central authorities came in conflict with the Gregorian calendar used in many rural areas in determining religious holidays.[38] The transformation of certain religious holidays—most prominently the Ascension, into national holidays—also brought about various reactions on the part of Orthodox believers. Although some participated compliantly in these commemorations, others paid only scant attention or ignored the Bucharest dictates regarding the specific types of ceremonies to be performed on that day.[39] Many saw the over-imposing of the Heroes Day onto the clearly demarcated holiday of the Ascension as simply an addition of names of those who fell in World War I to the commemoration of the local dead that was customary for Ascension. This minimized any significant change in ritual, religious practice, and signification of the holiday along any martial nationalist lines.[40] In other words, seeing the growing prominence of the Orthodox Church in state rituals is to some extent a misleading indicator of general changes in patterns of behavior among the population. There is more evidence of substantial impact among religious minorities, who were offended by this two-tiered relationship with the Romanian state and resented having to bow to the Orthodox Church through presumably secular rituals, such as Heroes Day. Catholics, Jews, and Protestants often resisted the Romanian state's mandates to treat what were Orthodox religious holidays as their own civic holidays, and often chose to simply ignore orders to participate in such rituals.[41]

Another equally important aspect of the relationship between the rising prominence of the Orthodox Church as a state-supported institution and religiosity among believers is church attendance. Before 1918, many Orthodox parishes in what would become Greater Romania did not have their own priests, especially in Transylvania, because the aggressive policy of Magyarization after 1867 made it difficult to sustain Orthodox churches and parishes, as they received no support (directly or through tax benefits) from the state, unlike all other religious denominations, including the Uniates.[42] Therefore, church attendance was somewhat of a mis-measure of religiosity, especially in areas where priests were not assigned to parishes. However, after 1918, the Romanian state spent a lot of money to begin building and restoring many churches and to train sufficient number of priests for the large Orthodox population in the country.[43] By 1940, even with over 10,000 churches and other places of worship in place, and with more than 8,500 priests and 10,500 cantors employed by the Church, church attendance among Orthodox believers was low. Only 10 percent of the flock attended church at least once a week.[44]

The meaning of this statistic is difficult to ascertain qualitatively (and even quantitatively) with any degree of nuance, as we are not privy to the methods employed to arrive at this number, nor the breakdown of the population along regional, gender, class, and rural/urban lines. However, based on my own research on the commemorative practices developed and continued during the interwar period, I posit that the rural population had a higher than 10 percent rate of attendance, for reasons that pertain to existing traditions, work patterns, and proximity of the place of worship. Women tended to attend church on Saturdays, for services dedicated to the memory of dead relatives, and for regular mass on Sunday. They also tended to participate in baptisms, weddings, and funerals, often for people who were not members of their own families. Men would more often attend Sunday mass and some of the rites of passage, but less often than women. Still, on average, male and female adults tended to go to church more than once a week. The proximity of places of worship to where most people lived also meant that the elderly could in fact attend church with less difficulty than in many urban settings.

In the meantime, urban dwellers, especially given their small percentage out of the total population (between 15 and 20 percent), had a significantly lower than 10 percent rate of attendance for reasons that have to deal with accessibility and comfort with specific settings. Many urban inhabitants were first generation and were more comfortable attending services that looked and felt like their rural homes. The Village Museum, inaugurated in Bucharest in the 1920s, served as an alternative for some newcomers to the metropolis. Young men and women could be seen on Sunday mornings making their way to the remote location of the museum, but this was not an option for those with small children and poor, as public transportation was for the relatively privileged in that era.[45] The urban rate can be gleaned anecdotally also from the scarce mention in newspapers and personal accounts of religious events.[46]

Another important difference in church attendance is along gender lines. It is difficult to assess precisely how much more often women attended church than men, but reports from religious war commemorations, pilgrimages, photos, as well as ongoing practices suggest that women attended church more often than men.[47] Women contributed actively (and essentially) to many commemorative rituals, especially regarding the cult of the dead, by preparing ritualistic meals and donations, as well as lists of the dead to be read in church by priests.[48] Considering the importance of the cult of the dead in Orthodoxy and the weekly Saturday liturgies dedicated to commemorating the dead, it is not far-fetched to claim that most active women Orthodox believers attended church at least twice a week—Sundays and often Saturdays, as well as on important holidays.[49] Based on these speculative calculations, I believe that by the end of the interwar period, the Orthodox population had become less religious in a differentiated manner, with urban populations and men in particular much further along the path toward secularization, while women, especially rural ones, continued religious practices much more in an unaltered fashion as they had done before 1918.

Historians of religion in Eastern Europe have generally been uninterested in the phenomenon of popular religion in terms of the gender dimensions of specific practices.[50] Evidence to consider this issue is sparse at best. Yet the issue of how particular traditions and the meanings ascribed to them became entrenched and continued to be so over a long period of time into the twentieth century is one that needs to be ad-dressed. We cannot assume that Orthodoxy existed as a living religion in the Romanian lands in the premodern periods simply because there was a Metropolitanate in Bucha-rest, Iasi, or Sibiu, and that a sizeable number of priests served in churches throughout the country. One has to look at the local level and in the area of material culture and ethnographic/folkloric evidence to understand this phenomenon of continuity.

The Orthodox Church under Communism (1945–1989)

The communist takeover has to be seen within the larger narrative of the modern period, in terms of both institutional relations with the state and of the differentiated declining religiosity evident at the beginning of World War II. The narrative of the Or-thodox Church under communism has varied greatly depending on who has crafted it and on whose behalf. Theologians and opponents of the communist regime in Roma-nia as a brutal atheist state have focused on the aggressive curbing of religious free-dom in Romania by looking at the imprisonment of large numbers of priests, monks, and nuns, as well as the infiltration of the clergy by the secret police and demolition of places of worship especially in the 1980s.[51] Critics of the Orthodox Church prefer to focus on the large degree of compliance on the part of priests through the Securitate. Other critical voices also focus on the takeover by the Orthodox Church of all Uniate assets after the latter was folded into the Orthodox Church in 1948.[52]

A more balanced view of the relationship between the Church hierarchy and the communist regime has begun to develop in recent years. In its 2006 Final Report, the Presidential Commission for the Study of the Crimes of Communism provides a narra-tive of religious oppression and compliance in communist Romania. The report high-lights the early abuses of the Romanian Communist Party against priests and religious institutions and also makes visible the extent to which the Romanian Orthodox Church was able to achieve a more autonomous and stable position by the 1960s, in clear con-trast to other religious denominations such as the Baptists or the Uniates.[53] More re-cently, Lucian Leuştean has offered an even-handed and well-documented analysis of how Orthodox Church leaders negotiated this position of autonomy.[54] Leuştean sees the techniques and results of the Orthodox Church in harmony and continuity with the tradition of *symphonia*, rather than as a break with the past.

Everyday Religiosity and Gender under Communism

Missing from these accounts and from much of the scholarly framing of studies on the fate of the Orthodox Church under communism is the question of what happened to the religiosity of Orthodox believers in the institutional context of early violence against outspoken believers or later in the context of compliance and even active work with the secret police. The Presidential Commission is silent on this matter, preferring to focus only on the exceptional fate of well-known priests and religious dissidents.[55] Leuştean frames his questions in terms of institutional relations between the Church and the communist regime. We are, therefore, left wondering what was happening on the ground.

Some evidence about religiosity among Orthodox Christians comes from anthropological research such as Gail Kligman's outstanding *Wedding of the Dead*. Her research shows a great degree of continuity (while acknowledging significant change) from before the communist period in terms of rural practices surrounding the cult of the dead among the Orthodox: "Although the state operates according to Marxist principles, village life continues according to traditional principles, among which religion is a guiding force."[56] Katherine Verdery's *Political Lives of Dead Bodies* also hints at continuities in religious practices.[57] The many Romanian ethnographers who wrote about "popular" traditions under the communist regime, as they could not write openly about popular Christianity, also attested to the continuity of such practices.[58] In addition, a growing oral history literature since 1990 has enriched the picture of localized religious practices.[59] The Museum of the Romanian Peasant, created in 1990, as well as the more traditional Village Museums in Bucharest, Sibiu, and Cluj are all products of the interwar period obsession with the "vitality" of peasant culture. They have focused significantly on religious rituals and practices in the countryside and have visually narrated these practices as continuously sustained—even during the communist period—through grassroots local traditions, rather than any specific institutional links with the Church or political regime.[60]

These studies and visual narratives suggest that the countryside remained somewhat removed from the struggles over authority and integrity that the Orthodox Church engaged in at the center of political power.[61] This is not to say that villagers were ignorant of or uncaring about, for instance, the imprisonment of priests or the infiltration of the clergy by the Securitate. In many informal conversations I have had over time with Orthodox believers, I heard strong opinions about the local priest, whom many suspected of being a collaborator of the Securitate, and thus avoided personally. But that seldom meant outright rejection of Orthodoxy.[62] Uniates who had been forced to go underground or, as an alternative, attend Orthodox Churches, were far more critical of the Church as an institution.[63] Yet they were not speaking on behalf of secularization, but rather on behalf of a different kind of religiosity. My own grandmother occasionally "shopped around" churches in Bucharest in search of priests who were less overtly compliant, but never quit attending church on the grounds of the priest's corruption by the communist regime.

In an oral history project conducted in 2009–2010, together with three other collaborators I interviewed over a hundred women from urban and rural backgrounds in the Hunedoara county. They ranged from in ages from mid-forties to mid-eighties and in educational/professional/economic backgrounds—from uneducated peasants to doctors, business entrepreneurs, and teachers. Of this cross-section of Romanian society, a majority not only self-identified religiously (overwhelmingly Orthodox), but they also expressed strong opinions about what the Church was and is or is not doing right. In other words, religiosity and the Orthodox Church were topics they wanted to engage with and they obviously thought about. Although these interviews took place twenty years after the end of communism, and it is obvious that these two decades have influenced the respondents' views on religiosity and the Orthodox Church in particular, some of their references suggested a longer process of self-identification with the Church:

> Back then [under communism] we weren't allowed but I made time, I made time to go [to church], my husband would take me, but somehow he wasn't really all that … religious, but he liked to take me; if I said "you know, I would like to go to Easter mass somewhere," he would get in the car and drive me, so he respected my religious beliefs … Back then we used to go in hiding … When I had to join the Party, we were forbidden to go to church, but if I felt like it, I would still go who knows where and I didn't care, I would go more seldom, but still went.[64]

In addition to underscoring the difficulty and fear associated with church attendance in relation to Communist Party membership, the quote above also exemplifies gender differences in church attendance. An interesting aspect of this narrative, which is not a-typical of other personal reflections on ways in which the Party attempted to curb religiosity, is that the description reflects a remembrance of fear of retribution, even as documentary evidence from Party archives suggests that such retribution was not as common from the 1960s on as it had been before. Most of the people we spoke with were in fact of generations that had not lived through the 1940s and early 1950s.

Other respondents also suggested interesting cross-religious attendance and traditions. A couple who were Protestant (he) and Catholic (she), spoke of keeping two Easters (Catholic and Orthodox) and participating in funerals and baptisms in the local Orthodox Church because their neighbors were Orthodox.[65] By contrast, one respondent spoke about self-identifying as an Orthodox, but attending a specific Catholic church in Hunedoara "because many years ago I was searching, I think I was sixteen [that would have been in 1985] and I was searching for a path and I wanted to find answers to some questions, and this was the church that gave me the answers I needed at that time."[66]

Overall, those who self-identified as regular churchgoers spoke of difficulties in practicing religious traditions in public under communism and of greater church attendance since then, but they also identified religious holidays (of which there are at least one per week in the Orthodox calendar) as important times when they used to go to church, in addition to Sundays. In terms of gendered involvement in religious holidays, they also identified men as being far less involved than women in keeping traditions alive both at home and in any kind of public fashion.

These observations help us better understand an apparently surprising statistic. In 1990, after half a century of tough atheistic communist dictatorship, church attendance of at least once a week among the Orthodox in Romania was at 20 percent, up 10 percent from 1940.[67] It is not clear what methods were used to measure attendance, and how to disaggregate this number. But even if we allow for a wide margin of error and for differences between the methods employed in the 1940 statistics and the 1990 ones, church attendance went up, not down, under an aggressively atheist regime that placed thousands of priests in prison.

Other important numbers can enhance our understanding of religiosity under the communist regime, with a pronounced gendered quality:

Table 1. Orthodox Monasteries in Romania (1938–1957)

year	1938	1949	1957
monks	1638	1528	1773
nuns	2549	3807	4041
monasteries monks	119	122	113
monasteries nuns	35	56	77

Source: Leuştean, *Orthodoxy*, 204.

This table highlights some little known developments that no scholar, to my knowledge, has tried to analyze in terms of gender differences. To begin with, most evidently, the number of people embracing religious orders grew significantly in the two decades from 1938 to 1957. At the height of the purges in the communist bloc, the number of monks and nuns allowed to take religious vows was growing. It is hard to analyze this data in terms of motivation of the people joining monasteries. But it is clear that, despite outspoken atheism and purges among the clergy, the Orthodox Church had a great deal of autonomy in replenishing the numbers of its dedicated clergy.

Second, and least known and analyzed of all, the number of nuns in Romanian monasteries grew much faster and was far larger than the number of monks. In addition, one is struck by how this significant disparity (a ratio of more than 2:1 nuns to monks by 1957) is reflected negatively in the number of monasteries dedicated to nuns versus monks. The number of monks stayed relatively flat over the two decades, as did the number of monasteries where they resided, so that the occupancy ratio goes from 14:1 to 16:1 between 1938 and 1957. This ratio also indicates that monks lived in relatively small monasteries and were likely assisted by the local population. By contrast, the ratio of nuns to monasteries declines from 73:1 to 53:1 during the same time period. Even with this decline, it is clear that far more nuns crowded into individual monasteries than monks.

The disparity in occupancy rate is so huge that it begs for an explanation. The size of monasteries may be an explanation, though not all or even a majority of nuns' monasteries were larger in size than those inhabited by monks. On the contrary, some of the largest and best-known monasteries in Romania are occupied by monks. Another possible explanation would be the inability of the Orthodox Church to obtain approval for building more women's monasteries to keep up with the growing number of women taking the veil.[68]

I would venture to suggest a few other important elements, all pointing toward the masculinist privileges that have defined many institutional practices of the Orthodox Church over time. Privacy has been deemed essential to monks' ability to focus on their religious practices of meditation, praying, and writing; however, the same has not been the case with nuns.[69] They are more often asked to reside in shared lodgings (several nuns in a cell) and their religious practice is more directly identified with active, public, and communal activities. Nuns are expected to prepare food, work in the fields, sew, and do artisanal work (especially textile and decorative arts, such as weaving, embroidery, and painting eggs, but rarely icon painting).[70] These may be important reasons why solitary living is not considered important for and by nuns, and why they tend to live in much larger communities than monks. Of course, the issue of personal safety, being protected from the threat of sexual assault, was also a prominent reason for nuns' placement in communal quarters.[71] However, a solid understanding of the reasons behind these gendered disparities demands more sustained ethnographic and sociological research.

The exponential growth of women taking the veil during the communist period may be viewed by some as a discontinuity. I choose to interpret it differently, and connect it to the existing religiosity among women especially in the countryside, in connection to some of the problems of adjustment of the rural population to the communist regime's economic and social policies. Obviously, a desire to serve the Church and live a religiously committed life was an important component. Yet other socioeconomic elements with pronounced gendered aspects played a role as well. Educational and economic opportunities did indeed open up a great deal for women under communism, but these opportunities were not always accessible (or perceived as such) to rural inhabitants. In order to pursue a high school or vocational school, rural children had to be removed from their environment and live in cities where they had no relatives or friends. Schools often had campus housing, but conditions were not appealing.[72] Although this is a conjectural link, I believe that the difficulties of such paths of empowerment for women in the countryside pushed them and their families toward making different decisions, especially for those who were deeply religious. With collectivization taking away one's means for subsistence in the countryside,[73] families with more than one child, and especially more than one daughter had to consider ways in which they could secure a future for their offspring. Joining a monastery was an option that suggested the lessening of financial burdens for the parents (dowries were and remain an important expensive custom in the countryside), as well as security for the young woman becoming a nun. Therefore, this phenomenon of growing numbers of nuns during communist Romania seems likely to be connected to both an ongoing religiosity among women in the countryside, as well as drastic changes brought about by the communist regime in especially the economy and education.

Conclusions

What do these numbers mean for our understanding of religion in Romania in the twentieth century? The most obvious observation to make initially is that the narrative of the institution of the Church (from underdog, to privileged state-supported religion, to censored and communist infiltrated institution) does not match the narrative of religiosity (from intense, to declining, to growing again under communism) among believers. On the contrary, one might surmise that in the modern period, when churches are not central political institutions able to closely control and regulate the lives of their members, religiosity is linked more to localized traditions and to socioeconomic conditions than to the power and visibility of the Church. Socio-economic adversity (both before 1918 and after 1945) seemed to have enhanced the desire or the need of many people to practice religious beliefs. Overall, it is clear that one cannot claim that communism destroyed or reduced religiosity among the Orthodox Christians in Romania. When speaking of "religion under communism" we need to pay greater attention to these important nuances, which suggest that continuity is the most important qualifier for describing religious practices, especially in the countryside.

A second and equally important conclusion is that gender norms and assumed identities have been crucial to how religiosity has developed among Orthodox populations in the modern period. Priests, as both representatives of the Church and embodiments of a particular ideal of masculine Christianity, have played a central gendered role in preserving certain dogmatic mandates of the institution and limits (for both women and men, yet in different ways) on performative aspects of religiosity, but women were also central to how religiosity developed during this period. Their role was far more informal, and thus it is less clearly evidenced in easy to understand traces. However, there is no doubt, based on both ethnographic evidence and also some of the statistics presented in this article, that women's greater religiosity and adherence to specific practices and rituals were essential in rendering Orthodoxy in the shape it exists today in Romania. The full story of this barely traceable force remains to be fully recovered and I hope this article has brought about questions among researchers of life under communism in Romania and among gender scholars, as this vast area of inquiry demands a multi-disciplinary long-term effort, in order to fully understand phenomena I was able to sketch out in broad and partial ways.

◊ Acknowledgments

I want to thank Kristen Ghodsee and Pamela Ballinger for their invitation to participate in the inspired and inspiring workshop "Spiritualities and Secularisms in Southeastern Europe: An Interdisciplinary Workshop" at Bowdoin College in October 2009. The experience would not have been what it was without the energy of all the participants involved—Milica Bakic-Hayden, Melissa Bokovoy, Keith Brown, Page Herrlinger, Emira Ibrahimpasic, and Mary Neuburger. Camp Balkans has remained a memorable bright moment of intellectual and personal fellowship. In addition, I am grateful to the five readers of the manuscript for their suggestions and criticisms. All remaining errors are my own.

◊ About the Author

Maria Bucur is John V. Hill Professor of East European History and director of the Russian and East European Institute at Indiana University. Her books include *Heroes and Victims: Remembering Romania's World Wars in the Twentieth Century* (Indiana University Press, 2009); *Gender and War in Twentieth Century Eastern Europe,* coedited with Nancy Wingfield (Indiana University Press, 2007); and *Eugenics and Modernization in Interwar Romania* (University of Pittsburgh Press, 2002). She is currently working on a project with Mihaela Miroiu, titled "Everyday Citizenship and Gender in Communist and Post-Communist Romania". Email: mbucur@indiana.edu

◊ Notes

1. *Coliva* is the Orthodox sweet dish cooked with boiled wheat, sugar, nuts, and spices, to embody and honor the links between the dearly departed and this world through a specifically Christian set of symbols—the wheat as resurrected life, the sweetness of the sugar as the sweetness of Christ's love. See Ofelia Văduva, *Steps towards the Sacred* (Bucharest: Editura Fundaţiei culturale române, 1999).

2. The shortest traditional version was *Bodaproste*, a literal translation of *Bogda prosit* from Russian—Thanks be to God.

3. Sf. Vineri was torn down after the 1977 earthquake, which provided a convenient reason (structural instability) to do away with quite a few such places of worship. On these demolitions, see Lidia Anania et al., *Bisericile osândite de Ceauşescu, 1977–1989* (The churches convicted by Ceauşescu, 1977–1989) (Bucharest: Editura Anastasia, 1995); Comisia Prezidenţială pentru Studiul Crimelor Comunismului (The Presidential Commission for the study of the crimes of communism), *Raport Final* (Final report) (Bucharest, 2006), 466–467.

4. Area of the church where women would be traditionally relegated to, its distance from the altar signifying the lowly status women had in the social order of Orthodox communities.

5. During the communist period women tended to retire ten years earlier than men. The retirement age for women was 50–55, and for men 60–65. This meant that in most cases, families in which both parents worked depended on either a kindergarten or grandmothers to tend to small children. Although available, daycare was often unreliable in terms of quality of staff, and parents preferred to leave small children with retired relatives (i.e., grandmothers or aunts). Therefore, it is safe to surmise that the impact of women of that generation, given their earlier retirement age, was significant in general for raising grandchildren. Since the generations I speak of are also those growing up in the interwar period (the grandmothers) and respectively in the 1970s (my generation), this further sheds light on the kind of information about religiosity that the older generation, rather than the children's parents, could pass on to their grandchildren.

6. This generalization is not based on the assumption of all people in Romania as Orthodox Christian, but rather on available statistics, which indicate that over 80 percent of Romania's population declared itself Orthodox during that period. This included most of those who had been part of the Greek Catholic Church that had been incorporated into the Orthodox Church in 1949, accounting for around 8 percent of the total population of the country. Currently, the proportion is at 87 percent. See Lucian Leuştean, *Orthodoxy and the Cold War* (London: Palgrave, 2009) and http://www.recensamant.ro (accessed 10 May 2010).

7. On the tradition of baptizing infants, see Vasile Răduca, *Ghidul creştinului ortodox de azi* (The guide for today's Orthodox Christian) (Bucharest: Humanitas, 1998), 141.

8. The Orthodox Church has its own television channel and radio station, both of them operating most hours of the day.

9. On the position of the Orthodox Church in Romania after 1990, including its presence in education, see Lavinia Stan and Lucian Turcescu, *Religion and Politics in Post-Communist Romania* (New York: Oxford University Press, 2007).

10. Ibid., 3–4.

11. For a larger perspective on the Orthodox Church under communism, see Sabrina Ramet, *Nihil Obstat: Religion, Politics, and Social Change in East-Central Europe and Russia* (Durham, NC: Duke University Press, 1998).

12. The question of how to characterize the development of religiosity among other significant denominations (Catholic, Lutheran, Unitarian, Calvinist, Jewish) falls outside the purview of this article, though it would be relevant in terms of the triangular relationship among religion, ethnicity, and political ideology/regime.

13. On secularization, see Comisia Prezidenţială, *Raport*, 25; Paul Caravia, Virgiliu Constantinescu, and Flori Stănescu, *Biserica întemniţată, România, 1944–1989* (The imprisoned church, Romania, 1944–1989) (Bucharest: INST, 1998); Cristina Păiuşan and Radu Ciuceanu, ed., *Biserica ortodoxă română sub regimul comunist, 1945–1958* (The Romanian Orthodox Church under the communist regime, 1945–1958), vol. 1 (Bucharest: INST, 2001).

14. In stating that the Orthodox Romanian Church is fundamentally misogynist I am not making any new claim. There has been an ongoing debate in the past decade over the dogmatic and consistent marginalization of women qua women from central ritualistic functions and from any discussion of opening priesthood to women. For more on this issue, see Mihaela Miroiu, "Fetzele patriarhatului" (The faces of patriarchy), *Journal for the Study of Religions and Ideologies*, no. 3 (Winter 2002): 207–226, http://www.jsri.ro/old/html%20version/index/no_3/mihaela_miroiu-articol.htm (accessed 18 January 2010); Miruna Munteanu, "Editorial feminist, frigid şi ateu" (Feminist, frigid, and atheist editorial), *Ziua*, (The day), no. 3452 (13 October 2005), http://www.ziua.net/display.php?data=2005-10-13&id=186511 (accessed 18 January 2010); Mihaela Miroiu, "Gâlceava danciachirilor cu demnitatea spirituală a femeilor" (The quarrel of the Danciachirs [reference to the author Dan Chiachir] with women's spiritual dignity), *Observator cultural* (The cultural observer), no. 35 (27 October–3 November 2005, http://www.romaniaculturala.ro/articol.php?cod=8277 (accessed 18 January 2010).

15. Christian symbols dating back to the second century A.D. have been found on what is today the territory of Romania, but even ardent nationalist scholars place the beginnings of (proto) Romanian Christianity a few centuries later, in the seventh century A.D. However, the first canonical recognition of a Metropolitanate in Wallachia and Moldavia is dated much later (1359 and 1401, respectively). See Mircea Păcurariu, *Istoria Bisericii ortodoxe române* (The history of the Romanian Orthodox Church), vol. 1 (Bucharest: Editura Institutului biblic şi de misiune al Bisericii ortodoxe române, 1992), 18.

16. See Păcurariu, *Istoria*; Elena Niculiţa-Voronca, *Datinele şi credinţele poporului român adunate şi asezate în ordine mitologică* (The traditions and beliefs of the Romanian people gathered and arranged in mythological order) (Iaşi: Polirom, [1903] 1998); Simion Mehedinţi, *Creştinismul românesc* (Romanian Christianity) (Bucharest: Fundaţia Anastasia, [1941] 1995); Simion Florea Marian, *Trilogia vieţii: Naşterea la români; Nunta la români; Inmormântarea la români* (The trilogy of life: Birth among the Romanians; weddings among the Romanians; burials among the Romanians) (Bucharest: Editura "Grai şi suflet—Cultural naţională," [1890–1902] 1995).

17. See Keith Hitchins, *Ortodoxie şi naţionalitate. Andrei Saguna şi românii din Transilvania. 1846–1873* (Orthodoxy and nationality. Andrei Saguna and the Romanians in Transylvania. 1846–1873) (Bucharest: Univers enciclopedic, 1995); Keith Hitchins, *Conştiinţă naţională şi acţiune politică la românii din Transilvania (1700–1868)* (National consciousness and political action among Romanians in Transylvania [1700–1868]) (Cluj: Dacia, 1987); Păcurariu, *Istoria*.

18. In a recent essay on the Greek-Orthodox/Uniate Church in Transylvania, Mircea Păcurariu, one of the most prominent historians of the Orthodox Church in Romania, underscores the claim of equating orthodoxy and Romanianness in his approval of Dumitru Stăniloae's article, "Rolul Ortodoxiei în formarea şi păstrarea fiinţei poporului român şi a unităţii naţionale" (The role of Orthodoxy in creating and preserving the identity of the Romanian people and national unity), *Ortodoxia* (Orthodoxy) 30, no. 4 (1979): 599: "This synthesis of Latinity and Orthodoxy, itself a miracle and a unique type of originality, has enabled the Romanian people to maintain its identity, through their Latinity that cannot be mistaken for Slavdom, and through their Orthodoxy that cannot be mistaken for the Catholicism of their western neighbors." Quoted in Mircea Păcurariu, "Pagini din istoria Bisericii româneşti. Consideraţii în legătură cu uniaţia în Transilvania" (Pages from the history of the Romanian church. Considerations regarding the Uniate [Church] in Transylvania), http://www.sfantuldaniilsihastrul

.ro/fisiere/uniatia.pdf (accessed 6 May 2010). Please note the title "Pages from the history of the Romanian church." There is in fact no such church. There are several denominations that also take on the "Romanian" adjective as part of their name, but do not claim to represent Romania or Romanianness in any official capacity (e.g., the Romanian Orthodox Church).

19. This is especially the case for the population in Transylvania and the Banat. Hitchins, *Ortodoxie;* idem, *Conştiinţă.*

20. Hitchins, *Ortodoxie.*

21. Marian, *Trilogia;* Stefania Cristescu-Golopenţia, *Gospodăria în credinţele şi riturile magice ale femilor din Drăguş (Făgăraş)* (Household activities in the beliefs and magic rituals of women from Drăguş [Făgăraş]) (Bucharest: Paideia, 2002); Stefan Dorondel, *Moartea şi apa. Ritualuri funerare, simbolism acvatic şi structura lumii de dincolo în imaginarul ţărănesc* (Death and water. Funerary rituals, aquatic symbolism and the structure of the world beyond in the peasant imaginary) (Bucharest: Paideia, 2004); Văduva, *Steps.*

22. Anna Careveli-Chaves, "Bridge between Worlds: The Greek Women's Lament as Communicative Event," *Journal of American Folklore* 93, no. 368 (1980): 129–157; Loring Danforth, *The Death Rituals of Rural Greece* (Princeton: Princeton University Press, 1982); Bette Denich, "Sex and Power in the Balkans," in *Woman, Culture, and Society,* ed. M. Rosaldo and L. Lamphere (Stanford: Stanford University Press, 1974), 243–262.

23. Văduva, *Steps;* Gail Kligman, *Wedding of the Dead. Ritual, Poetics, and Popular Culture in Transylvania* (Berkeley: University of California Press, 1988); Cristescu-Golopenţia, *Gospodăria.*

24. Marian, *Trilogia;* Văduva, *Steps.*

25. Marian, *Inmormântarea,* 79, 201.

26. For more on the issue of the gendered aspects of the cult of the dead in Romania, see Maria Bucur, *Heroes and Victims. Remembering Romanian's World Wars in the Twentieth Century* (Bloomington: Indiana University Press, 2009), ch. 1.

27. Nae Ionescu, *Roza vânturilor. Biserica ţăranilor* (The wind vane. The peasants' church) (Bucharest: Ed. Roza vânturilor, 1990). Identifying Ionescu as a "scholar" is not an endorsement by this author, but rather an acknowledgement of the reputation the fascist ideologue enjoyed during the interwar period, as professor of philosophy at Bucharest University, as well as the revival in his reputation among some post-communist intellectual elites.

28. Costion Nicolescu, *Elemente de teologie ţărănească* (Elements of peasant theology) (Bucharest: Editura vremea XXI, 2005), 10.

29. Dumitru Stăniloae, *Reflecţii despre spiritualitatea poporului român* (Reflections on the spirituality of the Romanian people) (Bucharest: Editura Elion, 2001).

30. See Leuştean, *Orthodoxy,* 17; Stan and Turcescu, *Religion,* ch. 2.

31. Lucian Predescu, *Enciclopedia României. Cugetarea* (The encyclopedia of Romania. The thought) (Bucharest: Editura Saeculum, 1999), 511; Marin Mihalache, *Cuza Vodă* (Vojvode Cuza) (Bucharest: Editura Tineretului, 1967). The process of transfer of these lands from the church to the state is generally referred to as "secularization" in Romanian. I prefer the term "nationalization," which more clearly identifies who had control over these lands after their confiscation from the church: the state.

32. This was reflected especially in the political alliance between the Conservative Party as a landholders' party and the Orthodox Church hierarchy.

33. On the subsequent woes of the rural population, see Phillip Eidelberg, *The Great Peasant Revolt of 1907* (Leiden: Brill, 1974).

34. Marian, *Trilogia.*

35. Leuştean, *Orthodoxy;* Păcurariu, *Istoria.*

36. Institutul Social Român, *Constituţia din 1923 în dezbaterile contemporanilor* (The 1923 Constitution in contemporary debates) (Bucharest: Humanitas, 1990).

37. Bucur, *Heroes.*

38. Ibid., 60.

39. Ibid., especially ch. 2.

40. In common Orthodox parlance, the Ascension is more often called "Ispas" or the "Easter of the Dead," or "Day of the Dead," not to be confused with the Catholic 1 November "Day of the Dead." See Simion Florea Marian, *Sărbătorile la români* (Holidays among the Romanians) (Bucharest: Editura Fundaţiei culturale române, 1994).

41. Bucur, *Heroes,* ch. 2.

42. Păcurariu, *Istoria;* Mircea Păcurariu, *Politica statului ungar faţă de Biserica românească din Transilvania în perioada dualismului, 1867–1918* (The politics of the Hungarian state towards the Romanian church in Transylvania during the period of dualism, 1867–1918) (Sibiu, 1986).

43. Dimitrie Gusti, ed., *Enciclopedia României* (The encyclopedia of Romania), vol. 2 (Bucharest: Fundaţia Regala Carol II, 1938–1940).

44. Leuştean, *Orthodoxy,* 47–48.

45. Adrian Majuru, *Bucureştii mahalalelor, sau periferia ca mod de existenţă* (Bucharest of the Mahalas, or perfidy as a mode of existence) (Bucharest: Compania, 2003); Ioana Pârvulescu, *Intoarcere în Bucureştiul interbelic* (Return to interwar Bucharest) (Bucharest: Humanitas, 2003).

46. *Universul* (The universe), *Curentul* (The current), *Porunca Vremii* (The command of the times), and other major newspapers often made mention of official commemorations at religious sites (churches, synagogues, and cemeteries), but did not make mention of cultural and other religious events at specific churches on a weekly basis.

47. A good source on the war commemoration pilgrimages is Societatea Ortodoxă Naţională a Femeilor Române (The National Orthodox Society of Romanian Women), whose entire archive, located in the National Central Archives in Bucharest, is replete with references to such events and the involvement of this organization in maintaining Orthodox traditions all over the country.

48. Bucur, *Heroes,* especially ch. 1.

49. On the multitude of religious holidays in the Orthodox Calendar, see http://www .goarch.org/en/chapel/calendar.asp (accessed 21 January 2010); on Romanian practices, see Irina Nicolau, *Ghidul sărbătorilor româneşti* (The guide of Romanian holidays) (Bucharest: Humanitas, 1998).

50. Eve Levin's classic book *Sex and Society in the World of the Orthodox Slavs, 900–1700* (Ithaca: Cornell University Press, 1989) is a noteworthy exception.

51. See, for instance, Caravia, Constantinescu, and Stănescu, *Biserica;* Păiuşan and Ciuceanu, ed., *Biserica ortodoxă;* Zosim Oancea, *Datoria de a mărturisi. Inchisorile unui preot ortodox* (The duty to confess. The prisons of an Orthodox priest) (Bucharest, Harisma, 1995); Nicolae Videnie, "Atitudinea anticomunistă, calvarul şi martiriul preotilor ortodocşi relfectate în presa exilului românesc (1945–1989)" (The anticommunist attitude, the calvary and martyrdom of Orthodox priests as reflected in the press of the Romanian exile [1945–1989]), in *Rezistenţa anticomunistă. Cercetare ştiinţifică şi valorificare muzeală* (Anticommunist resistance. Scientific research and museum valuation), ed. Cosmin Budeancă, Florentin Olteanu, and Iulia Pop, vol. 2 (Cluj: Argonaut, 2006), 32–51.

52. See Cristian Vasile, *Istoria Bisericii Greco-Catolice sub regimul comunist, 1945–1989* (The history of the Greek-Catholic church under the communist regime, 1945–1989) (Iaşi: Polirom, 2003).

53. Comisia Prezidenţială, *Raport,* 447–472.

54. Leuştean, *Orthodoxy.*

55. Comisia Prezidenţială, *Raport,* 447–472.

56. Kligman, *Wedding of the Dead,* 268.

57. Katherine Verdery, *The Political Lives of Dead Bodies. Reburial and Postsocialist Change* (New York: Columbia University Press, 1999).

58. Xenia Costa-Foru-Andreescu, *Cercetarea monografică a familiei: contribuție metodologică* (The monographic research of the family: Methodological contribution) (Bucharest: Institutul Social Român, 1945); T. Graur, "Predici rituale în structura și funcția ceremonialului de nuntă tradițională" (Ritualistic liturgies and the function of the traditional wedding ceremony), *Anuarul muzeului etnografic al Transivlaniei* (Yearbook of the Ethnographic Museum of Transylvania) (1976): 283–294; Mihai Pop, *Obiceiuri tradiționale românești* (Traditional peasant customs) (Bucharest: Consiliul Culturii și Educației Socialiste, 1976).

59. See, for instance, Aurora Liiceanu, *Nici alb, nici negru. Radiografia unui sat românesc, 1948–1998* (Neither white, nor black. Radiography of a Romanian village, 1948–1998) (Bucharest: Nemira, 2000); Zoltàn Rostas and Theodora-Eliza Văcărescu, eds., *Cealaltă jumătate a istoriei. Femei povestind* (The other half of history. Women narrating) (Bucharest: Curtea Veche, 2008).

60. On the Museum of the Romanian Peasant's central preoccupation with Orthodoxy, see http://www.muzeultaranuluiroman.ro/index.php?page=religioase (accessed 21 January 2010).

61. Monica Heintz, "Romanian Orthodoxy between the Urban and the Rural," Max Planck Institute for Social Anthropology Working Paper 67, Halle, 2004.

62. Fieldwork in Sighet, Ineu, and Trezenea, April–May 2000.

63. Vasile, *Istoria Bisericii;* I heard similar statements from Uniate believers during fieldwork in Cluj, March–April, 2000.

64. Interview with Osiceanu Elena, Hunedoara, 4 August 2009.

65. Interview with Viorica Vegh, Sîncrai, 6 August 2009.

66. Interview with Livia Laura Giurca, Hunedoara, 8 August 2009.

67. http://www.religioustolernace.org/rel_ratefor.htm (accessed 10 May 2010).

68. I can only speculate on this matter, as the Orthodox Church archives are not open to non-religious scholars like me, especially given my feminist agenda. The important question in examining this possible explanation is the extent to which the Church cared about this disparity in ratio and tried to do anything about it, and to what extent it was frustrated by the communist regime in these efforts. It would be easy for Church authorities to claim to have tried and not succeeded, given the reputation of the communist regime, but it is hard to prove that this was not the case without access to the Church's archives.

69. For descriptions and images of monastic life in twentieth-century Romania, see Dragoș Lumpan, *Chipuri de viață monahală* (Faces of monastic life) (Editura LiterNet, 2006), http://editura.liternet.ro/carte/218/Dragos-Lumpan/Chipuri-de-viata-monahala.html (accessed 21 January 2010); Ignatie Monahul, *Viața monahală în texte alese* (Monastic life in selected texts) (Bucharest: Lucman, 2006); George Enache and Adrian Nicolae Petcu, *Monahismul ortodox și puterea comunistă în România anilor '50* (Orthodox monasticism and communist power in Romania during the 1950s) (Bucharest: Partener, 2009). Much of the emphasis of the published materials is on monks' lives; it is difficult to find publications about nuns in wide circulation, although on occasion one can pick up self-published writings at women's monasteries.

70. Leuștean refers directly to these differential expectations of the Church along gender lines, by quoting a prelate who identified nuns as especially useful to the Church coffers through their artisanal work, which sold very well both locally and more broadly throughout the country and abroad. What is not clear at all from this book is whether these nuns see any direct returns on this type of arduous work, on a par with their greater contribution to the gross domestic product than that of priests and monks. See Leuștean, *Orthodoxy*, 135.

71. I thank one of the anonymous readers for pointing out this obvious oversight on my part.

72. Interview with E., Hunedoara, 8 August 2009.

73. Focus group Sîncrai, 5 August 2009.

Gendering Grief

Lamenting and Photographing the Dead
in Serbia, 1914–1941

Melissa Bokovoy

ABSTRACT

This article is part of a larger research project on the political, cultural, and social implications of interwar Yugoslavia's remembrance and mourning of its war dead. Eschewing a focus on state-centered commemorative practices, this article focuses on two types of sources, laments of Serbian women and photographs by Serbian military photographers, as entry points into understanding the private, cultural, and religious arenas of Serbian wartime and interwar remembrances. Drawing on research examining the political uses of lament and grief, the article considers the role Serbian women played in controlling and directing the "passion of grief and anger" within their communities as they remembered the dead. The photographic evidence reveals that traditional death rituals and laments were performed and that these rituals were significant socio-political spaces where women, families, and communities of soldiers advanced claims for recognition of their wartime experiences and memories. However, the photographs themselves are sites of memory and this article examines how military photographers, acting on behalf of the state, sought to control the representation of grief and by doing so politicized and secularized the way grief was expressed. Placing these sources side by side illustrates the intermingling of forms of mourning and remembrance that existed not only in the Balkans, but also in many other communities throughout Europe, especially among its rural inhabitants.

KEYWORDS: gender, laments, mourning, photography, Serbia, World War I

Somewhere in Serbia, six months after World War I began, Serbian photojournalist Rista Marjanović aimed his camera at two women, whose heads were bowed and covered and who sat beside a freshly dug grave in a cemetery. Marjanović framed his shot to capture the cemetery, its crosses, and the women kneeling beside the grave to inform the viewer that these women were engaged in the act of mourning (Figure 1). In

aspasia Volume 5, 2011: 46–69
doi:10.3167/asp.2011.050105

Figure 1. Untitled. Rista Marjanović, *Ratni Album,* 1912–1915 (War Album, 1912–1915). Internet edition. http://www.rastko.rs/fotografija/rmarjanovic/uvod_l.html. All photographs courtesy of the Vojni Muzej (Military Museum) Belgrade, Serbia.

1917, this same photograph appeared in the Serbian Princess Alexis Karadjordjević's memoir, *For the Better Hour,* which was published in Great Britain and written to expose the British to the wartime experiences of the Serbian population and its army.[1] Writing for a foreign audience, Karadjordjević briefly described a scene of mourning and a ritual for the dead, familiar to most Serbs but not at all familiar to her English-speaking audience. The book's publisher had chosen to illustrate Karadjordjević's description with Marjanović's photograph. A photograph of a similar scene appeared ten years later in a 448-page commemorative war album, *Ratni Album,* compiled and edited by Lieutenant-Colonel Andra Popović, the head of the Serbian Army's wartime photographic division. It was one photograph among thirty-seven images laid out in a seven-page spread that depicted the dead of the battle field and the wartime death rituals and burial practices of Serbs.[2]

Marjanović's photograph, Karadjordjević's description of women lamenting, and Popović's layouts in his monumental war album when placed side by side tell a story of death and remembrance among Serbs during wartime Serbia and afterward in the interwar Kingdom of Yugoslavia. The photographs and passage tapped into the cultural and religious knowledge of the Serbian viewer and the images related rituals familiar to them.[3] Over the centuries, Orthodox Slavs had developed a rich burial tradition, combining both pagan and Christian elements. These rituals and traditions became one of the many arenas for articulating war memories during and after World War I in the newly formed Kingdom of Serbs, Croats, and Slovenes.

The following discussion of photographs and descriptions of Serbs in mourning and then later, a discussion of women lamenting the war dead in interwar Serbia, highlight the struggle by different social actors and groups to articulate war memories and to modify or challenge existing and widely disseminated narratives. As noted by the authors of a recent collection of essays on memory and war, this struggle happens within "arenas of articulation" and "refers to those socio-political spaces within which social actors advance claims for the recognition of their specific war memories."[4] This article considers how traditional death rituals and laments performed during the war and afterward were significant socio-political spaces where women, families, and communities of soldiers advanced claims "for recognition" of their war memories and memories of the dead.

Studies of war commemoration often focus on public memorials and displays of mourning, which acknowledge a community's loss. Most memorials were built and supported by the state, veteran's organizations, or the military, and narrated a war-time story that sanctioned official versions of the past, utilizing secular discourses that spoke of duty and sacrifice to the nation, the homeland, or the state. Yet these studies tell us nothing about how people mourned or remembered away from these commemorative structures and spaces. Ignored by many scholars are the religious ceremonies and rituals that accompany public dedications of war memorials as well as the death rituals and traditions that occur beyond the purview of official and elite chroniclers of the war dead.

Central to the religious practices of innumerable cultures are rites about death. By studying these rituals in a specific historical context of burying and remembering World War I dead in Serbia, we can examine how narratives about the war drew on a wide array of discourses. Not only do the narratives that were created draw on secular discourses concerning national identity and citizenship, but the rituals being practiced also incorporated religious discourses familiar to orthodox Christians.[5] As one recent sociologist of religion observed, "Much of the core of the traditional Christian funeral rites is somehow related to story, the story of the deceased individual, the stories of the individual mourners, particularly family and above all, these stories are woven into and emerge from the story of the life, death, and resurrection of Christ."[6] Narrative, as a central expression of many of the world's religions, including Christianity, is often the vehicle for the communication of fundamental truths to participants and allows for interpretation and the inclusion of individuals and communities into the story. The inclusive nature of the narrative conveyed by the ritual gives them a timeless nature and invites others, through imagination, association, and identification, to make the stories their own.[7] Laments, as part of death rituals, often detail the life of the deceased but "are manipulated by the lamenter in order to benefit the living, that is, the female participants, and more indirectly, the community as a whole."[8]

Important for this theme on "Gendering the History of Spiritualities and Secularisms in Southeastern Europe" are the burial and mourning rituals practiced by Serbs in the first few decades of the twentieth century as a means to narrate and remember the dead of the Balkan Wars and World War I, not on parade fields or in public squares, but at the graveside and at the communal and familial feasts for the dead. In such rituals, women's laments express the pain of loss and their "emotional tone, dialectic exchange, and musical qualities act as a stimuli to memory" and "preserve the dead not only in individual but in communal memory."[9] By focusing on the spiritual and religious beliefs of one people, this article reminds scholars of gender, memory, and war of the centrality of death rituals and that religious discourses are significant socio-political arenas where collective remembrance takes place. Within these rituals, women often are able to control the articulation of grief, principally through lamentation.[10] By controlling one of the most "politically threatening of social emotions" women, especially during wartime, could instruct the community how to behave and act. The act of lamenting, as anthropologists have noted, is a most threatening social emotion to the state, religious authorities, and patriarchal structures because "voices raised in lament, rather than praise, acted as a sobering (and often subversive) critique of a polity and a culture which saw death as noble and heroic."[11]

When Marjanović,[12] as an agent of the Serbian military and state, turned his camera to the scene of women lamenting he erased the women's (social) memory of the dead, replacing the memory the women fixed and shaped through their words and performance with one framed and manipulated by the photographer. The photograph's static nature tamed the force of women's mourning and grief. Drawing on recent observations by Gail Holst-Warhaft and Catherine Moriarty, this article places these two seemingly disparate sources—photographs and laments—in the same historical space and investigates the process by which photographing laments, in other words, representations of lamentations and not the actual performances of the laments, began to assume the responsibility of remembering the dead and challenged women's traditional role as the ones who fixed and shaped the memory of the deceased.[13]

A Story of Mourning: Photographs

Marjanović most likely took his photograph during the period between December 1914 and October 1915 and somewhere close to Serbia's battle lines. Nothing is known about the circumstances of the death mourned; the women could have been burying a newborn infant, a child, a parent, a husband, or a near or distant relative, who had died of natural causes, of war related injuries, or a disease. We know nothing beyond what the image conveys and in the hands of an experienced photographer like Marjanović the indeterminate nature of the photograph allowed the viewer to inscribe his or her own meaning on the scene. By photographing the act of mourning, Marjanović captured the solemnity of the moment in order to convince his viewers that they were looking directly, without mediation, at a ritual for the dead.

Karadjordjević explained the circumstances of a ritual she observed:

In Serbia, Saturday is the day of the week usually set apart for visiting the cemeteries. From dawn to dark the roads are filled with women and girls carrying food and wine, which they set out, interspersed with lighted candles, on the family graves. Sitting about in groups, these women and girls work themselves up into hysterical wailing, which continues all day and fills the air for great distances, with the traditional lamentation.[14]

She then noted "the weird monotonous strains of their sad music, as it was wafted across the stream to me, hardly seemed to lend themselves to dancing, nor to suggest any form of gaiety; rather, they are the appropriate accompaniment to the national epic poems which immortalize the sufferings of this free spirited people when under the domination of the Turks."[15] Karadjordjević inscribed on the scene a heroic tale of sacrifice and suffering of the Serbian people who had been fighting to liberate themselves from foreign domination. Erased were the women's reasons for mourning.

Popović laid out, framed, reproduced, and captioned photographs of dead bodies and death rituals in two separate spreads in his commemorative war album. The first displayed twenty-seven photographs over five pages and depicted the wreckage, carnage, and detritus of the battlefield—abandoned wagons, carcasses of horses, corpses strewn about the battlefield or abandoned along roads and rivers and in fields, and the bodies of Serbian civilians executed by the Austro-Hungarians. Popović followed these images with a full-page photograph of twelve bodies collected and readied for burial. On the last page of this layout, he grouped together ten photographs on a page titled "Funerals of the Fallen" (Figure 2).[16] These photographs depicted bodies laid out

Figure 2. "Funerals of the Fallen." From Andra Popović, *Ratni Album, 1914–1918* (War Album, 1914–1918) (Belgrade: Uredništvo Ratnog albuma, Priv. izd., 1926).

to be buried, funeral processions, and grave side burials. The last photograph in the set captured a group of women, sitting among freshly dug graves marked by crosses, with baskets open beside them. The photograph, "Peasant Women Bringing Offerings to the Dead," acted as an end piece to this narrative arc (Figure 3). With the last two pages of his layout, Popović appeared to be rescuing the bodies from the profanity of detritus and restoring to them the dignity and decorum due to the dead by showing the viewer that soldiers had had a proper burial and had been respectfully mourned.[17]

Later in the album, Popović returned to images of the dead and displayed photographs in a two-page spread showing cemeteries near the front and women, children, and men "at the graves of the dearly beloved" (Figure 4). Particularly striking to Popović is the solicitude that Serbian soldiers displayed over graves. In the text at the bottom of the page titled, "Cemeteries near the Front," he noted, "The cemeteries for the fallen soldiers were kept in very good order by their comrades on festivals and whenever possible the priest held a service there." (Figure 5).[18] The photographs showed cemeteries along the front, neatly fenced, and the entrances clearly marked; some of the cemeteries even had grave stones with inscriptions already in place. Ten years earlier, Princess Karadjordjević had remarked on how the proprieties of death during the typhoid outbreak were being scrupulously observed. The bodies were given cultural value by placement in a cemetery and through the lamentation of women.

The pretty little cemetery with its daily increasing number of graves, some marked only by a plain wooden cross, some by crosses of stone, and others

Figure 3. "Peasant women bringing offerings to the dead." From Popović, *Ratni Album.*

Figure 4. "At the graves of the dearly beloved." From Popović, *Ratni Album.*

again by tattered banners, had become almost the center of the village life. One heard constantly the wail of *kuku mene* (woe is me), a characteristic expression of grief. There is a pretty legend which connects the *kuku* with the death of Lazar. After the battle of Kosovo in 1389, when Lazar was captured and beheaded by the Turks, the souls of his daughters were permitted by the pitying gods to enter the bodies of cuckoos, whose melancholy notes has ever since embodied the soul of grief.[19]

Karadjordjević's description of women lamenting and Marjanović's photograph are both generic and specific representations of mourning. Through their representation of mourning, they appropriated these women's grief by presenting them as timeless and spaceless symbols of suffering and mourning, but also situated them within the context of Serbian history and myth.

Marjanović intentionally framed his shot of the mourning women to tell both a local and a universal story of life and death.[20] As an internationally recognized photojournalist, Marjanović knew that the photographs he shot had the possibility of being published abroad and that women mourning at a grave would have a widespread appeal. By allowing this photograph to be published in a memoir narrating the triumphs, setbacks, and tragedies of Serbia's war with Austro-Hungary and Bulgaria, the photograph narrated a story of life and death under wartime conditions, conditions with

Figure 5. "Cemeteries near the Front." From Popović, *Ratni Album.*

which many in Europe could identify. The publication of such photographs was based on the "European ideal of moral universalism which posits that compassion is a natural emotional response to the universal suffering of the Other and that publicizing misery is sufficient to stimulate the consciences of the public and motivate citizens to acts of benevolence or political action."[21] For some, such as South Slavs who experienced the war, the photograph became a substitute for memory, its effect contingent on the associations the observer brought to it. However, for the historian the photograph demonstrates the multiplicity of social actors engaged in remembering the dead—the women, the photographer, and the viewer—each advancing a claim for the recognition of their memories at different points in time and space.

In the commemorative war album, Popović chose photographs that represented the different stages of the death ritual—bodies laid out; open coffins draped with white cloths and laurels; a processional for the dead; a priest standing over an open coffin, possibly outside a church; soldiers gathered at a grave side burial; and a group of women mourning (Figure 6). To the viewer, these photographs could stand in for the actual event or person.[22] For many in interwar Serbia, these photographs possibly offered a glimpse or imagined sighting of a loved one buried at the front. The photographed scenes, which were displayed and captioned, narrated the sequence of Orthodox Christian stages and rituals associated with the dead. The first stage is home (or wherever the person has died or is laid out). The priest arrived in vestments, bear-

Figure 6. Cover page. From Popović, *Ratni Album.*

ing incense, and conducted a service that contained many ancient Christian liturgical texts and chants and focused on commending the dead to God.[23] Participation in the psalms, hymns, intercessory prayers, and scripture readings were shared among the gathered, usually the family, the priest, and community, representing the totality of the church. When ready, the body was taken from the home (or where it was first laid out) to the church in a procession. In the church, the open coffin was placed on a raised platform in the center front of the church.

The second portion of the ritual then began. The priest read a number of liturgical and biblical texts and prayers were said. Included was most likely the prayer, "God of all spirits and flesh."[24] Usually this prayer was said twice, during the laying out of the body and then again during the burial service. The service concluded with the "last kiss" or "farewell"; all present were invited to give a farewell kiss to the deceased. Procession to the place of burial followed. Popović displayed a photograph of a processional in the village of Klenje (Figure 7). There was someone in front carrying the cross and a banner and it is unclear if someone was carrying *koljivo* (dish made of boiled wheat). The priest was followed by those who were singing the *zaupokojene* (passing away) songs and bearing the coffin; the family and other participants trailed behind.[25] At the grave site, other rites were performed. From Popović's photographs, there appeared to be a conjoining or an abridgment of the rituals given the circumstances and locations of death. This photograph and the photograph of the bodies laid out, priest with bible in hand, and soldiers standing with bowed heads demonstrated to the observer that the death rituals were being followed in the most significant ways.

Laying out of the body prepared the person to part from this world. Following death, it was necessary to prepare the corpse. Particular care was paid to the body because it was viewed as the temple of the soul. On judgment day, the physical body of the deceased would be resurrected, and the dead would rise from their graves. Consequently, the coffin and the graveyard were a "home" for the dead, and the deceased's relatives had a duty to maintain the graveyard. The corpse was to be placed in the coffin in a dignified state; clothing was very carefully chosen because it was believed that what the person was wearing and what they looked like when they were buried was the way they would appear in the other world. Photographs of the dead in Popović's album showed bodies prepared for burial, fully clothed, footwear on, and covered with shrouds.

The Orthodox Church required an open casket with the corpse present during the funeral service. The coffin was only closed when the deceased was transported to the graveyard. The bereaved often placed items in the grave, such as money, personal belongings, food, and sacred objects such as small icons of saints, church candles, and prayer books. Before the coffin was placed in the ground, the soil at the bottom of the grave was sprinkled with holy water, and the grave was filled with the smoke of burning incense. The burial served as a ritual farewell to the relatives and friends of

Figure 7. "The Village of Klenje." From Popović, *Ratni Album.*

the deceased and was seen by them as a journey into the afterlife. The final part of the ritual occurred back at the house, where family and community members took part in the popular religious practice of the funeral meal. An extended and obvious period of mourning existed, which included the feasting of the dead seven days, forty days, a half year, and a year after death.

Much care was taken, when possible, to find an appropriate final resting place, even under wartime conditions. Preferable sites were those next to water. Orthodox Slavs thought that running water would protect those who had come into contact with death. Cemeteries and graves could be found next to some sort of stream or other water because spirits and demons could not go through water and hurt the living. Returning from the cemetery, one might wash their hands with water and before entering the deceased's house, everyone washed and cleaned with water. Those who had prepared the bodies for burial or even those who were in the same room or participated in the burial were considered unclean and had to cleanse themselves. In Serbian folk religion the means for such a cleansing were not only water but also wind, air, the rising sun, and fire.[26]

Several photographs taken high atop the Moglena Mountains during the late summer of 1918 offensive of the Serbs and their allies against the Bulgarians demonstrate the adherence to such rituals, even under the most difficult circumstances of war. An unnamed photographer captured on film a member of the Drina artillery division, Živadin Marković, sitting next to a spring, water cascading down the rocks, and a spigot designed to funnel the water below. Over the spigot was an etching of man on a cross with an angel looking down on him and read: "To the fallen soldiers who stood on the threshold of endearing themselves to the Serbian homeland."[27] Another photographer, positioned behind the line at Miletina Kosa, photographed a fountain built by a unit of the Fourth Regiment of the Timok Division. The photograph framed the fountain's plaque that read: "The years and centuries will pass by you, fountain, no one will come and no one will appreciate the fallen comrades, but this rippling water, the howling wolves, the thunder, this blooming grove will sing to the glory of the Serbian people"[28] (Figure 8). Unknown to the viewer is where the bodies of these "fallen soldiers" were buried or what had been the circumstances of their burial. The photographer, the photographed subject, and those who displayed the photographs sought to suggest associations that were tied to traditional death rituals.

By choosing to build fountains, use water, and invoke nature to mourn the dead, the soldiers' commemorative art borrowed death motifs from Serbian death and burial rituals. The inscriptions on the fountains demonstrated the soldiers' awareness that they were fighting for the homeland and the "glory of the Serbia people," yet the physical sites conveyed a meaning that was neither heroic nor mythical. The running water conjured up older motifs of bereavement and mourning. That is not to say that older motifs did not take on new meanings and forms, but it does demonstrate that the political meaning of their death did not erase the individual or familial significance.

The rituals photographed and described obscured a much more disturbing scene associated with death in wartime Serbia: the abandonment of bodies during battle and retreat, the hasty disposal of over 150,000 corpses during the 1915 typhus outbreak,[29] and then again an unknown number of corpses of soldiers and civilians who had sur-

Figure 8. "Vojnik Drinskog Haubičkog diviziona Živadin Marković, pokraj česme koju je sagra-dio na Jelaku 1917" (Soldier of the Drina Howitzer Division Živadin Marković beside a fountain which was built at Jelak in 1917). From Radovan Blagojević, ed., *Spomenici i groblja ratova Srbije 1912–1918* (Monuments and cemeteries from Serbia's wars, 1912–1918) (Belgrade: Beogradski izdavačko-grafički zavod, 1976).

vived the retreat through the frozen and snow covered mountains of Albania in late 1915 and succumbed to disease and starvation in the weeks following their evacuation to the Greek islands of Corfu and Vido. Beginning in 1916, the waters around Vido, a small Ionian isle at the mouth of Corfu's bay where the sick had been quarantined, were christened Plava Grobnica (Blue Tomb). Here French and British sailors uncer-emoniously piled corpses onto the decks of the very boats that had just evacuated the retreating Serbs, motored off the coast of Vido, and threw corpses, one by one, into the sea (Figure 9). A French nurse recalled the scene, "Corpses, piled up like planks of wood one on top of the other ... four in a row, sometimes six. ... An arm, a leg, or a convulsed face stuck here and there."[30]

Milutin Bojić, a soldier and survivor of the retreat who eventually succumbed to typhus in late 1917, attempted to recover the humanity of these bodies and give them

Figure 9. "The dead are thrown into the sea." From Popović, *Ratni Album.*

"sacred reverence and care."[31] In the first stanza of his poem, "Plava Grobnica" (Blue tomb), Bojić portrayed these waters as a sacred and consecrated burial ground:

> There on the bottom [of this holy water], where shells slumber
> And seaweed gently falls on the dead,
> Lie the graves of the brave, brother lying next to brother
> Prometheuses of hope, apostles of tragedy
> Don't you feel how the sea gently swells
> So it doesn't disturb the rest of the fallen company?[32]

Countering the image of corpses piled high that had erased the meaning of individual life, Bojić's requiem honored his comrades and restored their dignity amid the chaos and necessity of disposal.

The chaos of mass death and the expediency of burial meant that bodies of men, women, boys, and girls could not and would not be identified and brought back home. The absence of bodies meant that Serbian families relied on traditional religious death rituals to mourn the missing. In the Orthodox tradition, symbolic burial was required when the body was unavailable. In these cases, one publicly lamented about the person's clothing or other things that belonged to him or her. The clothing would be laid on the earth, on a table, or bed, so that it would appear as if a person was present. Then the clothes were buried with the customary ceremony. It was believed that without the funeral rite the spirit could not pass into the other world. A collection of laments compiled by the Serbian ethnographer Novica Šaulić described the mother of a soldier lost during World War I who mourned him in absentia by laying out his clothes and asking the priest to perform the death ritual.[33] In other cases, a family might erect a tombstone above an empty grave. In the village of Slatina in the region of Bora, the family of Jovan Simonović, a soldier who died in 1916, buried his clothing and followed the funerary rites. His sister erected a tombstone above the empty grave and carried out the feast days for the dead as well as *zadušnica* (memorial).[34] In some villages where several families were unable to recover the bodies of loved ones, tombstones would be erected in the same place, side by side. In western Serbia these types of monuments were called *krajputaši* (roadside memorial) because most often they were placed next to roads or paths. In the village of Samaila near Kraljevo in central Serbia, the peasants had been erecting *krajputaši* since the nineteenth century but the largest number of them commemorated World War I casualties.[35]

A Story of Mourning: Laments

The responsibility of remembering the dead was commonly assigned to women as performers of laments. Marjanović's photograph of mourning women and Popović's decision to place photographs of women lamenting in his war album displayed this part of the mourning ritual and highlighted laments as "an indispensable part of the death rituals."[36] Women mourned men everywhere. Mothers mourned sons, sisters—brothers, and wives—husbands. They appealed to the deceased, asking how they would live without him/her. It was considered a friendly obligation that each woman who came to mourn lamented over the body, during which time the lamenter often asked the recently deceased to carry a message to their dead loved ones.[37] Women carried gifts in a small basket: candles, a glass to drink, and flowers. The photographer whose image was featured in Popović's album cropped the image in order to include baskets sitting alongside the mourning women at the grave (Figure 3). When women arrived at a gravesite, they would take out the provisions and offer them to those present, and then they would place the flowers on the grave and light candles. In one of Popović's photographs, those who laid out the body of an officer ringed his body with a laurel of flowers. The flowers might possibly have been basil, which was very important as a medicinal flower and was prescribed for its mystical power. Basil followed people from birth to the grave. The deceased was sprinkled with it and basil was planted on the grave.[38] Integral to these rituals—which included the laying out of the dead, the preparation and distribution of ritual foods, the lighting of candles, and the placing of flowers on the graves—was lamentation.

Before the twentieth century, Serbian men and women participated equally in the creation, preservation, and transmission of the collective memories that recalled the events, heroes, martyrs, and notables of the Battle of Kosovo (1389).[39] Throughout the Balkan Peninsula, including Serbia, women were the principal transmitters of the song tradition. Because women's work often occupied their hands with weaving, knitting, carrying water and firewood, and cooking tasks, song in its vocally presented medium rather than instrumental music has been their main musical outlet.[40] Women's songs spoke of their love of homeland, family, home, children, and loss. Men dominated the singing when the subject turned to narrating war and heroism; these songs spoke of men's heroism, sacrifice, loyalty, and courage. However, in their songs men did not express the grief and sorrow of death that is often featured in women's singing.[41] Šaulić wrote, "women's emotions are deeply felt, created from sorrow, and borne of tears." He observed that Serbian women in the 1920s were carrying on the tradition bequeathed to them from earlier generations and learned through the recitation of two poems from the Kosovo epic cycle: "The Kosovo Maiden" and "The Death of the Mother of the Jugovići."[42]

A mother's grief was presented in a series of powerful images in the song, "The Death of the Mother of the Jugovići." Unable to wait for the news about the battle, the Mother of the Jugovići asked God to reveal to her the outcome of the battle. God granted her swan's wings, which enabled her to fly to Kosovo, and a falcon's eyes, which allowed her to see the battlefield. Upon seeing her husband and her nine sons dead, she did not cry out, but remained stalwart. The Mother of the Jugovići returned home with their horses, dogs, and falcons and was greeted by the other soldiers' grieving widows and orphans. Holding back her tears, she detailed the battle. However, when two ravens dropped the severed arm of her youngest son, she no longer repressed her grief. Holding the severed hand and whispering to it, she began a lament:

> My hand, O green apple,
> Where did you grow, where were you ripped off?
> You grow upon this mother's lap
> You were torn away on the Kosovo plain.[43]

At the end of the song, the mother's heart broke with pain and death came as a relief.

In the poem "Kosovo Maiden," a young woman looks for her betrothed among the wounded and the dead on a bright Sunday morning following the Battle of Kosovo. In her hands she carries bread and two gold jugs—one filled with cold water, the other with red wine. As she searches for her fiancée and his two blood brothers, she turns bodies over in order to identify her loved ones. In the midst of the fallen, she comes upon one of the poem's heroes, Pavle Orlović,[44] who is in his death throes. He asks her, "Maid of Kosovo, my dearest sister, what misfortune leads you to this plain, to turn the warriors over in their blood? Whom can you be looking for out here? Have you lost a brother or a nephew? Have you lost perhaps an aging father?"[45] She replies by telling him she is looking for her betrothed and his blood brothers, all of whom she had promised to remember and as she tells the warrior, "And so I search upon this field of slaughter." Orlović responds thus,

> O my dearest sister, Maid of Kosovo!
> Do you see, dear soul, those battle-lances
> Where they're piled the highest over there?

That is where the blood of heroes flowed
In pools higher than the flanks of horses,
Higher even than the horses' saddles—
right up to the riders' silken waistbands.
Those you came to find have fallen there;
Go back, maiden, to your white-walled dwelling.
Do not stain your skirt and sleeves with blood."
When she has heard the wounded hero's words
She weeps, and tears flow down her pale face;
She leaves the plain of Kosovo and walks
To her white village wailing, crying out:
"O pity, pity! I am cursed so utterly
That if I touched a greenly leafing tree
it would dry and wither, blighted and defiled."[46]

The maiden's pain and her wailing, sobbing, and grief as well as the grief of the Mother of the Jugovići are fully on display in the last lines of both poems. Both of these poems direct the listener to the women's responsibility of mourning and lamenting the dead.[47]

Šaulić drew a direct line between the Kosovo Maiden and the Mother of the Jugovići to Serbian women mourning the dead from Serbia's wars of liberation. In his research, he concluded, "Serbian women, mothers, and sisters guard from forgetting [the soldiers] and remember warriors in their heartfelt songs, elevating them to immortal glory and everlasting life. Through the misty acknowledgment of life after death, heroes are brought back to life by grace, love, and imagination. … These women as mothers and sisters and as lamenters contribute moral strength and they create a powerful bond of a people to the land."[48] Šaulić firmly rooted his lamenters in the full national pathos of the Kosovo legend and the laments he collected and framed stipulated how women should remember the dead. By collecting and presenting women's laments, Šaulić channeled women remembering into a nationalist and heroic framework.

The laments that Šaulić recorded included not only laments heard at funerals but also laments heard on days of the dead, annual memorial services, All Soul's day, or days reserved for remembrance in the first year of a loved one's death. The songs relevant for this study are from women whom Šaulić recorded chanting about those who died during the Balkan Wars or World War I. He observed and recorded them on days of mourning. His observations yielded three basic elements found in the performance and structure of laments: lyrical, dramatic, and epic elements.[49]

Šaulić noted that the epic element was the dominant form sung most often when the deceased died in battle. He observed, "The addressing of those present and the way in which one lamenter takes over from another lends the laments a dramatic element. One leads, the others in turn repeating her words, exactly or with slight modification, or in great grief."[50] What is important to note from this description is that the process was a social one, the grief of an individual became common grief and the praise of the departed in the presence of a large gathering gave permanence to the memory of the deceased. These laments when chanted among a group of people were not as a rule free improvisations—women may have agreed in advance what part each one was to perform. The drama of family members, community members gathered and listening to descriptions of the deceased's personality and merits, and the lyrical elements of the lament tied to other oral traditions stimulated memory and stitched the deceased's life into the fabric of the community.[51] In the case of those who died in battle, the lamenter's grief and the preservation of the memory of the departed's heroism coexisted as seen in this lament titled, "A mother for her dead sons." This mother had lost a number of sons in battle and she chanted:

> Oh my children, oh my fortune
> Never will I grieve for you enough
> In glory vanished from my life
> Where our heroes soar
> Where our ancestors are avenged.[52]

In a different lament, another mother sang:

> With heroes did you go to war
> With heroes did you fall in battle
> Vanished from this life with honor.[53]

And in the same lament, she noted other members of her community who died a heroic death.

> Loudly rang the cry of [our] bravest men:
> Now who will be the first to charge?
> And leap into the enemy trench?
> A dozen men chose this
> Each one a plucky Bukilić.[54]

Šaulić collected these laments during the first decade following the end of World War I and what struck another contemporary ethnographer was how much they were "like heroic ballads" and that the "lamenting women's tone ceases to be lyrical and becomes epic."[55] Inserted into some of the laments was the lamenter's intimate knowledge of the battles or circumstances of the soldiers' death. Laments in Šaulić's collection featured descriptions and details from battles of the Balkan Wars such as those at Kumanovo, Prilep, and Monastir/Bitolj; and World War I battles such as Cer, Drina, Kolubara, the defense of Belgrade, and Dobro Polje. He noted that some of the chants had the rhythms and sentiments of heroic epics. In a lament that he titled "Mother for her son and to all heroes," the mother set the scene for her son's death at the battle of Kumanovo (1912):

> When that army set out to fight
> It was like a storm blocking the sky
> The Army was like rolling clouds

The officers like soaring eagles
And they attacked (fell upon) the Turkish horde
Brandishing their nimble guns
In powerful and heroic hands
And at their waists hung razor swords
On tightened belts about their waists
Unhesitating they charged the foe
In hordes that swarmed like hungry wolves
Who see white lambs upon the slopes.[56]

What is obvious is that the women knew first hand or from descriptions of others what were the facts of the battles and the topography of battlefields. There were others mentioned who fell in the same battle (often neighbors) or there was even mention of those who had fallen in earlier conflicts and whom lamenters knew only from tradition. In the same lament, the women, noting the heroism of these men, also encouraged others to sacrifice for their community and for their children:

But I would like to call upon you
All to be courageous,
All to be of lordly lineage ...
May all your children grow up
Nurture them and bring them up
When we no longer live in shame
Under the tyranny of alien rulers
So at last our children sing
Freely in our schools and churches.[57]

Another feature of the laments was commentary on political and social conditions about a current enemy. Referencing the Second Balkan War, a woman noted "the faithless Bulgarian cousins." Another sang, "Austria that brings ill fortune" and "Hungary's bloodthirsty soul," which referred to Serbia's enemies during World War I.[58] In these laments, Šaulić presented a central trope that runs through many of the laments in his collection, namely how Serbian peasants suffered from having alien overlords, especially the Turks. The language of oppression and struggle for "national liberty" was now applied to their recent enemies—the Bulgarians, Austrians, and Hungarians.

Šaulić referred to Serbian women's laments as women's poetry. John Berger has commented on the cathartic role that poetry can play, stating how all human pain, apart from physical pain, is caused by "one form or another of separation." He writes, "Poetry can repair no loss, but it defies the space which separates. And it does this by its continual labor of reassembling what has been scattered."[59] One can argue that this is the process that took place during laments; women gathered the "truth" about the deceased and through their private and personal grief and remembrances, they gave meaning to the individual life.

The laments cited here demonstrate how "the expression of grief represents power" and the role women played in controlling and directing the "passion of grief and anger" within their communities.[60] Prior to the twentieth century, Serbian women's role in the rituals of mourning went unchallenged and, as can be seen through the photographic record, was a role that those photographing and displaying war believed to be of cultural, religious, and political significance. Serbian women, through their laments, were able to weave into their songs history of earlier struggles against oppressors and the wars that Serbia fought between 1912 and 1918 and direct grief outward. Others, such as official photographers and representatives of the state, appropriated and represented the rituals and laments for political purposes. As noted by Gail Holst-Warhaft in *The Cue for Passion: Grief and Its Political Uses,* as communities and states expanded and mobilized men to serve in the military, the state began to "take control of mourning rituals, monopolizing and frequently politicizing the way grief was expressed."[61]

Conclusion: Gendering Grief

Until the mid-twentieth century in the Balkans, the responsibility for remembering the dead was commonly assigned to women, who were performers of laments. As this tradition began to wane, other ways of remembering the dead supplanted or were privileged over the lament and women's role in death rituals. As has been detailed in a significant body of literature on war and memory, states in the twentieth century built monuments to their war dead, erected plaques, created official days for remembering the dead, and held ceremonies and dedications to honor the dead. Most of these sites of memory were designed to stimulate a feeling, an emotion, or a memory that all members of the state were expected to share. Such commemorative activity did not abrogate other individual, communal, or religious rituals for the dead, but competed with these other forms to tell the "truth" about the deceased.

Photographs emerged as a new medium, which were credited with preserving and stimulating memory.[62] The photograph appeared to be able to capture a moment of an individual's existence and became irrefutable evidence of the existence of a dearly loved one. Thus photographs took on a claim of truth, a truth that did not come from religious or communal traditions but instead came from the photographer and his camera, which provided an objective and dispassionate truth. Few commemorative forms were able to conjure up the individual as a photograph could. The only exception, as Holst-Warhaft has pointed out, are laments. Singers of laments and their audiences believed that laments told the truth about the dead. The photographs of women in mourning provide evidence of the coexistence of these two commemorative acts—lamentation and photography—and how the male photographer's gaze, through framing, cropping, display, caption, text, and distribution, fixed the meaning of the photographed object and privileged a particular one. Photographs of women mourning might prompt a memory of the viewer as to the act of mourning. Displayed as these were in a Serbian war album, the photographs are both generic and specific. Their power to stimulate memory derives from what the tragic figure of the women in expressive grief shares with other depictions of lamentation over the dead, such as paintings, poem, sculptures, or similar artistic expressions of grief.[63] The photographs also became timeless and spaceless symbols of suffering and mourning. The actual person mourned and the circumstances of his death disappeared from the frame and became a secularized symbol of heroic sacrifice. Women's mourning was represented but the photographer silenced their voices in the ritual of death.

Thus there are important differences between laments and photographs, not only with regard to who performs or produces them, but also in the substance and performance of these memorial arts. "The lament's truth is based on the aesthetic of pain. When performed, the lament, because of its emotional tone, dialectic exchange, and its musical quality, stimulates memory and has the power to evoke a memory."[64] The performance of the lament and its ability to engage the audience stimulated memory but did not fix a specific meaning or privilege a particular one as photographs did.

The photographs in these war albums had an ambiguous relationship to memory. Memory was actually fixed by the photographer, the photographer whose glimpse through the lens captured the subject. The viewer of the photograph most likely was not present at the moment the subject passed in front of the lens and thus the viewer could not know the truth of the moment but instead brought to the photograph his or her own associations. A lament structures and evokes a memory because it is a communal memorial structured for the living. The laments brought in individual, familial, and communal references in order to fix the soldiers' life in the public memory of his community. He is more than a soldier: he is a husband, brother, son, and member of the community. Lamentation evokes multiple memories and associations and "establish[s] the memory of the individual and place it where it can be accessed in the future through performance. … By contrast, the photograph requires no involvement of the community and despite its incontestability, often fails to conjure the presence of the dead."[65] In the case of Serbia, women's ability to shape these memories and associations gave them substantial authority within a community. Laments not only praised the heroism of those who went to war but also acted as a critique of existing conditions.

In cultures where laments and photographs began to coexist, such as Montenegro, Herzegovina, and parts of Serbia prior to World War II, photographs may have acted as stimuli to lament. Šaulić in his ethnography described a scene at some point in the 1920s where the mother of a deceased soldier from the Balkan Wars clutched a framed photograph to her chest at a gravesite as she uttered cries of dismay for her dead son. Photographs, framed and placed conspicuously in the home, embedded or placed on tombs, placed in family albums, found in a commemorative, photographic war album, or a photographic exhibition, coexisted with the lament in this scene. Placed in this context one can see how a photograph, clutched by the mother, provoked the lament but it was through the act of lamenting that the mother recalled memories of the deceased. The photograph could not speak. Such images fascinated or suggested associations of the deceased but they did not remember by themselves. As we have seen with laments, their lyrical, dramatic, and heroic elements placed the private memory of the dead into a communal web of geographic, familial, and social relationships. Women played a vital role in shaping and defining the meaning of death and they did so by conversing with the dead and remembering them on the community's behalf.

◊ About the Author

Melissa Bokovoy is an associate professor of history and Regents' Lecturer at the University of New Mexico. Her current research examines the culture and politics of World War I commemoration in interwar Serbia. She is the author of *Sharing the Stage: Biography and Gender in Western Civilization* (2003), with Jane Slaughter, and *Sharing the World Stage: Biography and Gender in World History* (2009), with Jane Slaughter, Patricia Risso, Ping Yao, and Patricia Romero. Her first historical monograph, *Peasants and Communists: Politics and Ideology in the Yugoslav Countryside* (1998) won the AAASS Barbara Jelavich Prize. Email: mbokovoy@unm.edu.

◊ Notes

1. Princess Alexis Karageorgevitch [Karadjordjević], *For the Better Hour* (London: Constable and Company, 1917). 45.

2. Popović's was a "monumental and voluminous" album of 448 pages and 1,500 pictures; "a photographic collection of the personalities and events of the war in all countries." Throughout the album, there are at least another dozen photographs of women in mourning. In this album, Popović gathered the photographs of the best wartime photographers in order to create "a token of gratitude and glory for all the living and the dead participants in the Great War, to our own people, as well as to our Allies, the known and the unknown, to all those who, fighting for honour, justice, and Freedom, have with the help of God, contributed toward the realization of the Liberation and the Unity of our nation." Andra Popović, *Ratni Album, 1914–1918* (War album, 1914–1918) (Belgrade: Uredništvo Ratnog Albuma, 1924), 4.

3. As anthropologists engaged in the study of death rituals have noted that ritual is defined by its formality and routine, resembling a recipe, a fixed program, or a book of etiquette. Richard Huntington and Peter Metcalf, *Celebrations of Death: The Anthropology of Mortuary Ritual* (Cambridge: Cambridge University Press, 1991). For Serbian death rituals, see Slobodan Zečević, *Kult mrtvih kod Srba* (The cult of death among the Serbs) (Belgrade: Etnografski muzej, 1982).

4. T. G. Ashplant, Graham Dawson, and Michael Roper, eds., *The Politics of War Memory and Commemoration* (New York: Routledge, 2000), 18. This work provides a nuanced framework for thinking about the political and cultural processes involved in the struggle to articulate war memory. The editors identify three parts necessary for understanding this struggle: narratives of articulation; arenas of articulation; and agencies of articulation, which refer to those institutions through which social actors seek to promote and secure recognition for their war memories."

5. There exists a substantial body of literature on death, dying, and bereavement. Philippe Aries's works remain the foundational texts: *The Hour of Our Death* (New York: Knopf, 1981); and *Western Attitudes Toward Death: From the Middle Ages to the Present* (Baltimore: Johns Hopkins University Press, 1974). For texts dealing specifically with death, mourning, and war, see Jay Winter, *Sites of Memory, Sites of Mourning: The Great War in European Cultural History* (Cambridge: Cambridge University Press, 1998); Catherine Merridale, *Night of Stone: Death and Memory in Russia* (London: Granta, 2000); and Drew Gilpin Faust, *This Republic of Suffering: Death and the American Civil War* (New York: Alfred A. Knopf, 2008).

6. Kathleen Garces-Foley, *Death and Religion in a Changing World* (Armonk, NY: M.E. Sharpe, 2005), 93–94.

7. Ibid.

8. Anna Caraveli-Chaves, "Bridge between Worlds: The Greek Women's Laments as Communicative Event," *Journal of American Folklore* 93 (April–June 1980): 129–130.

9. Gail Holst-Warhaft, "Remembering the Dead: Laments and Photographs," *Comparative Studies of South Asia, Africa and the Middle East* 25, no. 1 (2005): 152.

10. For reading on death and laments in the Balkans, especially Greek lamentation, see Margaret Alexiou, *The Ritual Lament in the Greek Tradition* (Cambridge: Cambridge University Press, 1974); Loring Danforth, *The Death Rituals of Greece* (Princeton: Princeton University Press, 1982); Gail Holst-Warhaft, *Dangerous Voices: Women's Laments and Greek Literature* (London: Routledge, 1992); and Gail Holst-Warhaft, *The Cue for Passion* (Oxford: Oxford University Press, 2000). Gail Kligman, *The Wedding of the Dead: Ritual, Poetics, and Popular Culture in Transylvania* (Berkeley: University of California Press, 1988) is still the only English-language monograph dealing with Slavic lamentation and death rituals in the Balkans.

11. Parita Mukta, "Lament and Power: The Subversion and Appropriation of Grief," *Studies in History* 13, no. 2 (1997): 209–246.

12. In 1912, Dragutin Dimitrijević-Apis, the chief of Serbian Intelligence and head of the secret society Black Hand, which was implicated in the assassination of Franz Ferdinand, invited Rista Marjanović to document the experiences of Serbian soldiers during the First Balkan War. The photographer took a leave of absence from his job as illustrations editor at the European edition of the *New York Herald* (Paris). When World War I broke out, he returned to Serbia and became one of the most prolific of Serbia's wartime photographers and a favorite of the royal government and court.

13. In the last decade, feminist and gender scholars, especially anthropologists and sociologists, have begun to examine the relationship between photographic representations of suffering and grief. For an example of this literature, see Holst-Warhaft, "Remembering the Dead: Laments and Photographs," 152–160. She places these two "arts" in the context of telling the "truth" about the dead. Also see Christina Konstantinidou, "Death, Lamentation and the Photographic Representation of the Other during the Second Iraq War in the Greek Newspapers," *International Journal of Cultural Studies* 10, no. 2 (June 2007): 147–165; and Catherine Moriarty, "Though in a Picture Only: Portrait Photography and the Commemoration of the First World War," in *Evidence, History and the Great War*, ed. Gail Braybon (New York: Berghahn Books, 2005), 30-47.

14. Karageorgevitch, *For the Better Hour*, 45.

15. Ibid., 46.

16. Popović, *Ratni Album*, 85–91.

17. Drew Gilpin Faust notes how both Confederate and Union soldiers and their families feared that they would not receive a "decent" burial or rituals if they died on the battlefield. See Gilpin Faust's chapter on "Burying," in *This Republic of Suffering*, 61–101. During World War I, burial services were usually "brutally short" with rituals briefly observed, sometimes bodies were simply left unburied as battles continued or armies, especially along the Eastern and Serbian fronts, retreated. For a recent discussion of the dead and missing along the Western Front, see Neil Hanson, *Unknown Soldiers: The Story of the Missing of the First World War* (New York: Vintage, 2006); and Jay Winter, *Sites of Memory, Sites of Mourning* (Cambridge: Cambridge University Press, 1998), 15–53.

18. Popović, *Ratni Album*, 334–335.

19. Karageorgevitch, *For the Better Hour*, 47.

20. Marjanović was part of an international group of photographers who began to photograph war. Beginning with the wars of the early twentieth century in Europe, a specific visual vocabulary of modern war photography developed that contained a central core of symbols that come together around at least three themes: (1) the conflict and the technology (fighting soldiers, combat, battlefields, air raids); (2) civilian life or wartime city; and (3) human casualties and their physical, psychological or emotional effects (war victims, mourning, refugees).

For a discussion of this trend, see Catherine Brothers, *War and Photography: A Cultural History* (London: Routledge, 1997), 201.

21. K. Rozario, "Delicious Horrors: Mass Culture, the Red Cross and the Appeal of Modern American Humanitarianism," *American Quarterly* 55, no. 3 (2003): 417–455, cited in Konstantinidou, "Death, Lamentation and the Photographic Representation of the Other," 156.

22. John Berger, *Ways of Seeing* (New York: Penguin, 1990), 10.

23. The following description is summarized from Zečević, *Kult mrtvih kod Srba*.

24. The prayer is as follows: O God of spirits and of all flesh, Who has conquered death and destroyed Satan and Who has granted life to Your world: O Lord, rest the soul(s) of Your departed servant(s) (name) in a serene, luxuriant, and peaceful place where all pain and sorrow and lamentation are absent. As a gracious God Who loves mankind, forgive all transgressions committed by (him, her, them) by word or deed or thought, for there is no one living who does not sin; You alone are without sin, Your truth is truth in eternity and Your word is truth. For You are the resurrection, life and repose of Your departed servant(s) (name) O Christ our God, and we give glory to You and to Your eternal Father and You most holy, gracious and life-giving Spirit, now and ever and forever. Amen.

This prayer is taken from a seventeenth- or eighteenth-century manuscript from the Serbian Hilandar monastery on Mt. Athos. They were translated from the Church Slavonic text, but are close to the Greek prototypes that appear in other old manuscripts and were published in the nineteenth century. See "Orthodox Prayers," http://troparion.com/various.htm (accessed 16 July 2010).

25. Zečević, *Kult mrtvih kod Srba*, 47.

26. Ibid.

27. Radovan Blagojević, *Spomenici i groblja iz ratova Srbije 1912–1918* (Monuments and cemeteries from Serbia's wars, 1912–1918) (Belgrade: Beogradski izdavačko-grafički zavod, 1976), 68.

28. Ibid.

29. The typhoid epidemic is estimated to have caused over 150,000 deaths during 1915. Henry Oliver Lancaster, *Expectations of Life: A Study in the Demography, Statistics, and History of World Mortality* (New York: Springer Verlag, 1990), 333.

30. P. de Mondesire, *Albanska Golgota* (Albanian Golgotha) (Belgrade: Prosveta, 1936), 61. Cited in Andrej Mitrović, *Serbia's Great War, 1914–1918* (West Lafayette, IN: Purdue University Press, 2007), 160.

31. Gilpin Faust, *This Republic of Suffering*, 62.

32. Milutin Bojić, *Plava Grobnica* (Blue tomb), in *Antologija srpske književnosti* (Anthology of Serbian literature), http://www.antologijasrpskeknjizevnosti.rs/ASK_sr_projekat.aspx (accessed 20 May 2010).

33. Novica Šaulić, *Srpske narodne tužbalice* (Serbian National Laments) (Belgrade: Grafički institut "Narodne misao," 1929), 19.

34. On four Saturdays of the year, families and loved ones visit the grave site and hold a feast for the dead.

35. Zečević, *Kult mrtvih kod Srba*, 57.

36. Ibid., 47.

37. Description from Šaulić, *Srpske narodne tužbalice*, 21.

38. Ibid.

39. This discussion about the tradition of Serbian women's laments is taken from Melissa Bokovoy, "Kosovo Maidens: Serbian Women Commemorating the Wars of National Liberation, 1912-1918," in *Women and War in Twentieth Century Eastern Europe*, ed. Nancy Wingfield and Maria Bucur (Bloomington: Indiana University Press, 2006), 157–170.

40. Timothy Rice, "A Macedonian *Sobor:* Anatomy of a Celebration," *Journal of American Folklore* 93 (April–June 1980): 121.

41. Patricia K. Shehan, "Balkan Women as the Preservers of Traditional Music and Culture," in *Women and Music in Cross Cultural Perspective,* ed. Ellen Koskoff (Urbana: University of Illinois Press, 1989), 47.

42. Šaulić, *Srpske narodne tužbalice,* 19.

43. Anne Pennington and Peter Levi, *Marko the Prince: Serbo-Croat Heroic Songs* (London: Palgrave Macmillan, 1984), 17.

44. This was the young man who carried Prince Lazar's battle-standard.

45. *The Battle of Kosovo,* trans. John Matthias and Vladeta Vučković (Athens, OH: Swallow Press, 1987), 70.

46. Ibid., 74–75.

47. Svetlana Slapšak, "Women's Memory in the Balkans: The Alternative Kosovo Myth," *Anthropological Journal on European Cultures* 14 (2005): 95–111, discusses the nationalist appropriation of women's grief and interpretations of the epic poetry of the Battle of Kosovo. She argues that the voices of women in the epics not only express the misery and grief of death but are also expressions of reproach to men who decided about the collective's future.

48. Šaulić, *Srpske narodne tužbalice,* 26.

49. Ibid.

50. Ibid.

51. Ibid.

52. Ibid., 40. Translations modified by the author from J. Šaulić, "The Oral Women Poets of the Serbs," *Slavonic and East European Review,* 42, no. 98 (December 1963): 161–183.

53. Šaulić, *Srpske narodne tužbalice,,* 42.

54. Ibid.

55. P. Slijepević, "Prilozi narodnoj metrici" (Contributions to the oral poetic tradition), in *Godišnjak filosofskog fakulteta* (Yearbook of the philosophy department) (Skoplje: Južna Srbija, 1930), 29. Cited in Šaulić, "The Oral Women Poets of the Serbs," 161–183.

56. Šaulić, *Srpske narodne tužbalice,* 35.

57. J. Šaulić, "The Oral Women Poets of the Serbs," 172.

58. Novica Šaulić, "Šala i podrugivanje u tužbalicama" (Jest and mockery in laments), in *Iz naše književnosti* (From our literature) (Belgrade: Rotacija, 1938), 31. Cited in Šaulić, "Oral Women Poets," 170.

59. John Berger, "The Hour of Poetry," in *The White Bird,* ed. Lloyd Spencer (London: Hogarth Press, 1988), 249. Cited in Moriarty, "Though in a Picture Only," 40.

60. Holst-Warhaft, *The Cue for Passion: Grief and Its Political Uses,* 9.

61. Ibid.

62. The premise that photographs preserve memory has been challenged, among others, by Roland Barthes, John Berger, and Susan Sontag. See, for example, Roland Barthes, *Camera Lucida: Reflections on Photography* (New York: Hill and Wang, 1982); John Berger, *About Looking* (New York: Vintage, 1992), and Susan Sontag, *On Photography* (New York: Picador, 2001).

63. Byzantine artists were among the first, in the ninth century, to paint Christ's disciples taking him down from the cross. By the eleventh century, artists gave a more prominent position to Mary who either cradles Christ's body or he is draped across her lap. In a very early Byzantine fresco found in the Church of St. Panteleimon (Nerezi) in Macedonia, the artist changed the site of mourning from that of below the cross to outside the mouth of the tomb. Gertrud Schiller, *Iconography of Christian Art,* vol. 2 (London: Lund Humphries, 1972), 164–181.

64. Holst-Warhaft, "Remembering the Dead," 152.

65. Ibid., 155.

The *Krŭchma*, the *Kafene*, and the Orient Express

Tobacco, Alcohol, and the Gender of Sacred and Secular Restraint in Bulgaria, 1856–1939

Mary Neuburger

Abstract

This article explores shifts in patterns of consumption of alcohol and tobacco in Bulgaria, with a focus on public establishments in the nineteenth and first half of the twentieth century. In exploring both the gender dimension of such shifts and its religious implications, the article argues that public consumption of tobacco in particular both reflected and was constitutive of dramatic historical change. At the same time, the increased consumption of such culturally fraught substances provoked an increase in both religious and secular campaigns of "restraint," in which gender played a key role.

KEYWORDS: abstinence, Bulgaria, café, communism, consumption, gender, religion, Protestantism

In the early years of the twentieth century a young Bulgarian woman who was traveling alone on the Orient Express almost set the train on fire. Raina Kostentseva, who later related this tale proudly, recorded neither her age nor the exact year of the incident, but the details seemed burned into her memory. At a prolonged stop, Raina had stepped out of her compartment and onto the platform for a breath of fresh air. Among other passengers milling about, two "foreigners" engaged her in conversation and offered her a cigarette. Although ready and willing to smoke, Raina suggested that they adjourn to her compartment where she had special Bulgarian cigarettes and, more exotically—a special lighting kit involving flint, steel, and tinder that would have been "completely unknown to them." As Raina related, she lit the fire in front of the curious foreigners who used it to light up the Bulgarian cigarettes she offered. She, in turn, smoked one of theirs. After they finished smoking amid lively conversation, they

aspasia Volume 5, 2011: 70–91
doi:10.3167/asp.2011.050106

parted with warm goodbyes. Raina extinguished the fire and put the paraphernalia back in her bag. She lay down to sleep but soon smelled smoke, rapidly growing in intensity. With her quick wits about her, Raina immediately threw the bag, now smoldering, out the window, saving the train and herself from catastrophe.[1]

On this occasion the Orient Express was heading north and west, away from the Orient. Indeed, Bulgaria was but a stop on the Orient Express, one that was already in the "Orient" from the perspective of many travelers. Even Raina herself served up a bit of the exotic to the "foreigners," who were presumably on their way home, with her Bulgarian cigarettes and lighting kit. At the same time, in her behavior, her traveling alone, her smoking with foreigners—presumably men, though perhaps a couple—Raina was violating many of the behavioral conventions for Bulgarian women of the time. Such conventions were in gradual flux at the turn of the century, and more rapidly after World War I, Raina, a girl from a rather "modern" family in Sofia—the capital of the Bulgarian principality since 1878—was still surely ahead of her time. And she was headed, so to speak, both literally and figuratively westward in her bearing and smoking indulgence. Still her scandalous incident on the Orient Express was one of the few moments of relative social abandon in Raina's memoirs, which detail her life in Sofia from the turn of the century through the interwar years. For even in Sofia, a rapidly growing and rather cosmopolitan city by this time, the social world of this relatively "modern" girl was deeply embedded in a matrix of tightly intertwined ethnic and religious conventions, in which gendered patterns of sociability and consumption were rigidly circumscribed.

In Bulgarian history, as globally, shifting patterns of gendered consumption—particularly of such culturally loaded substances as alcohol and tobacco—both reflected a host of social and cultural changes and were constitutive of such changes.[2] Shifts in gendered patterns of tobacco and alcohol consumption, particularly public consumption, are intertwined with the profound social changes that characterized late-Ottoman and post-Ottoman Bulgaria. In Bulgaria, as globally, the late nineteenth and early twentieth century were characterized by the rapid rise in consumption of alcohol and tobacco, which along with coffee had become the global drug of choice in the modern period. New modes and a greater scale of consumption of these intoxicants, and indeed the participation of women in such consumption, have been linked by numerous scholars to global urbanization, modernization, and the effects of the World Wars.[3] For each, alone or in combination, a range of "social lives" developed with very context-specific and historically meaningful forms and consequences.[4] In Bulgaria these social lives had their own highly gendered trajectories and outcomes, embedded in traditional mores, but also shaped by a variety of new influences, religious as well as secular. Bulgarian[5] men had traditionally gathered in the alcohol-imbibed *krŭchma* (tavern), but over the course of the nineteenth century they also entered into the "sober" social life of the *kafene* (coffeehouse), where coffee and smoking became part of the rituals of public male sociability.[6] The *kafene*, though already a centuries-old Balkan Muslim tradition, was a new phenomenon in nineteenth-century male Balkan Christian life, arguably part of their political coming of age that culminated in self rule in 1878. For Bulgarian women, their entrance into the Bulgarian *kafene* as well as the *krŭchma*, beer-hall, and other public consumption venues was also a kind of a social coming of age. But it was one that began only slowly, as leisure venues became more varied and prevalent in fin de siècle Sofia. Accelerating rapidly after World War I, the pattern of women entering public establishments was certainly a general European, if not global trend, but there was something very specific about the Bulgarian experience of this transformation of social life. For women such changes were in the shadow of a process of Europeanization of these very local institutions and so were connected at least figuratively with a kind of boarding of the Orient Express, on its "return" to Europe. But this, significantly, did not lessen the social alarm at such changes, and indeed may have actually enhanced them.

In the interwar period when women began to enter into a rapidly changing field of public alcohol and tobacco consumption in significant numbers, a Bulgarian abstinence movement rapidly mushroomed. That is not to say that women were the explicit focus of the movement, and indeed men were still to a large degree the primary problem in the eyes of abstinence advocates. Still disparate visions and solutions to the perceived blight of alcohol and tobacco that arose and evolved in interwar Bulgaria both focused on and deeply involved women in complicated ways. These voices of restraint were not entirely new in Bulgaria, but they took on ever more radical forms in the face of what seemed like the moral crisis of the period. Most abstinence advocates prior to World War I had been connected to the active Protestant missionary milieu, and abstinence had been a distinguishing factor between Protestant converts and the local Orthodox Christian populations, just as sobriety distinguished Muslims from the alcohol-imbibed Orthodox Christian community.[7] Interestingly, as a home-rown non-Protestant abstinence movement began to take hold in Bulgaria in the early twentieth century and particularly after World War I, it was primarily Bulgarian communists who founded and directed the new movement. Ironically, the communists' abstinent vision of the future shared much with the Protestant one. Though philosophical opposites in a sense, both Protestants and at least a segment of socialists developed similar concerns about the medical, but more pointedly moral and social effects of alcohol and tobacco in Bulgaria. Of course, unlike Protestant religious underpinnings, it was on the soil of radical secularism that the communist movement—though far from unified on the abstinence question—built its variety of solutions to Bulgaria's particular social ills. Te similarities in these voices of interwar "restraint" were striking. Both targeted drinking and smoking by modern pleasure seekers and connected them specifically to the ill effects of Western penetration. But both also looked forward, advocating a deeply gendered "moral uplift" to counter "backward" and traditional patterns of *krŭchma* drunkenness. Both communists and Protestants expected women, as presumed "natural abstainers," to be at the forefront of the movement, enforcing abstinence at home and playing an important role in bringing about communism and creating believers worthy of the kingdom of heaven, respectively. But both also saw the drinking and smoking "fallen woman"—an ambiguous term at best—as a threat to dreams of a bright future. In short, from the streets of Sofia to the Orient Express, "modern girls" were both the hope and potential threat to both religious and secular visions of a brighter future.[8] Indeed, the "modern girl" herself became both a participant and an object of scrutiny, as Bulgaria's future rapidly unfolded.

From the *Krŭchma* to the *Kafene*

In the revered pages of Ivan Vazov's *Pod igoto* (Under the yoke), published in 1889, the *kafene* is iconic.[9] The novel, arguably the most famous and widely read in Bulgarian history, is set in a central Balkan mountain town during the period just prior to the April Uprising of 1876. Ganko's *kafene* is the central hub of town, crowded with archetypal characters from Ottoman Bulgaria who drink bitter coffee, ruminate and debate, laugh, and observe within a "dense fog of tobacco smoke."[10] In *Pod igoto kafene* sociability, immersed in coffee and smoke, seems to possess an almost primordial connection to Bulgarian rural life. But the novel by no means portrays a timeless or static Ottoman past. Instead, on the set of the *kafene* Vazov skillfully stages a late Ottoman social landscape rife with change: "Ganko's *kafene* was, as usual, filled with noise and smoke. It was the meeting place of old and young alike, where public matters were discussed, and the Eastern Question too, as well as all the domestic and foreign policy of Europe. A miniature parliament one might say."[11]

In its small town setting, Ganko's *kafene* offered a forum for interaction and debate among the various social strata (and to a lesser extent ethnic communities) that were present during the so-called Bulgarian national revival period (1762–1878).[12] Indeed, *kafene* life itself was indicative of, if not essential to, the social, cultural, and eventual political change of the period, and it was the move from the *krŭchma* to the *kafene* that both characterized and was arguably critical to the changes of the "revival period." Smoking and sipping coffee in the Ottoman *kafene* (and later pan-European café),[13] in fact, became intimately connected to upward mobility among Bulgarian men and their increased authority in Ottoman villages, towns, and cities, in short, a national and political "awakening."[14]

Not that the *krŭchma* was left behind. The consumption of alcohol had dominated male Balkan Christian social life since the Middle Ages and the *krŭchma* in fact remained a central rural and urban social institution for Balkan Christian men, a "parallel reality" to the primarily Muslim *kafene*.[15] Alcohol had been produced in the region since antiquity, and the Christian Medieval Bulgarian state had encouraged alcohol production, even among priests and monks, who historically had traded in wine.[16] This continued after the Ottoman conquest in the fourteenth century, as Islamic law—which made the sale and consumption of alcohol illegal—applied only to Muslim populations, including migrants and converts in the Eastern Balkan region.[17] For Muslims, alcohol was banned but even coffee and tobacco were initially scrutinized, questioned, and even banned by the Ottoman Ulema. By the seventeenth century, however, both substances were deemed socially acceptable offerings in the ubiquitous Muslim coffeehouse, the sober foil to the Christian *krŭchma*.

That is not to say that abstinence was totally unknown, or not valued, in the Christian Orthodox world. It was quite common for nineteenth-century Bulgarian writers, for example, to employ the trope of the *krŭchma*, as opposed to Vazov's productive *kafene*, as a source of malaise for the rural peasant.[18] This by no means meant an embrace of abstinence, but examples of abstinent living were not totally unknown. Indeed, the most well-known case of total abstinence was associated with the revered, if not sanctified Vasil Levski, the most legendary of all Bulgarian nationalist revolutionaries from the national revival period.[19] In a colorful biography of Levski, Mercia MacDermott makes this point repeatedly, Levski himself rigorously abstained from both alcohol and tobacco, both of which he considered to be harmful to the individual and society. He was a firm believer in personal as well as organizational discipline."[20] Similar statements about the "purity" of Levski, while certainly prone to hero worship and myth making, permeate sources from his contemporaries as well.[21] Nikola Obretenov, a revolutionary comrade in arms, asserts in his memoirs that Levski "would not touch alcoholic drinks of any kind, and did not smoke; he drank only tea."[22] But the fact that his abstinence was noteworthy points to a more general trend toward drinking and smoking that pervaded Bulgarian male society, including its revolutionaries, during this period. Indeed, the average Bulgarian revolutionaries of the 1860s were very much urban animals, with all the vices of the *kafene* and *kruchma* where they routinely networked.[23] Levski, in contrast, continued to play the role of the monk-like, indeed saint-like, figure to respect and worship. A former Orthodox Christian deacon, Levski was mythologized to such a degree that his moralistic practices could be justifiably ignored by mere mortals.[24]

Very few could actually follow in his footsteps. Even Orthodox Christian priests were routinely involved in the production, sale, and consumption of alcohol in villages and towns. According to Protestant missionary observers of Bulgarian religious and social life there was much need of temperance in a society where "the village priest was commonly accorded the privilege of acting as the local liquor dealer."[25] Protestant missionaries—though over-simplifying the matter—often noted that Bulgarians were not receptive to Protestantism precisely because of its abstinence requirement. And most reports make it clear that the few who accepted Protestantism and/or abstinence were exceptions in a sea of drunkenness and even hostility.[26]

At the same time, it is worth mentioning that contrary to Protestant observations, Orthodox practices were not entirely devoid of codes of conduct and restraint when it came to alcohol and tobacco consumption. Although more in word than practice, critiques of drunkenness can be found in traditional church writings, namely the *damaskin* (Orthodox texts used primarily for sermons).[27] More pointedly, Orthodox practices held Bulgarian women to a higher standard than men in terms of moderation or abstention, especially in public settings. That is not to say that women did not smoke or drink in nineteenth-century Ottoman-Bulgarian society. Numerous Western travelers to the region commented on the prevalence of drinking and smoking among women of all religions, but mostly in private homes and gardens, in the baths, or at fashionable parties in elite Ottoman or foreign homes.[28] Indeed, the fact that tobacco crossed the gender divide actually seemed to impress some Western women observers, like Mary Patrick Mills, the first director of the Protestant missionary American Women's College, who marveled that "During the whole period of Turkish history good tobacco was one of the luxuries of life, enjoyed equally by men and women alike."[29] This speaks to the fact that a separate but significant (and largely unexplored) world of female sociability—that also included smoking—existed outside the male-dominated coffeehouse.[30] Still, to the extent that Bulgarian women smoked—and evidence of such a fact is rather limited—it was an activity that generally took place behind closed doors or high garden walls, as in the case of other Ottoman women.[31] In fact, one of the most famous women of the Bulgarian national revival, Nedelia Petkova—remembered as Baba (grandmother) Nedelia—was known to have scandalized Bulgarian sensibilities with her public smoking habit. Baba Nedelia, famous as the founder of women's education in Sofia, was one of the first Bulgarian women teachers of the era. But in many of the places she taught Baba Nedelia was treated with open hostility, as an "emancipated and independent woman."[32] According to one source, Nedelia was fired from a job in the town of Prilep in Macedonia "because of her extremely free behavior, clothing, and her smoking."[33] Although Bulgarian women did find various ways to participate in the national revival that did not include smoking and sociability, it is telling that one of the most famous of the "emancipated" female figures from the revival was also a known smoker. But even more telling is that this practice was openly censured and cause for professional difficulties.

For Bulgarian men, in contrast, the smoking habit, along with drinking, became a given and was intimately tied to changing notions of masculinity, which included new levels of participation in an expanding leisure and "productive" *kafene* sociability. In the course of the nineteenth century, patterns of consumption and sociability changed dramatically, along with commercial revival, urbanization, and other phenomena that accompanied Ottoman reform and Bulgarian "national revival." Initially elite Bulgarian men—generally local notables or itinerant merchants—began to enter the Muslim *kafene* for commercial or administrative purposes. The reforms of the Ottoman Tanzimat (1839–1876), increased the profile of Balkan Christians and also their need to enter the sober and, arguably, "productive" coffeehouse institution, where local administration and commerce were conducted.[34] But the *kafene* was productive in other ways, as a critical mass of Slavs formed urban colonies at home and abroad where they began to discover and invent "Bulgarianness," among other places, amid *kafene* conviviality.[35] This shift in social interaction amid consumption practices from the drunken *krŭchma* to sober *kafene*, with all of the pharmacological effects implied, is a compelling way to rethink social and cultural change in the nineteenth-century Balkans. At the same time, just as Bulgarians began to enter the *kafene* the line between *kafene* and *krŭchma* was often blurred, especially in small town settings. In fact, in some cases, the same institution would be a *kafene* by day and a *krŭchma* by night; as the sun went down the coffee would be replaced with wine or *rakia* (plum brandy), all consumed with the ubiquitous tobacco.[36] Hence sober pursuits that arguably helped fuel the revival, were always accompanied by familiar "drunken" forms of social interaction, which persisted and even flourished. This was particularly true given the late-Ottoman and post-Ottoman urbanization of Bulgarian populations and Europeanization of Bulgarian urban life. Both factors brought alcohol consumption into the heart of Bulgarian cities, from the village or from abroad, respectively.

Still the seemingly sobering shift in Balkan practices was viewed with approval by many Western observers from the time, particularly by Protestant missionaries who had a significant presence and influence by the mid-nineteenth century.[37] Admittedly for numerous Western observers, the Ottoman coffeehouse became emblematic of an Oriental penchant for pleasure and idleness and so Ottoman decadence and decline.[38] But at the same time, Protestant observers, most famously St. Clair and Brophy, were decidedly more aghast at the drinking practices of the proverbial Balkan Christian man who was "morally and physically degraded and idly lies dead drunk upon a dung-heap."[39] This image of the drunken Bulgarian man was one of many that justified, in a sense, British support for the maintenance of the Ottoman political status quo, the sober Muslim ruling over the drunken Christian.[40] Along these lines, as British traveler Fanny Blunt was critical of the "lazy and sensuous Turk," she also recognized that coffee and tobacco were at least nonalcoholic and hence sober pursuits.[41] In fact, she described with approval the coffeehouse culture that had also developed among Bulgarian shopkeepers and guilds, who met before and after work for coffee and a *chibouk* (long pipe) full of tobacco. Hence, she claimed, "Bulgarians are not all drunks as most assume."[42] Recast as sober and positive, for Blunt the smoking Turk became a veritable role model for the Bulgarian Christian, as the coffeehouse was recast a temple of sobriety and productive interchange. Indeed, for Bulgarian men in the mid-late nineteenth century the *kafene* apparently served just such a function. Still, notions of idleness and moral degradation connected to the *kafene* and *krŭchma*—along with an increasingly blurred line between sober and intoxicated pursuits—were never far from the surface. And the West itself was paradoxically both the cause and solution.

Sacred and Secular Abstinences

Throughout Bulgaria, a brisk Europeanization, institutional and cultural, followed the attainment of autonomy following the 1877–1878 Russo-Turkish War. Among other things this meant the rapid influx of European-style cafés and new leisure venues and aesthetic forms in the last decade of the nineteenth century.[43] Sofia in particular underwent a major transformation in the years following autonomy in 1878, when foreign ideas, bodies, and commerce flooded the new capital. By the 1880s and 1890s Sofia had developed a lively and assertive café set, with its own diverse political and intellectual ferment. Bulgarian *kafene* and urban leisure culture blossomed, particularly after the extreme social changes wrought by World War I. But these changes, far from universally welcomed, for many Bulgarians were seen as deleterious to native culture, harbingers of physical and moral degradation. With this in mind various quarters of Bulgarian society sought renewal in a variety of ways, and women were invariably seen as critical to such visions.

Protestant missionaries and local Bulgarian converts were one of the first and remained one of the most critical advocates of Bulgarian "moral" renewal throughout this period. The 1848 establishment of Robert College in Istanbul, on the picturesque banks of the Bosporus Straits, was indicative of the American commitment to educational and "moral uplift" in the region.[44] Robert College graduated more Bulgarians than any other ethnic group, and a women's college was also set up in Istanbul that attracted numerous Bulgarian women.[45] Anglo-American Protestant missionaries established missionary bases and schools for boys and girls throughout Ottoman and post-Ottoman Bulgaria, penetrating and contributing to early Bulgarian educational and cultural life as with no other Balkan people.[46] From the very beginning the Protestants focused on abstinence as a way to counter traditional moral ills, as well as newly proliferating "immoral pursuits" coming from the West.[47] As Cyrus Hamlin, Robert College's founder and president until 1877 bemoaned, "Steam made Constantinople a commercial city and brought civilization, the arts, and the vices, of the West and the East together in the Ottoman capital."[48] Mary Patrick Mills, who was president of the American College for Girls in Istanbul from 1889 to 1923, described the city as a "labyrinth of temptation."[49] Bulgarian students continued to be educated in the diverse and debauched milieu of Istanbul, while Sofia underwent a far-reaching Europeanization after 1878, which brought a wide range of new leisure venues including cabarets, taverns, and beer halls. In part as a direct reaction, by the 1880s Protestant organizers were distributing some 100,000 tracts annually in the new Bulgarian principality.[50] Protestant women associated with the international Women's Christian Temperance Union were actively involved in the host of measures that included teaching abstinence in Protestant schools, disseminating tracts, and posting columns in the Protestant publications.[51] By the early twentieth century even the sober *kafene* raised the ire of the Protestant abstinence movement, because of the apparent mixing of alcohol with tobacco, gambling, and presumed sexual debauchery. This potent brew seemed to ferment unfettered in Bulgaria's new public establishments, and Protestant abstinence movements increased in intensity along with this perceived "moral freefall" of Bulgarian society.[52]

The Protestants were not the only ones to take note and react to these apparent changes. Bulgarian socialist activists, under the influence of select European socialists, were also increasingly focused on Bulgarian "vices" in their visions of brighter socialist future. By the early twentieth century, a colony of left-leaning Bulgarian students studying in Geneva, with the ringleader Khristo Dimchev, became the core of a socialist-oriented abstinence movement that slowly began to proliferate through student and teacher organizations in Bulgaria beginning in 1906.[53] For most Bulgarians, as Dimchev acknowledges, Protestantism and abstinence were synonymous before World War I, and it would have been "incomprehensible for a healthy person to abstain, if they were not an evangelist."[54] With this in mind socialists were not highly successful in this prewar period, and the Protestants were by far the most numerous and organized forces of abstinence. Significantly, Dimchev admitted that he was influenced by Protestant thought on abstinence, though his primary inspiration was the well-known Swiss psychologist and outspoken socialist and temperance advocate August Forel, whose ideas and personage were central to the Bulgarian socialist (later communist) movement.[55] Forel, notably, was a former Protestant who had turned away from religion and toward socialism in the course of his intellectual and political development, his utopian visions of a bright future shifting from heaven to earth.[56]

As strange as it seems, there was a certain natural symbiosis between the activities and paradigms of these two very different sets of social actors. In a certain sense, both socialists and Protestants had laid deep roots in nineteenth-century Bulgarian intellectual soil. In addition to the authority of Robert College, many figures from the Bulgarian national revival, most famously Petko Slaveikov, were also influenced by Protestant visions of moral uplift and restraint. At the same time most Bulgarian nineteenth-century intellectuals were attracted to some form of socialist thought, which in part explains the prevalent social critique of the *krŭchma* as a source of "oppression"

for both the rural peasant and urban working-class man.[57] In addition, there was notable Protestant-socialist cross-breeding. Dimitur Blagoev (1856–1924), for example, was
the product of a Protestant upbringing and went on to become the founding father of
Bulgarian socialism, establishing the Social Democratic Workers Party in 1891 and its
left-wing precursor to the Communist Party in 1903.[58] Other notable socialists were
brought up in Protestant households, like Nikola Vaptsarov (1909–1942), revolutionary, poet, and vanguard of the abstinence movement in the Pirin region.[59] Protestantism, in a sense, was the first step to radical socialism, representing as it did a break
from Bulgarian religious (and hence even national) tradition. Raina Kostentseva—our
intrepid traveler who almost ignited the Orient Express—concurred: "There is something revolutionary in the activities of the Protestants that that attracts many people."[60] Indeed, by the early twentieth century, at least a segment of Bulgarian socialists
seemed to echo Protestants in their call for a revolutionary change in Bulgarian consumption habits, that is, the advocacy of total abstinence.

But as Bulgarian socialism matured it was also deeply divided along a number
of lines, including attitudes toward abstinence. For Protestant missionaries temperance was a defining feature of their community, but for socialists only a segment of
the movement rejected alcohol and tobacco as capitalist vices. The rest were mired in
smoke at Sofia's *kafene* Sredets or Tsar Osvoboditel as well as other cafés across Bulgaria
or Europe. Or they were drinking and smoking with their fellow working-class or rural "toilers" at the local *krŭchma*.[61] Indeed the *krŭchma* became a major organizational
point for socialists, especially in rural areas, as the urban *kafene* was both a meeting
place and a workshop for leftist intellectuals in Bulgaria and abroad.[62] Drinking, and
particularly smoking, was by no means taboo among Party operatives, men or women.
Vela Blagoeva (1858–1921), the founder and one of the leaders of the socialist women's
movement in Bulgaria and the wife of Dimitur Blagoev, was a well-known smoker and
one of Bulgaria's first "modern girls."[63] Even in the interwar period as the communist
abstinence movement gained momentum, the head of the Party Georgi Dimitrov was
still a well-known public smoker, and a "passionate" one.[64] For men their consumption practices were by no means a departure from established traditions of Bulgarian
male sociability, but for women in the movement such practices marked them as political and social radicals. For women in the Party and its left-leaning supporters the
radical ideology of communism provided a breaking point from traditional Bulgarian
society, an ideology of emancipation and release. And certainly public smoking, as
drinking, was not taboo for women in the movement. In fact, politically left-leaning
women were among the few who braved the still male dominated public spaces, even
in the interwar period.

For most women, Orthodox strictures, perhaps mixed with newer notions of middle-class propriety, were still in force before and after World War I. According to Raina,
the Orthodox notion of "shame" and "sin" kept women in this period from smoking
or drinking in public:[65]

> It was shameful for women to cut their hair, to smoke, to go on walks alone, to
> wear a dress without a jacket, to go alone into a *kafene* or *krŭchma* … in a word
> the understandings of the time of shame and sin were to a large degree a prod-

uct of [Orthodox Christian] religious fanaticism and religious superstitions ... the biggest victim of the superstitions and the stupidity of the guild class was women, treated for centuries as a lower being than men.[66]

Raina noted that the "modern girls"—drinking, smoking and sporting the *bubikopf* (German for short haircut)—were still quite unusual and scandalous in interwar Bulgaria.[67] Although an increasing number of venues were open to women in this period, namely Sofia's new "family" establishments, even there women could only appear accompanied by husbands or fathers without a loss of propriety. Raina, in fact, visited many such venues, but however liberated she might have been, she still lamented not being able to visit a certain "family" *biraria* (beer hall) because "no one would accompany her."[68] As she frankly admitted, only Orthodox women with a "certain past" could be seen smoking in public, or indeed appearing in traditionally male establishments, or virtually any establishment unaccompanied, in this period.[69] Some of the first women to frequent public establishments in postliberation Bulgaria, she claimed, were prostitutes, singers, or dancers—in short, either foreign women or local women of questionable morals.[70] Indeed, this specter of the "fallen woman" kept women like Raina continually on the outside looking in.

But of course not all women who took their place in Bulgaria's rapidly expanded interwar leisure possibilities were of the "fallen" variety. Gradually newly educated and engaged women intellectuals began to appear in the gleaming European-style café, clubs, and cabarets of Sofia, some without accompanying men. The city—its anonymity, chaos, and for many a loss of religious or ethnic-family strictures—meant a new freedom of movement, an intoxicating escape, especially after World War I. For Bulgaria as elsewhere, the war had a cataclysmic effect on public life. Mobilized on the front and home front, a whole new range of social actors, namely peasants, workers, and women, had heightened expectations for political and social change in the wake of war. At the end of World War I, Europe was shaken by the Russian Revolution followed by leftist revolts in defeated Germany and Hungary; Bulgaria also ended the war in defeat and a near revolutionary situation that swept the Agrarian Party to power. The Bulgarian Communist Party was at the peak of its popularity in the immediate postwar years, garnering the second highest number of votes in the national assembly and taking power in a number of municipalities. Along with such political changes, Bulgarian social life evolved rapidly with the penetration of a number of new direct outside influences—namely the occupying Allied troops and large numbers of Russian Whites fleeing revolution. In tandem with such transformations, new clubs, cabarets, cafes, and other consumption venues expanded and smoking and drinking became palpably more widespread, among men as well as women of various classes.[71]

Bulgarian observers explained these changes as a reaction to the devastation and humiliation of war, which propelled this surge in escapist modes of consumption. As one article in the abstinence newspaper *Trezvenost* noted, in Bulgaria the war took 100,000 lives, injured 220,000, called for the equivalent of 450 million dollars in reparations, and caused the number of taverns in Bulgaria to mushroom to 20,000. The abstinence advocates called this the "unstoppable march of the *krŭchmi*" that allowed for one tavern for every 245 people.[72] In the burgeoning abstinence literature, these changes were described as part and parcel of a larger moral freefall caused by both the ill-effects of war and the rapid influx of Western material culture and vice.[73] Of course, for Bulgaria as elsewhere, men supplied with cigarettes at the front, and a variety of new roles and responsibilities for women (and their attendant emancipation) on the home front, were responsible for the rise in smoking. After all, for these reasons smoking and drinking in ever-greater amounts was a global phenomenon in this period, in states that won or lost the Great War.[74] Or perhaps it was less the loss of war than the need to forget that drove postwar societies into their own versions of the American Roaring Twenties." As Bulgarian journalist and keen observer of interwar Sofia Khristo Brŭzitsov posited, "maybe those who were saved from the war just wanted to live … a second life!"[75]

But it was not just broken men returning from the front, but also women in the cities of postwar Bulgaria, who entered beer halls, coffeehouses, and other public smoking (and drinking) venues and upset traditions of male-dominated leisure.[76] In the villages, where religious and social stricture stayed firmly in place, the village *kafene* and *krŭchma* remained an exclusively male space until after World War II.[77] But in Sofia as other Bulgarian cities, American films that featured and popularized glamorous smoking ingenues—for example, Greta Garbo and Marlene Dietrich—began to be widely shown. At the same time the Bulgarian cigarette industry began to market women's cigarettes, with pictures of glamorous smoking women on the front or with pictures of Hollywood stars inside. By the 1930s new brands of women's cigarettes had emerged, fighting for sales among Bulgarian women, like Zora with its Hollywood-like ingenue on the pack. Zora also featured collectable pictures of Hollywood stars, as well as Miss Bulgaria or other glamorous beauty pageant participants.[78] Although working-class women began to smoke in a more limited way, among the growing ranks of the female bourgeoisie and intelligentsia smoking became a common social practice, an indicator of emancipation, glamour, and cosmopolitan urban culture. As an article in *Trezvenost* noted, "And in Bulgaria everyone smokes, even children, students, and grownups, old people, and recently it is considered a sign of special breeding and an expression of 'good taste;' of high intelligence."[79] But it was also, for observers, a specter of rebellion, of communism, of prostitution, of a threat to traditional values. As smoking men in this period occupied a central role in Bulgarian political and cultural life, the emergence of smoking Bulgarian women, particularly in public, was far from mainstream; indeed it was still slightly scandalous, brazen, and outside the social norm.

A number of well-known Bulgarian women intellectuals and/or performance artists, generally associated with the political left, became visible regulars in the Bulgarian café scene.[80] Most famously the poetesses Elizaveta Bagriana and Dora Gabe—along with other writers from the journal *Zlatorog* and its attendant intellectual circle by the same name—frequented the upscale café Tsar Osvoboditel as well as other social venues where intellectuals and other Bulgarian elites congregated on a daily basis.[81] This was seemingly progress, but of the 106 regulars whom Alexander Dobrinov immortalized in his famous 1935 caricature-painting of Tsar Osvoboditel only four were women—all of them at the same table.[82] Undoubtedly, not all women at the Tsar Osvoboditel would have been considered regulars, let alone been worthy of a place in

Dobrinov's painting. Some women came in as admirers, or to be admired, taking their place in the picture window to be seen by all that were inside as well as by passersby.[83] As Dimo Kazasov—a prominent leftist cultural and political figure from the period— pointed out, Tsar Osvoboditel was the only café where women could go in this period, and even this was decidedly *muzhko* (male) in the morning; women only made their appearance in the afternoon and evening.[84] This was surely an exaggeration, as other sources mention the presence of women in other *kafenes*, for example the Café Royale and the Sredets, and Bagriana was known to frequent the Armensko, where she met left-leaning intellectuals of the time.[85] Although the communist movement was forced underground by the 1923 rightist Sgovor coup that was staged by a collection of rightist political and military forces, men and women of the Bulgarian left retained a visible place at the central cafes of Sofia, as café culture continued to be central to Party organization and to intellectual labor.

Of course for much of the Party this kind of public visibility became impossible after 1923. For example, for Tsola Dragoicheva—the most famous and politically committed communist woman from the period—a life of daily smokes and conversation at the Tsar Osvoboditel would have been out of the question.[86] Dragoicheva essentially lived underground, unregistered, with a variety of pseudonyms, hiding out in apartments arranged by other Party operatives. In the early postwar years, Dragoicheva had frequented the rather modest Party club in Sofia,[87] but for most of the interwar years the Sofia *kafene* scene was far too dangerous for her. In fact, Dragoicheva, as other leftists of the period, was openly critical in her memoirs about Sofia's "highlife" and the "snobbism" of the rich elite on their way to the "Union Club," while "unseen plebians with empty pockets" filled the city.[88] Dragoicheva, though, was never openly critical of leftist intellectuals who gathered in such venues, and certainly not Party elites, like Georgi Dimitrov and others who met primarily in the cafes and restaurants of Berlin.[89] Still, for Dragoicheva, as for many other communists (male and female), such decadent—even capitalist—behavior seemed to conflict with their deep commitment to social change and political revolution. But she, and perhaps even the more hard-core ascetic communists of the period, was cognizant of the need to tolerate the smoking and drinking that was so prevalent among many of its thinkers, supporters, and even leadership.[90]

The communist-led abstinence movement was rapidly coalescing in this period, especially after the repression, restrictions, and eventual banning of the Bulgarian Communist Party from 1923 to 1925. Increasingly, abstinence associations became the only political platform for Party work on the local level and across Bulgaria they became front organizations in which the Party cooperated with noncommunist elements, including Protestants.[91] For communists, abstinence was deemed "revolutionary," a "revolution in morals … of the soul and the mind, without barricades."[92] This cooperation was in fact by Comintern (the international communist organization in Moscow) dictate, the so-called united front policy of the 1920s, and popular front (or second united front) policy (1935–1943). Left-leaning teachers and students founded or infiltrated abstinence organizations, using them to push forth the broader Party agenda in one way or another. In part they did this by advocating abstinence as a "social question" connected to malaise among the working class and even the "exploitation" of the worker and peasant by the *krŭchmar* (tavern owner).[93] During this time, communist interests strangely dovetailed with Protestant ones and both worked together—sometimes uneasily—in various "neutral abstinence Unions."[94] Admittedly, for communists capitalism and imperialist war were the culprits and communism the cure, while for Protestants Godlessness was cause and true religion the cure, but for both the "afflicted" and the most immediate solution was abstinence.

A philosophical synergy between the two movements that were in no way political allies developed more than at any other point in history precisely because of shared notions of abstinence and social regeneration. Protestants and at least a segment of Bulgarian communists developed similar concerns about the medical, but more pointedly moral and social effects of alcohol and tobacco use and abuse in Bulgaria. Both targeted drinking and smoking by modern "pleasure" seekers in the Bulgarian urban *kafene* or cabaret, which they connected specially to the ill effects of Western penetration. Interestingly, for both, modernity was simultaneously the problem and the solution, as modern notions of hygiene and discipline were to supplant backward and/or newly minted practices of decadent and nonproductive leisure. As an article from a leftist abstinence publication explained:

> The abstinence movement, as a social phenomenon, was the child of modern society. It arose as a reaction against certain social ills: alcoholism, tobacco smoking, licentiousness, gambling and prostitution, which are inseparably connected to and a product of the modern social economic structure of society, which especially during its decline in the current period have increased markedly, strengthening the processes of demoralization and degeneration.[95]

This resonated with similar claims in Protestant publications from the period that seemed to agree with communists that Western "vices" and even social malaise were a central part of the problem. But both wings of the abstinence movement also believed that modern man should be able to use reason and science to "reign in inimical passions."[96]

In addition, for both, there was also a clear overlap in the gender dimension to their abstinence message, in terms of both the diagnosis and cure. Notably, across the movement activists advocated a deeply gendered "moral uplift" to counter "backward" patterns of *krŭchma* drunkenness as well as modern *kafene* and *krŭchma* debauchery in the city as well as village. Working-class or rural women, in particular, were seen as the express "victims" of alcoholic and smoking men, who spent sparse family resources on unnecessary intoxicants. Working-class and peasant men were also "victims," in a sense, of the scourge of alcohol and tobacco (and also capitalism), at the same time, as the primary consumers of alcohol and tobacco, they were also perpetrators of misery and violence against their wives and children.[97] Numerous images and morality tales from the abstinence literature depicted working-class men drinking and smoking in neighborhood bars, while their wives (and children) literally starved to death at home in cold unheated flats.[98] This is not surprising given that men not only consumed alcohol and tobacco in much greater quantities than women, but they did this in public venues where they could be openly judged by society at large. Furthermore, according to abstinent observers, worker and peasant women who took care of the household in addition to working in the fields and factory seemed to *work* much harder than men who squandered their free time and precious money in the tavern.[99] The social implications of these analyses were clear: alcohol and tobacco were killing the Bulgarian family, if not the nation, and women were bearing the brunt of the ruinous practices.

Women were expected, to do more than men, and were beseeched in abstinence publications to control men's behavior and make men stop their rampant consumption of alcohol and tobacco.[100] But women were also implored to become actively involved in abstinence organizations.[101] And women did participate in increasing numbers, as the cause appealed to ever larger numbers of Bulgarian women.[102] A number of well-educated and articulate women, like Vera Zlatareva and Ianka Tosheva, gave their names, energy, and intelligence to the cause, contributing regularly to abstinence publications.[103] But even when women were not actively involved in the movement, they were cited as the major audience for musical evenings, lectures, and other cultural events organized by abstinence associations. Women, together with local Turks, were generally seen as well as avid supporters of referendums to close local *krŭchmi* that were continuously waged by abstinence organizations, although with only limited success.[104] But despite all efforts and achievements, abstinence movements continually lamented the lack of participation among women in this period, which seemed to indicate that women "accepted" the "vulgar" consumption habits of their men.[105]

The specter of the "fallen woman," smoking, drunk, and often selling herself for sex, was a central figure in both radical religious and secular concerns.[106] In general, smoking and drinking women were a threat not only to "traditional" mores—now tainted by "Western vice"—they were also a threat to visions of an abstinent future. This reflected not only the rise in male drinking and smoking—as opposed to the abstinent women foil—but also deep-seated concerns about the growing visibility of smoking and drinking women.[107] It was the "modern girl," in fact, and not just the prostitute, who embodied this threat, the moral freefall that would doom the nation. And it was not just men who were alarmed by these implications, as an article by Ianka Tosheva—an educated and vocal advocate of abstinence—amply shows. Ruminating on just these issues, Tosheva explores how Bulgarian women had been emancipated in a variety of ways, but seemed to be squandering the opportunity.[108] To illustrate, Tosheva relates a probably all too familiar anecdote about a rural young woman who comes to Sofia to study medicine at the university. Soon after the young woman in question begins to attend soirées and other such events, she falls into "modern company," with whom she begins to "smoke, drink and flirt." As Tosheva narrated with a clear sense of irony, eventually the girl returns home to the provinces completely "modern." According to Tosheva, the girl-student's understanding of "modern" was highly misguided and needed to be redirected toward "a true understanding of science and health." If the girl was a truly modern emancipated woman, she should have returned home with her knowledge of medicine and used it for the common good. As a "good citizen," she would not smoke or drink and would demand that her future spouse and children also be abstinent.[109] But alas, as Tosheva lamented, this was by no means the norm.[110] In fact, as Tosheva complained, only 3 percent of women university students in Bulgaria were members of abstinence associations, which they shunned because of their own "egoism" and the idea that such activities were a "waste of time."[111] Tosheva continued that most "modern smokers" felt that abstainers were merely "sectarians" [Protestants] who wanted to "limit their freedom."[112] In the end, for Tosheva the misguided "modern girl" needed to be redirected so that another brand of modern woman could emerge, one that was abstinent and socially conscious. Interestingly, this vision of the abstinent modern woman propagated by the left was not so far afield from the "new Bulgarian" women of the Protestants, or even right-wing Bulgarians, that is, she was pure, productive, and managing herself and her children in service of the nation.

Ultimately, whatever goals Protestant and communist abstinence movements shared, their marriage of interest was doomed by politics. Evangelical groups were tense and suspicious about the communist "reddening of the movement" in the interwar period and its use for political ends.[113] They repeatedly and futilely called for political neutrality to be maintained, as communist youth, they complained, would show up at meetings with red banners, attracting the attention of local police and provoking arrests.[114] Protestant anxieties were surely heightened by the increased suspicion of Bulgarian authorities toward these abstinence associations, which by 1927 or 1928 had led to the shutting down of student congresses and organizations.[115] Although revived in the early 1930s, by the end of the decade the movement was again under direct fire because of its clear communist dimension. Meetings were watched and broken up if seen as political, and finally in 1938 all abstinence associations and publications were dissolved, only to re-emerge (in a purely communist guise) after World War II.[116] In the end, those who saw drinking and smoking Bulgarian men and women as threats were defined as threats themselves. It was only after the dramatic change in power after World War II, that the communist abstinence message would re-emerge, though strongly mitigated by communists for whom smoking and drinking was an inseparable part of their utopian future.

Conclusion

For both Bulgarian women and men, changing patterns of consumption in the nineteenth and early twentieth century were tightly intertwined with dramatic changes in political participation, patterns of social interaction, and religious orientation. For Bulgarian men, the move from the *krŭchma* to the *kafene* became a critical step in participation in Ottoman and European social practices and patterns, hearkening a new era of commercial revival, secularization, urbanization, and sociability. At the same time, until early in the twentieth century women were expressly excluded from both the sober *kafene* and the traditional drunken *krŭchma*. But as changes ensued for both men and women, a range of concerns emerged, in part as a result of the transgression of boundaries that marked the twentieth century—between the sober *kafene* and the drunken *krŭchma*, between traditional public male and private female sociability, and finally between East and West. As a result, radical models of morality and sobriety, religious (Protestant) and secular (socialist), developed and spread as Bulgarians groped for grounding in a rapidly changing landscape of pleasure and propriety. Perhaps things would have been different if Raina had actually set the Orient Express on fire?

◊ Acknowledgments

I thank the International Research Exchanges Board and the National Council for East European and Eurasian Research for funding much of this research on this project. I like thank Kristen Ghodsee and Pam Ballinger for organizing the very fruitful symposium on secularisms and religiosities in Southeastern Europe in the fall of 2009. The comments of all present were highly stimulating and useful for rethinking this article. Finally I thank the *Aspasia* readers and especially editors, Sharon Kowalsky and Krassimira Daskalova, for their thorough editing and insightful comments.

◊ About the Author

Mary Neuburger is an associate professor in the department of history at the University of Texas in Austin. She is the author of *The Orient Within: Muslim Minorities and the Negotiation of Nationhood in Modern Bulgaria* (Cornell University Press, 2007). Her current book project, with the working title *Inhaling Modernity*, explores tobacco commerce, production, and consumption in Bulgaria in the modern period. In addition, Neuburger is coediting a book with Paulina Bren, *Communism Unwrapped*, on consumption under communism in Eastern Europe. Email: burgerm@austin.utexas.edu

◊ Notes

1. See Raina Kostentseva, *Moiat roden grad Sofiia v kraia na XIX–nachalo na XX vek i sled tova* (My native city Sofia in the end of the nineteenth–beginning of the twentieth century and after that) (Sofia: "Riva," 2008), 80–81. All translations from Bulgarian are by the author unless otherwise noted.

2. Scholars of consumption have long explored how patterns of consumption are embedded in and foster social change. On gender and consumption, see Victoria de Grazia and Ellen Furlough eds., *The Sex of Things: Gender and Consumption in Historical Perspective* (Berkeley: University of California Press, 1996).

3. Tobacco has been dubbed one of the "big three" along with alcohol and caffeine in the global "psychoactive revolution" in which "soldiers, sailors, and students were the advance guard." David Courtright, *Forces of Habit: Drugs and the Making of the Modern World* (Cambridge: Harvard University Press, 2001), 9. On the "social lives" of commodities see Arjun Appadurai, "Introduction: Commodities and the Politics of Value," in *The Social Life of Things: Commodities in Cultural Perspective*, ed. Arjun Appadurai (New York: Cambridge University Press, 1992), 3–63.

4. See, for example, Susanna Barrows and Robin Room, eds., *Drinking: Behavior and Belief in Social History* (Berkeley: University of California Press, 1991).

5. Here I use the term "Bulgarian" for Slavic-speaking Christians, but with the assumption that much of this population did not have a well-defined national consciousness until well into the twentieth century.

6. I put "sober" in quotes here because in many coffeehouses opium and hashish were also routinely consumed. On tobacco and the Ottoman coffeehouse, see James Grehan, "Smoking and 'Early Modern' Sociability: The Great Tobacco Debate in the Ottoman Middle East, Seventeenth to Eighteenth Centuries," *American Historical Review* 3, no. 5 (2006): 1352–1376. On coffee, see Antony Wild, *Coffee: A Dark History* (New York: W.W. Norton, 2005), 52–53.

7. Khristo Kulichev, *Zaslugite na Protestantite za Bŭlgarskiia narod* (Protestant contributions to the Bulgarian nation) (Sofia: Universitetsko Izdatelstvo "Sveti Kliment Okhridski," 2008), 267.

8. I use the term "modern girl" to refer to the recognized global phenomenon of the inter-war period—of the young woman as emancipated producer-consumer. For an insightful study of its global reach and variations see Alys Eve Weinbaum et al., eds., *The Modern Girl Around the World: Consumption, Modernity, Globalization* (Durham: Duke University Press, 2008).

9. In the Ivan Vazov novella, *Chichovtsi: Galeriia ot tipove i nravi Bŭlgarski v Tursko vreme* (Uncles: A gallery of Bulgarian types and morals in Turkish times), satire mixes with the burlesque in cutting representations of Bulgaria's persistent and newly emergent social types in this period centered around the *kafene,* in all of its ethnic and social complexity.

10. Ivan Vazov, *Under the Yoke* (New York: Twayne, 1971), 163.

11. Ibid., 110. It is unclear whether Vazov was familiar with Balzac's famous novel *The Peasants* in which he called the café the "parliament of the people," but it is entirely likely that he would have been exposed to the work of the French writer.

12. These dates have been hotly debated in Bulgarian historiography. For a recent monograph on the historiography of the period, see Rumen Daskalov, *The Making of a Nation in the Balkans: Historiography of the Bulgarian Revival* (Budapest: Central European University Press, 2004).

13. The European café was originally an adaptation of the Ottoman coffeehouse, where coffee and coffeehouse sociability was first discovered by Europeans. See, for example, Danilo Reato, *The Coffee-House: Venetian Coffee-Houses from the 18th to 20th century* (Venice: Arsenale, 1991), 13; and Markman Ellis, *The Coffee-House: A Cultural History* (London: Weidenfeld and Nicholson, 2004), 75.

14. I am not using the term awakening in the traditional, primordialist sense of awakening pre-existing national identities, but rather in the modernist sense of the imagining or constructing of national identities by national elites. For the concise discussion of the modernist-primordial debate, see Anthony D. Smith, *The Ethnic Origins of Nations* (Oxford: Blackwell, 1986), 6–18.

15. Katiia Kuzmova-Zografova, "Ot imaniarski legendi do mornata utopiia 'Viensko kafene'," (From the treasure-hunters' legends to the girl's utopia 'Vienna Café') in *Kafene Evropa* (Café Europe), ed. Raia Zaimova (Sofia: Izdatelstvo Damian Iakov, 2007), 158.

16. See Yassen Borislavov, *Bulgarian Wine Book: History, Culture, Cellars, Wines* (Sofia: Trud, 2007), 74–79.

17. Relli Shechter, *Smoking, Culture and Economy in the Middle East: The Egyptian Tobacco Market, 1850–2000* (New York: I.B. Tauris, 2006), 17.

18. Khristo Nedialkov, *Bŭlgarski kulturni deitsi za trezvenostta* (Bulgarian cultural activists for abstinence) (Sofia: Meditsina i fizkultura, 1977), 15–26.

19. Kristo Stoianov, *Dvizheniie za trezvenost v razgradski okrŭg* (The abstinence movement in the Razgrad region) (Razgrad: Okrŭzhen Komitet za Trezvenost, 1983), 6.

20. Mercia MacDermott, *Apostle of Freedom: A Portrait of Vasil Levsky against a Background of Nineteenth Century Bulgaria* (South Brunswick, NJ: A.S. Barnes, 1969), 258.

21. A Communist period proponent of abstinence even claims that Levski organized the first abstinence organization in Karlovo in 1869, though this claims seems to be entirely spurious. See Pavel Petkov, *Borbata za trezvenost vŭv vrachanski okrŭg, 1920–1980* (The struggle for abstinence in the Vratsa region, 1920–1980) (Sofia: Izdatelstvo na Otechestveniia Front, 1982).

22. As cited in MacDermott, *Apostle of Freedom*, 269.

23. Raina Gavrilova, *Kolelota na zhivota: Vsekidnevieto na Bŭlgarskiia vŭzrozhdenski grad* (The wheel of life: The everyday in the Bulgarian revival city) (Sofia: Universitetsko Izdatelstvo

"Sv. Kliment Okhridski," 1999), 45–46; and Petur Karaivanov, *Vasil Levski po spomeni na Vasil Karaivanov* (Vasil Levski according to the memoirs of Vasil Karaivanov) (Sofia: Izdatelstvo na Otechestveniia Front, 1987), 138. See also Nikola Alvadzhiev, *Plovdivska khronika* (Plovdiv chronicle) (Plovdiv: Izdatelstvo na "Khristo G. Danov," 1971), 336.

24. On the cult of Levski in Bulgarian history, see Maria Todorova, *Bones of Contention: The Living Archive of Vasil Levski and the Making of Bulgaria's National Hero* (Budapest: Central European University Press, 2009). For earlier work on Levski as hero see Nikoli Genchev, *Vasil Levski* (Sofia: Voenno Izdatelstvo, 1987).

25. William Webster, *Puritans in the Balkans: The American Board Mission in Bulgaria, 1878–1918: A Study in Purpose and Procedure* (Sofia: Studia Historico-Philologica Serdicensia, 1938), 88.

26. Protestant educational and other efforts were sometimes welcomed, but it is also important to note the stonings of local missionaries, the ostracization of cooperating or converted Bulgarians, and even the kidnapping of American missionaries. Conversely, one source boasted of a temperance rally attended by 400 people in 1884. See, for example, James Clarke, *Bulgaria and Salonica in Macedonia* (Boston: American Board of Commissioners for Foreign Missionaries, 1885), 5; and Webster, *Puritans in the Balkans*, 22–23.

27. Nikola Aretov, "Kafeneta, krŭchmi, saloni, i khanove v Bŭlgarskata literatura ot vtoriia polovina na XIX vek" (Cafés, taverns, salons, and hans (inns) in Bulgarian literature from the second half of the nineteenth century), in Zaimova, *Kafene Evropa*, 60.

28. Grehan, "Smoking," 9–10. Numerous nineteenth-century travelers and diplomats comment on the prevalence of smoking among women of all religions, but again not in the coffeehouse. See for example, the writings of diplomat William Miller from the 1890s. William Miller, *Travels and Politics in the Near East* (New York: Arno Press, 1971), 418.

29. Mary Patrick Mills, *Under Five Sultans* (London: Century, 1929), 239.

30. There has been some work done on elite Ottoman women's social world. See, for example, Suraiya Faroqhi, *Subjects of the Sultan: Culture and Daily Life in the Ottoman Empire* (London: I.B. Tauris, 2005), 101–123.

31. As Raina Gavrilova argues, it was not unusual for Bulgarian women to smoke and even Bulgarian nuns were known to "smoke the *chibouk* (pipe) and drink *rakia* (brandy)." Gavrilova, *Koleloto na zhivota*, 308.

32. Georgi Traĭchev, *Grad Prilep: Istoriko-geografsko i stopansko pregled* (The city Prilep: A historical-geographical and economic review) (Sofia: Pechatnitsa "Fotinov," 1925), 137.

33. Ibid.

34. It is worth noting that newer literature on non-European societies, including the Ottoman empire, has decidedly decentered the Habermasian notion of the "public sphere" as a uniquely Western European phenomenon., that is from Jürgen Habermas, *The Structural Transformation of the Public Sphere: An Inquiry into a Category of Bourgeois Society* (Boston: MIT Press, 1992). On non-Western perspectives see, for example, William Rowe, "The Public Sphere in Modern China," *Modern China* 16 (1990): 309–329; and Srirupa Roy, "Seeing a State: National Commemorations and the Public Sphere in India and Turkey," *Comparative Studies in Society and History* 48 (2006): 200–232. On the Ottoman empire, see Cengiz Kirli, "The Struggle over Space: Coffeehouses of Ottoman Istanbul, 1780–1845" (PhD diss., State University of New York at Binghamton, 2001); Cengiz Kirli, "Coffeehouses: Public Opinion in the Nineteenth Century Ottoman Empire," in *Public Islam and the Common Good*, ed. Armando Salvatore and Dale F. Eickelman (Boston: Brill, 2004), 75–97; and Selma Özkoçak, "Coffee Houses: Rethinking the Public and Private in Early Modern Istanbul," *Journal Of Urban History* 33 (2007): 965–986.

35. Recent Bulgarian historiography has begun to draw a direct connection between the *kafene* and Bulgarian urbanization, cultural revival, "political emancipation," and the Bulgarian

"public sphere." See, for example, Raina Gavrilova, "Kafeneto: Vreme i miasto," (The café: time and place) in *Kafeneto kato diskurs* (The café as discourse), ed. Georg Kraev (Sofia: Nov Bŭlgarski Universitet, 2005), 94–95.

36. Khristo Danov, *Za teb mili rode* (For you my dear homeland) (Plovdiv: Izdatelstvo "Khristo G. Danov," 1978), 113.

37. There is a rich literature that analyzes the agendas and implications of Western travel literature on the Balkans. See Maria Todorova, *Imagining the Balkans* (New York: Oxford University Press, 1997); John Allcock and Antonia Young, eds., *Black Lambs and Grey Falcons: Women Travelers in the Balkans* (Bradford, UK: Penguin Classics, 1995), 170–191; and Bozidar Jezernik, *Wild Europe: The Balkans in the Gaze of Western Travelers* (London: Saqi Books, 2004).

38. Earlier accounts of Ottoman leisure, such as the famous letters of Lady Montagu, reflect more fascination than judgment in relation to the smoking men and harem women of the empire. See Mary Wortley Montagu, *Turkish Embassy Letters* (Athens: University of Georgia Press, 1993). See also Stanislas St. Clair and Charles Brophy, *Residence in Bulgaria; Or, Notes on the Resources and Administration of Turkey: The Condition and Character, Manners, Customs, and Language of the Christian and Musselman Populations, with Reference to the Eastern Question* (London: J. Murray, 1869), 116.

39. St. Clair and Brophy, *Residence in Bulgaria,* 408. Brophy was quite shocked that his Bulgarian hosts, who gave him shelter in a remote village, offered him wine in the morning before travel instead of coffee, which would be offered by Muslim hosts (ibid., 22).

40. As Todorova points out, such imagery was guided by class bias and echoed British writings on their own lower classes of the period. Todorova, *Imagining the Balkans,* 102.

41. Fanny Blunt, *My Reminiscences* (London: J. Murray, 1918), 27.

42. Ibid., 231.

43. See, for example, Petŭr Tashev, *Sofiia arkhitekturno gradoustroĭstveno razvitie: Etapi, postizheniia i problemi* (Sofia architectural and urban development: Stages, achievements, and problems) (Sofia: "Tekhnika," 1972); and Khristo Ganchev, Grigor Doĭnchev, and Ivana Stoianova, *Bŭlgariia 1900: Evropeŭskata vliianiia v Bŭlgarskoto gradoustroŭstvo, arkitektura, parkove, i gradini, 1878–1918* (Bulgaria 1900: European influences in Bulgarian urban development, architecture, parks and gardens, 1878–1918) (Sofia: Izkustvo i arkitektura, 2002).

44. As George Washburn, the college's second but longest reigning president, articulated in his memoirs, the goals of Robert College were to inculcate "moral strength" into young men from the region. Alcohol was strictly forbidden at Robert College, and numerous students were notably "publicly thrashed" or expelled for coming back to the college dormitories drunk. George Washburn, *Fifty Years in Constantinople and Recollections of Robert College* (New York: Houghton Mifflin 1909), 298, 70.

45. Five graduates of Robert College went on to become prime ministers, and many others became ambassadors, industrialists, and other people of import in post-1878 autonomous Bulgaria. Washburn, *Fifty Years,* 95.

46. As Ilchev argues, American missionaries lived in more direct and intimate contact with Bulgarians than any other foreigners during the nineteenth century. Ivan Ilchev and Plamen Mitev, eds., *Bŭlgaro-Amerikanski kulturni i politicheski vrŭzki prez XIX–pŭrvata polovina na XX v.* (Bulgarian-American cultural and political ties in the nineteenth and first half of the twentieth centuries) (Sofia: Universitetsko Izdatelstvo "Sv. Kliment Okhridski," 2004), 12. See also James Clarke, *Bible Societies, American Missionaries, and the National Revival of Bulgaria* (New York: Arno Press: 1971).

47. Washburn, *Fifty Years,* 298.

48. Cyrus Hamlin, *My Life and Times* (Boston: Congregational Sunday School and Publishing Society, 1893), 230.

49. Patrick Mills, *Under Five Sultans,* 170.

50. With cooperation of the Bulgarian Ministry of Education, American reverend James Clark produced and distributed approximately 350,000 temperance tracts throughout Bulgaria from 1907 to 1910. James F. Clarke, *Temperance Work in Bulgaria: Its Successes* (Samokov, Bulgaria: Evangelical School Press, 1909), 3. See also Edward Haskell, *American Influence in Bulgaria* (New York: American Board of Commissioners for Foreign Missionaries, 1913), 7.

51. On the Women's Christian Temperance Union see Ian Tyrell, *Women's World, Women's Empire: The Women's Christian Temperance Union in International Perspective* (Chapel Hill: University of North Carolina press, 1991).

52. See, for example, Kulichev, *Zaslugi na Protestantite,* 273.

53. Khristo Stoikov, *Dvizheniete na tresvenost v Razgradski okrŭg, 1920–1982* (The abstinence movement in the Razgrad region, 1920–1982) (Sofia: Okrŭzhen komitet za trezvenost, 1983), 7.

54. Khristo Dimchev, "Pŭrvi Iskri" (First sparks), in *Iubileen Nauchen Sbornik, 1922–37* (Anniversary scholarly collection, 1922–37), ed. Krum Akhchiiski and Ianka Tosheva (Sofia: Studentska vŭzdŭrzhatelno druzhestvo, 1937), 10.

55. Petkov, *Borbata za trezvenost,* 21.

56. *Trezvenost,* April 1922, 7.

57. Although such figures were later claimed as abstinence "activists," their messages were not about total abstinence, but rather moderation and critique mostly aimed at lower classes, for which the consumption of intoxicants was presumed to be financially devastating. For an example of later (communist period) appropriation of such figures for the abstinence movement see Nedialkov, *Bŭlgarski kulturni deitsi,* 15–26.

58. Significantly, at least some American Protestants in Bulgaria were quite approving of the "essentially Christian virtues" of early Bulgarian socialists in this period. Haskell, *American Influence in Bulgaria,* 7.

59. See Boika Vaptsarova, *Nikola Vaptsarov; Letopis za zhivota i tvorchestvoto mu* (Nikola Vaptsarov; Chronicle of his life and work) (Sofia: Izdatelstvo na Akademiia na Naukite, 1978), 12–13, 34.

60. Kostentseva, *Moiat roden grad Sofiia,* 180.

61. Lora Shumkova, "Mechtaniia za kafene sredets," in Kraev, ed., *Kafeneto kato diskurs,* 68.

62. In Vaptsarov's hometown of Bansko, for example, the local chapter of the "narrow socialist" or later Bulgarian Communist Party was formed at a local *krŭchma.* Vaptsarova, *Nikola Vaptsarov,* 22.

63. See Elena Bogdanova, *Vela Blagoeva: Biografichen ocherk* (Vela Blagoeva: Biographical essay) (Sofia: Izdatelstvo na BKP, 1969). For a recent presentation of Vela Blagoeva, see Krassimira Daskalova, "Vela Blagoeva," in *A Biographical Dictionary of Women's Movements and Feminisms. Central, Eastern, and South Eastern Europe, 19th and 20th Centuries,* ed. Francisca de Haan, Krassimira Daskalova and Anna Loutfi (Budapest and New York: Central European University Press, 2006), 62–65.

64. Paun Genov, *S fakela na trezvenostta: Momenti ot borbata protiv pianstvoto i tiutiunopusheneto pres 1300-godishnata istoriia na Bŭlgariia* (With the torch of abstinence: Moments from the struggle against drunkenness and smoking in the 1300-year history of Bulgaria) (Sofia: Natsionalen Komitet za Trezvenost, Meditsina i fizkultura, 1980), 43.

65. Kostentseva, *Moiat roden grad Sofiia,* 172.

66. Ibid., 173.

67. For an exploration of the *bubikopf* and emancipated woman phenomenon in interwar Yugoslavia, see Marina Vujnovic, *Forging the Bubikopf Nation: Journalism, Gender and Modernity in Interwar Yugoslavia* (New York: Peter Lang, 2009).

68. Kostentseva, *Moiat roden grad Sofiia,* 223.

69. Ibid., 80.

70. Petŭr Mirchev, *Kipezhŭt: Kniga za Sofiia, 1878–1884* (Ferment: The book of Sofia, 1878–1884) (Sofia: Otechestven front, 1971), 39-44.

71. See Rumen Daskalov, *Bŭlgarsko obshtetsvo, vol 2, Naselenie, obshtestvo, kultura* (Bulgarian society, vol. 2, Population, society, culture) (Sofia: IK "Gutenberg," 2005), 152–156.

72. *Trezvenost* (Sofia), May 1922, 7.

73. Ibid., 11.

74. In the United States, under pressure of the temperance movement alcohol was made illegal on 16 January 1919. This day became celebrated as "abstinence day" among all abstinence groups in Bulgaria. The United States remained always at the forefront of global abstinence efforts, whether religious or secular, even after prohibition was abolished in 1933. On American temperance see, for example, John Burnham, *Bad Habits: Drinking, Smoking, Taking Drugs, Gambling and Sexual Misbehavior in American History* (New York: New York University Press, 1993), 1–47. At the same time, the Bolsheviks made alcohol illegal in the newly formed Soviet Union. On temperance in the Soviet Union, see Laura Phillips, *Bolsheviks and the Bottle: Drink and Worker Culture in St. Petersburg, 1900–1929* (DeKalb: Northern Illinois University Press, 2000); and Kate Transchel, *Under the Influence: Working-Class Drinking, Temperance, and Cultural Revolution in Russia, 1895–1932* (Pittsburgh: University of Pittsburgh Press, 2006).

75. Khristo Bruzitsov, *Niakoga v Sofiia: Spomeni 1913–1944* (Once in Sofia: Memoirs 1913–1944) (Sofia: Bŭlgarski pisatel, 1970), 67–68.

76. Raina Gavrilova, *Bulgarian Urban Culture in the Eighteenth and Nineteenth Centuries* (Cranbury, NJ: Susquehanna University Press, 1999), 149.

77. Dobrinka Parusehva, "Politicheska kultura na kafeneta" (The political culture of the café), in Zaimova, ed., *Kafene Evropa*, 85.

78. Rosen Takhov, *Golemite Bŭlgarski senzatsii* (Big Bulgarian sensations) (Sofia: Izdatelstvo "Iztok-Zapad," 2005), 233.

79. *Trezvenost* (Ruse), April 1922, 5.

80. Bruzitsov, *Niakoga v Sofiia*, 69; and Konstantin Konstatinov, *Pŭtuvane kŭm vŭrkhovete: Portreti, spomeni, eseta* (Journeys to the peaks: Portraits, memoirs, essays) (Varna: Knigodatelstvo "Georgi Bakalov," 1976), 144.

81. Konstatinov, *Pŭtuvane kŭm vŭrkhovete*, 115.

82. For the image and a list of all of the figures in the painting, see Aleksandŭr Dobrinov and Adelina Fileva, *Alexander Dobrinov, 1898–1958* (Sofia: Grafik "Amat," 1998), 40. The women represented in the painting are Fani Popova-Muftafova, Elizaveta Bagriana, Dora Gabe, and Iana Iazova. See also Konstatinov, *Pŭtuvane kŭm vŭrkhovete*, 115.

83. Konstatinov, *Pŭtuvane kŭm vŭrkhovete*, 113.

84. Dimo Kazasov, *Vidiano i prezhiviano, 1891–1944* (Seen and experienced, 1891–1944) (Sofia: Otechestven Front, 1969), 111–114, 127.

85. Dimitrova and Vasilev, ed., *Mladostta na Bagriana*, 162; and Nedialkov, *Bŭlgarski kulturni deitsi*, 50–52.

86. In one case she did meet with operatives in a luxury café in Ruse but her discomfort was palpable in the text. Tsola Dragoicheva, *Poveliia na dŭlga: Spomeni razmisli* (The call of duty: Memoirs and reflections), vol. 1 (Sofia: Profizdat, 1975), 129.

87. Ibid., 160.

88. Ibid., 153.

89. Dragoicheva, *Poveliia na dŭlga: Shtŭrmŭt* (The call of Duty: The assault), vol. 2 (Sofia: Profizdat, 1975), 110–111.

90. Nedialkov, *Bŭlgarski kulturni deitsi*, 14–22.

91. Petkov, *Borbata za trezvenost*, 9.

92. *Trezvenost* (Ruse), 1922, 1.

93. Petkov, *Borbata za trezvenost*, 8.

94. Ibid., 73.

95. Titko Chernokolev, "Kharakter i rolia na vŭzdŭrzhatelnoto dvizheniie" (The character and role of the abstinence movement), in Akhchiiski and Tosheva, ed., *Iubileen Nauchen Sbornik, 1922–37*, 65.

96. *Trezvenost* (Ruse), April 1922, 7.

97. *Trezvenost* (Sofia), February 1924, 23.

98. *Trezvenost* (Sofia), April 1922, 14–15.

99. See, for example, *Trezvenost* (Sofia), October–November 1921, 8–9.

100. *Trezvenost* (Ruse), June 1922, 3.

101. *Trezvenost* (Sofia), August 1924, 23.

102. Stoikov, *Dvizheniete na tresvenost*, 31, 43. See also Petkov, *Borbata za trezvenost*, 126.

103. See, for example, their articles in Akhchiiski and Tosheva, ed., *Iubileen nauchen sbornik, 1922–37*. Vera Zlatareva was for many years President of the *Bŭlgarski neutralen vuzdurzhatelen suiuz* (Bulgarian Independent Temperance Union). On Zlatareva, see Krassimira Daskalova, "Vera Zlatareva," in *A Biographical Dictionary of Women's Movements and Feminism*, De Haan, Daskalova, and Loutfi, ed., 620–623.

104. Stoikov, *Dvizheniete na tresvenost*, 13–14, 26, 42–43.

105. *Trezvenost*, October–November 1921, 8.

106. The "morality" tale of the women who has "fallen" victim to drink and tobacco was a common trope in the abstinence literature. *Trezvenost* (Sofia), October–November, 1921, 15; Sofia, Bulgaria, Central State Archive, f. 1027k, op. 1, a.e. 22, l. 9. See also, Stoianov, *Dvizheniie za trezvenost*, 26.

107. *Trezvenost* (Ruse), April 1922, 39.

108. Ianka Tosheva, "Roliata na studentkata v vŭzdŭrzhatelno dvizhenie" (The role of the student in the abstinence movement), in Akhchiiski and Tosheva, ed., *Iubileen nauchen sbornik, 1922–37*, 36.

109. Tosheva, "Roliata na studentkata," 38–40.

110. Ibid., 40.

111. Ibid.

112. Ibid., 37.

113. Sofia, Bulgaria, Central State Archive, f. 1027k, op. 1, a.e. 16, l. 194.
 Sofia, Bulgaria, Central State Archive, f. 1027k, op. 1, a.e. 12, l. 16.

114. Sofia, Bulgaria, Central State Archive, f. 1027k, op. 1, a.e. 16, l. 194.

115. Petkov, *Borbata za trezvenost*, 15.

116. These scant sources were in direct reference to the Nazi antismoking movement, literature, and congresses on the subject. See, for example, *Zdrave i zhivot* (Health and life), 3 March 1940, 2.

Reconstructing the History of the Constitutional Era in Ottoman Turkey through Women's Periodicals

Serpil Atamaz-Hazar

ABSTRACT

This article discusses the historical value of Ottoman women's periodicals published in the aftermath of the 1908 Revolution, which marked the beginning of the Constitutional Era (1908–1918). Through specific examples of women's writings in the press, it illustrates how these periodicals can shed light on the previously unexplored aspects of this period. The article argues that women's journals allow scholars both to recover the identities and stories of hundreds of women, which would have been lost otherwise, and to challenge the mainstream historiography, which has traditionally presented a one-dimensional portrayal of the Constitutional Era by privileging men's voices and experiences over women's. It demonstrates that women's journals not only reveal a dynamic, flexible, and complex milieu, in which women could and did act as agents of both social and political change, but also signify the multifaceted transformation the Revolution of 1908 caused in Ottoman society in the early twentieth century.

KEYWORDS: Constitutional Era, journals, Ottoman Empire, press, Revolution of 1908, Turkey, women

In this article I explore the different ways to employ women's journals, published in the early twentieth-century Ottoman Empire, in reconstructing the history of the Constitutional Era (1908–1918). I argue that women's journals, which should be treated both as valuable sources of information and as agents of change, are essential to gaining a better understanding of the Constitutional Era, because they present unique perspectives on the hitherto ignored social and cultural transformation that the Ottoman society underwent after the 1908 Revolution. In the following pages, I briefly discuss the Constitutional Era and its significance, offer a short critique of the existing literature on the period, provide some observations about the history and the characteristics of Ottoman women's press based on original research, explain the potential of women's

aspasia Volume 5, 2011: 92–111
doi:10.3167/asp.2011.050107

journals as a historical source, and finally employ specific examples from these journals to demonstrate how revolutionary the Revolution of 1908 really was. Through
this article I hope to raise interest in this relatively unknown set of sources by and
about Ottoman women and encourage further scholarly research that will ultimately
help us understand the ways the revolutionary process and women influenced each
other in the late Ottoman Empire.

The Constitutional Era and Its Significance

In July 1908, Ottoman constitutionalists—also known as the Young Turks—who had
been trying to put an end to the autocratic rule of Sultan Abdülhamid II for decades,
finally reached their goal and declared the establishment of a parliamentary regime.
This event, which came to be known as the Second Constitutional Movement in Turkish historiography and as the 1908 Revolution in the Western world, was the end product of a long process of reform that aimed to strengthen the Ottoman Empire as well
as the culmination of an ideological debate about the nature and extent of this reform.[1]
Behind this movement stood the *İttihat ve Terakki Cemiyeti* (Committee of Union and
Progress [CUP]), an umbrella organization that brought together a large number of
intellectuals, bureaucrats, and military officers who believed that they could solve the
problems of the empire and prevent its disintegration by establishing a constitutional
regime.[2]

   After having to share power with other political groups for five years, the CUP
finally launched a coup d'état on 23 January 1913 to take full control of the state and
started to implement a series of centralist and nationalist policies, aimed at transforming the political, social, and economic structure of the empire.[3] Even though they based
their struggle against the sultan on the principle of liberty, the Unionists—once in
charge—ended the multiparty system, imposed censorship over the press, and abolished all the political organizations to suppress the opposition. Establishing a military
triumvirate, Enver, Talat, and Cemal pashas began ruling the empire with an iron fist
and ironically brought its end as they allied themselves with Germany and dragged
the country into World War I (1914–1918).[4]

   The period that started with the Constitutional Revolution of 1908 and ended with
the collapse of the Ottoman Empire in 1918 was noteworthy because it resulted in the
establishment of a multiparty regime and the emergence of an active political life that
Turkish society had never witnessed before.[5] Moreover, this period played an influential role in shaping the political and economic structure of the Turkish Republic, as
it set the background for the reforms implemented by Mustafa Kemal Atatürk in the
1920s and 1930s.[6] The seeds of the nation-state were also planted in the years following
the Revolution of 1908. Because of the Balkan Wars (1912–1913) and the policies of the
CUP, the Ottoman state rapidly transformed from an empire with a multiethnic and
multireligious population into a nation-state based on Turkish and Muslim identity.[7]
In short, the late Ottoman Empire and the early Turkish Republic were closely connected through institutions, organizations, people, and policies, despite the difference
between their political regimes.[8]

Women in the Constitutional Era

Furthermore, the Constitutional Era witnessed major changes in systems of justice, education, transportation, and communication, which led to the rearrangement of the social order, which in turn led to the transformation of gender roles and relations. The lives of Ottoman women from the upper and upper middle classes were already changing in the nineteenth century, albeit slowly, as a result of the growing influence of Western ideas on reformist statesmen and intellectuals. The Constitutional Movement accelerated this process in two major ways. First, it brought to power the CUP, which, in accordance with its liberal stance on the "woman question," implemented a series of policies that both improved women's status and encouraged them to play a more active role in public life.[9] These policies included facilitating women's involvement in politics,[10] expanding women's opportunities for higher education[11] and professional employment,[12] and introducing penal and civil codes that advanced women's legal standing vis-à-vis men.[13] Second, curtailing the powers of the monarch and lifting the censorship on the press, the 1908 Revolution allowed individuals from different segments of society to voice their opinions freely through the press and to establish social organizations, which were particularly important for women's rights advocates. These periodicals and organizations played a vital role in the formation of the Ottoman women's movement by bringing together women from different ideologies and classes, enabling them to get to know each other on a personal level and providing them with a public forum through which they could disseminate their ideas to thousands of people and display their willingness and ability to solve both women's and the nation's problems.

The Historiography of the Constitutional Era

In accordance with its significance, the Constitutional Movement of 1908 has been one of the most heavily researched topics in Ottoman history. Scholars such as Ernest Ramsaur, Feroz Ahmad, Erik Jan Zürcher, Sina Akşin, Şükrü Hanioğlu, and Aykut Kansu have thoroughly debated the circumstances in which the Constitutional Movement emerged and analyzed its causes and consequences.[14] Relying on traditional sources—principally parliament proceedings, government documents, official correspondences, daily newspapers, historical biographies, and memoirs of prominent figures from the era—historians of the Constitutional Era have successfully demonstrated how the 1908 Revolution changed the structure and style of politics in the Ottoman Empire. However, because they limited their analyses to the political scene and based their studies on male-centered sources, these scholars have consistently left women out of the story of the Constitutional Movement. Although women are missing from the general histories of this period, they have appeared as passive objects in the works of those who examined the ideological debates on the "woman question" in the nineteenth and early twentieth centuries, as the attention was on how male intellectuals conceived of women's roles and rights rather than on how women themselves evaluated and transformed their own position in society.[15]

Women's Journals

The silence of the scholarly literature on women certainly was not the result of women's absence from the public sphere or a lack of sources, for women were quite active and productive in the postrevolutionary period, as illustrated by the female press that flourished between 1908 and 1918. The growing influence of the feminist movement on Turkish scholarship in the 1990s has finally brought to light the writings and activities of women in the late Ottoman Empire. Aynur Demirdirek and Şefika Kurnaz have offered general overviews of the women's movement in the last years of the Ottoman Empire, and Serpil Çakır has focused on a women's journal, *Kadınlar Dünyası* (Women's world), in her monograph.[16] Concerned with tracing the origins of Turkish feminism, these scholars have skillfully used women's journals to find out about the literary, social, and occasionally political writings and activities of early twentieth-century female activists. Yet they have not probed how these writings and activities reflected the changes triggered by the Revolution of 1908, because they were not interested in examining the constitutional process from the perspective of women and gender.

There are several reasons why I have chosen to focus on women's journals. Historians studying the Constitutional Era have closely examined the newspapers and periodicals published by men, but they have hitherto ignored the women's press. Also, despite constituting the largest collection of sources produced by women in the late Ottoman Empire, women's journals have been the least visible in historical studies of this period. These journals are all the more important, because the other primary sources we have by women are limited in both number and scope. There are hardly any women's memoirs and autobiographies from this time period,[17] and the few published literary works present only the voices of a small group of female intellectuals and do not provide as much diverse information on women and society in the revolutionary era as women's journals do. Containing news stories about political, social, and cultural events regarding women, opinion pieces by various female columnists, and readers' letters from all over the Ottoman Empire, these periodicals allow us both to find out what women from different social, economic, and cultural backgrounds did and thought, and to identify who these women were. In fact, our knowledge about most of the female activists of this era is based on their writings in the press. In other words, it would have been impossible without the women's journals to recover the names, voices, and stories of hundreds of women and integrate them into the history of the Constitutional Era. These journals also offer new and unique perspectives on the 1908 Revolution by revealing the specific changes it generated in women's lives and gender relations, something that the male-centered press fails to do. That is to say, establishing women's long-denied agency as well as illustrating the multifaceted transformation the revolutionary process caused in society, women's periodicals can help scholars reconstruct the history of the late Ottoman Empire.

The History of the Women's Press in the Ottoman Empire

Before I discuss some of the ways we can benefit from these journals, I would like to provide a brief history of the Ottoman women's press to highlight their value as both historical sources and agents of change. It was during the Tanzimat Era (the Period of Reorganization, 1839–1876), which brought about significant changes in social and cultural life, that the women's journals surfaced. The first journal to be published in Ottoman Turkish was *Terakki-i Muhadderat* (Progress of virtuous women), which came out in 1868 as a supplement of the newspaper *Terakki* (Progress). It was soon followed by others such as *Vakit* (Time), *Mürebbi-i Muhadderat* (Educator of virtuous women, first published in 1875), *Ayine* (Mirror, 1875), *Aile* (Family, 1880), *İnsaniyet* (Humanity, 1883), *Hanımlar* (Ladies, 1883), *Şükufezar* (Garden of flowers, 1886), *Mürüvvet* (Benevolence, 1888), *Parça Bohçası* (Bundle of pieces, 1889), *Hanımlara Mahsus Gazete* (Ladies' own gazette, 1895), *Hanımlara Mahsus Malumat* (Information special to ladies, 1895), and *Alem-i Nisvan* (World of women, 1906).[18] *Vakit*, *Mürüvvet*, and *Hanımlara Mahsus Malumat* were supplements of prominent newspapers, whose primary audience was men. All the periodicals—except *Ayine* and *Alem-i Nisvan*, which were based in Thessalonikki and the Crimean Peninsula, respectively—were based in Istanbul, the center of publishing in the Ottoman Empire. Although it is possible to encounter letters sent by female readers as well as some articles written by women in earlier journals, *Şükufezar* in 1886 was the first journal to be owned and published entirely by women.[19] Arife Hanım, who was the owner of *Şükufezar*, told her readers that the mission of her journal was to disprove the statement often used by men to ridicule women— "women are long-haired and short-brained."[20] The owners of *Parça Bohçası* and the editor of *Alem-i Nisvan* were also women.[21]

Among the prerevolutionary women's journals, *Hanımlara Mahsus Gazete* deserves the most attention, because it was the longest lasting journal in the history of the women's press in the Ottoman Empire, running for more than six hundred issues from 1895 to 1908. In addition to surviving for more than a decade under the rule of Abdülhamid, who was known for his tight measures on the press, this journal was able to publish its own auxiliary for girls and a separate gazette for children.[22] Declaring its goals of turning women into good mothers, good wives, and good Muslims, *Hanımlara Mahsus Gazete* concentrated on such topics as house management, childcare, family life, fashion, health, and women's education.

Although women's journals emerged in the late nineteenth century, they flourished in the aftermath of the 1908 Revolution, due to the atmosphere of freedom it created in the empire. *Demet* (Bouquet), which appeared as a weekly in September 1908 and ran for seven issues, was the first in a series of women's periodicals published in the postrevolutionary period. *Demet* was owned and edited by a poet Celal Sahir (1883–1935), and featured literary selections as well as articles on women's education, morality, health, child raising, and fashion. Despite the fact that its columnists were mostly male intellectuals,[23] there were prominent women who contributed to *Demet*, among whom were literary figures and activists such as Halide Salih (later Edip),[24] İsmet Hakkı, Nigar Bint-i Osman,[25] Ulviye Bint-i Asım, Fatma Müzehher and Hadiye Ebüzziya, under the pseudonym Ruhsan Nevvare.[26] *Demet* was soon followed by another journal, named *Mehasin* (Virtues). Even though they were similar in terms of their content, contributors, and the date of publication, *Mehasin*, a monthly that ran for twelve issues, carried more articles by women and included more contributions on social and political issues than *Demet*. Often decorated with pictures of European women in low-neck clothes, *Mehasin* tried to promote a new and Western-inspired conception of beauty. Among its female columnists were writers, poets, and public figures, including İsmet Hakkı, Emine Semiye,[27] Münevver Asım, Evliyazade Makbule, Fatma Sabiha, Şükufe Nihal,[28] Zühre Hanım, Fatma Aliye,[29] and Nigar Bint-i Osman.

The next women's journal to surface in the postrevolutionary period was *Kadın* (Woman, 1908), which was different from *Demet* and *Mehasin* in many ways. First, it was based in Thessalonikki, the center of revolutionary activity. Second, it was openly supported by the CUP and promoted in the committee's official publication, *Tanin* (Echo). Third, lasting for thirty issues as a weekly, *Kadın* proved to be more enduring than most of the women's journals of the time. Despite being owned and edited by men like the other two journals, *Kadın* contained a considerable number of articles by women, dealing with a wider range of issues, including marriage, divorce, women's role and status in the society, women's nature, and national progress. The women writers whose names appeared in *Kadın* were Ayşe İsmet, Cavide Peyker, Fatma Seniye, Zekiye Hanım, Pakize Seni, Fatma Bint-i Haşim, Nakiye,[30] Emine Lütfiye, and Refia Şükran.

Another women's periodical worth being discussed in detail is *Kadınlar Dünyası* (Women's world), which ran for more than two hundred issues between 1913 and 1921, becoming one of the longest running journals of the postrevolutionary era, second only to *Hanımlar alemi* (Ladies' world, 1914–1928).[31] As the journal of the *Osmanlı Müdafaa-i Hukuk-u Nisvan Cemiyeti* (Ottoman Society for the Defense of Women's Rights), *Kadınlar Dünyası* was owned by Ulviye Mevlan,[32] the founder and president of this association. *Kadınlar Dünyası*, whose goal was to defend the rights and interests of Ottoman women regardless of their ethnicity and religion, had only published articles written by women.[33]

The decision to publish only women authors was explained as follows: "*Kadınlar Dünyası* will not open its pages to men, unless our rights are regarded as part of universal rights and women as well as men can participate in all sorts of activities."[34] The demand for full equality, repeated many times by different women in the pages of *Kadınlar Dünyası*, was one of the several things that set this journal apart from most others. The editors of the journal also stated that even though they were thankful to those who defended womanhood, they—as Ottoman women—could protect their own rights, using their own methods. "They should leave us alone," they said, "How can we deign to accept men's benevolence to end the suffering we endure because of them?"[35] Discussing women's emancipation in both private and public spheres, defending women's social, economic, legal, and political rights, questioning established gender norms, and challenging patriarchy in different ways, *Kadınlar Dünyası* became the first women's journal in the Ottoman Empire to adopt an explicitly feminist agenda.[36] *Kadınlar Dünyası* was also the first journal to publish photographs of Muslim women.

Kadınlar Dünyası had a substantial number of contributors, among whom were prominent female columnists, educators, and activists such as Ulviye Mevlan, Aziz Haydar,[37] Emine Seher Ali, Mükerrem Belkıs, Atiye Şükran, Belkıs Şevket,[38] Yaşar Nezihe,[39] Nimet Cemil, Meliha Cenan, Azize, Feride Servet, Pakize Sadri, Naciye Şerif, Bint-ül Halim Seyhan, Mesudet Bedirhan, Feride İzzet Selim, Meliha Zekeriya, Bedia

Kamran, and Fatma Zerrin. In addition to its regular columnists, European journalists such as Odette Feldmann (*Berliner Tageblatt*) and Grace Ellison (*Times*)[40] contributed articles and letters to *Kadınlar Dünyası,* as did its readers.

Other women's journals that were published after the Revolution of 1908 included *Musavver Kadın* (Imagined woman, 1911), *Kadın* (Woman, 1911), *Kadınlık Hayatı* (Life of womanhood, 1913), *Seyyale* (Stream, 1914), *Siyaset* (Politics, 1914), *Kadınlık* (Womanhood, 1914), *Hanımlar Alemi* (Ladies' world, 1914–1928), *Osmanlı Kadınlar Alemi* (Ottoman women's world, 1914), and *Bilgi Yurdu Işığı* (The Light of the homeland of knowledge, 1916). Decorated with pictures of Western women and illustrated with advertisements, many of them contained large sections on beauty, fashion, health, and cooking. They also featured literary works such as poems and serialized stories and covered regional and international news concerning women. Some of them, *Kadınlık* and *Kadınlar Alemi* in particular, addressed more serious topics as well, carrying opinion pieces on women's clothing, marriage practices, the elevation of women's status, feminism, nationalism, and women's participation in social and economic life. Unfortunately, with the exception of *Hanımlar Alemi* all of these journals were short-lived, none of them being able to survive the first year.

Prerevolutionary versus Postrevolutionary Women's Journals

Women's journals of the postrevolutionary period were different from those published before 1908 in many ways. Unlike the ones in the prerevolutionary period, women's journals of the Constitutional Era were run mostly by women. Both the owners and editors of such journals as *Kadınlar Dünyası, Siyaset, Kadınlık,* and *Seyyale* were women; others such as *Kadınlık Hayatı* and *Kadınlar Alemi* were owned by men but had female editors. The number of women working as columnists and typographers also increased after the Revolution of 1908. Accordingly, women came to have more influence in the press, which helped them take matters into their own hands and shape the public discourse on women in the direction they wanted. Although their opinions on certain issues varied, they all emphasized the need to redefine gender roles and gender relations in Ottoman society and made it clear in their writings that they no longer wanted to leave this matter in the hands of men; they wanted to protect their own rights, using their own methods.

Operating under a liberal regime that officially lifted the censorship on the press, women's periodicals allowed women to express their opinions freely and engage in lively discussions. As political issues became debatable, Ottoman women, for the first time in history, found the chance to demonstrate their interest in and knowledge about the subject openly through their writings. Accordingly, women's journals published in the postrevolutionary period carried fewer literary works and more position pieces dealing with issues regarding the nation as well as the womanhood.

Women in the press not only made their voices heard, but also, different from the prerevolutionary era, made themselves and other Muslim women visible to their readers by publishing their pictures. This change, initiated by *Kadınlar Dünyası* in 1913, reflected the social and political transformation that the Ottoman society was undergoing at the time. Through education, employment, and a variety of activities under different organizations, more women were getting involved in social and economic life, becoming more visible in public. Moreover, subscribing to the ideology of Turkish nationalism, which gained momentum during the Balkan Wars, many intellectuals, including the publishers of *Kadınlar Dünyası,* began using national rather than Western symbols, images, and references to call for change in the gender system. Consequently, women's periodicals increasingly portrayed Turkish instead of European figures as role models for other women to follow. Among these were Belkıs Şevket and Aziz Haydar, members of the Ottoman Society for the Defense of Women's Rights, and the seven women hired by the French phone company—Bedra Osman, Bedia Şekib, Nezihe Mustafa, Hamiyet Derviş, Mediha, Refika Mustafa, Seniha Hikmet—who were also the first Turkish women whose pictures appeared in the press.[41]

Women's increasing influence, combined with the atmosphere of freedom following the 1908 Revolution, helped change the mission of women's journals. The main goal of the earlier journals was to turn women into ideal mothers and homemakers through domestic instruction, but most postrevolutionary periodicals aimed at putting women's issues and demands on the public agenda as well as promoting and defending women's rights.[42] Therefore, rather than having to reiterate the official ideology and to serve as propaganda tool in the hands of the state, as they did under Abdülhamid's rule because of the strict censorship, women's journals in the Constitutional Era acted independently and pushed for a more egalitarian society by referring to the principles the 1908 Revolution was based on: freedom, equality, and progress.[43]

Finally, postrevolutionary women's journals suffered from financial difficulties just like those published before the Revolution, resulting in the closure of most after a year. However, the postrevolutionary journals lasted longer than their predecessors, an indication of an increase in the size of the readership. Although we do not have definite numbers and most figures are speculative, we know that the copies sold through individual sales and subscriptions were in the lower thousands.[44] More important, the fact that readers were sending letters from different cities, demonstrates that the demand for and the accessibility of women's journals in the postrevolutionary period was not restricted to urban areas in the Western part of the Ottoman Empire.[45]

The Social and Political Transformation after 1908 Exemplified in Three Points

As should be clear by now, women's periodicals have a lot of potential and can be employed by scholars in different ways. Here I focus on three issues that can help us better understand the social transformation that the Ottoman society underwent following the Revolution of 1908: women's criticism of the government; women's ability to change the direction of the debate on women and gender; and the concrete changes women's journals helped to implement.

Women's Criticism of the Government

Taking advantage of the liberties that they gained following the revolution, Ottoman women often used their pens to question the sincerity of their leaders about constitutionalist principles and to criticize the government for not paying attention to women. In an article titled "Erkekler Hakikaten Hürriyetperver Midirler? Kadınlar Ne İstiyor?" (Are men really advocates of freedom? What do women want?), columnist Naciye Hanım praised men's struggle for freedom over the centuries against clergy and kings, but mentioned that although men were able to gain their freedom with great revolutions, there was still a large group of people whose freedom had not been ensured.[46] "That was us; pitiful women," she wrote, "although men seem to be advocates of freedom, in reality they are nothing but little despots. Even when they were drowning large continents in blood for freedom, they ignored the world of women, which was much greater and more important than theirs."[47] Naciye Hanım accused men of being hypocrites, because even though they were successful in establishing a constitutional regime, they had not yet given women their basic rights, let alone political rights.

In another article published in *Kadınlar Dünyası*, women continued to criticize the politicians and the government.[48] They wrote that despite declarations such as "Ottomans gained their freedom," "people could make their own laws," and "despotism was removed," women's rights were not guaranteed by the state. After "waiting for men for five years to recover from the intoxication of freedom" and to do something about women, they felt the need to remind the government that women, too, were Ottomans, paying taxes, and affected by the decisions the government made. They said: "Our patience was in vain. They excluded women from the principle that 'people are born equal' … This is outright cruelty, an attack against womanhood. Our situation has become unbearable. It is essential for the salvation of society to eliminate the social circumstances that have turned us into victims and slaves."[49]

Sharing the same sentiments, activist Emine Semiye compared the statesmen of the Constitutional Era to those of the repressive Hamidian era. In her words, "[f]ollowing the declaration of freedom our statesmen have ignored us, just like they did in the era of despotism."[50] To prove her point, she narrated a conversation that she had had with a "progressive" and "trustworthy" man, who had performed and was still performing "great services for his homeland." Before the Revolution, Emine Semiye had requested that after the establishment of freedom the government would give women their due and this individual had given her his assurance. After the establishment of the constitutional regime, she felt the need to remind him of women's rights, which she thought were forgotten. He responded with: "We were able to save freedom, but could remove only one piece from the castle of fanaticism. Please wait; the time for women's progress will come, too. People like you, who love their country and nation, cannot lose hope and should work without falling into despair."[51] Emine Semiye shared this conversation with her readers, because she wanted them to realize that the politicians currently were not concerned with women's problems and would not be so for quite a while. Thus it was up to women, those who understood that being a woman did not have to mean being deprived of the rights and virtues of humanity, to make sure that the next generation of women would be raised in accordance with the necessities of the civilized world.[52]

Other articles also voiced women's discontent with the government and made specific demands. Columnist Mükerrem Belkıs, for example, wanted the government to take steps to reform family life,[53] while Ulviye Mevlan complained that, although after the 1908 Revolution people could talk openly about the necessity to educate women, to increase their cultural level, and to provide them with scientific knowledge, the government made no serious attempt to implement any of these changes.[54]

Women's complaints about and demands from the government reflected the transformation of state-society relations as well as women's claim to equal citizenship, both of which directly resulted from the Revolution of 1908. The revolution created an environment that allowed individuals to participate in the political process and to play a role in setting agendas and formulating policies through the press, associations, and elections. For the first time in the history of the Ottoman Empire, people had a say in how they were governed and the government had responsibilities to its people. That was why it was possible for the members of society, including women, to criticize the government and to petition for specific changes. Also, the revolution gave women both the opportunity and the justification to demand equal rights by defining themselves as part of the nation. Women could openly and rightfully claim equal citizenship, because the constitutional regime was established on principles such as freedom, equality, and progress. In other words, women were simply asking for what was already promised.

Women Changing the Direction of the Debate on Women and Gender

Since the Tanzimat, men had discussed in detail how they wanted women to be and how they expected them to behave, as they considered the "woman question" central to the construction of a new society and its progress. Many of them even blamed the problems of the society on women's shortcomings. However, the Constitutional Era gave women the chance to turn the tables on men and to change both the issues at hand and the language that was used. Aziz Haydar was one of the most outspoken women on this issue. In "Yalnız Kadınlar mı Islaha Muhtaç?" (Is it only women who need to be reformed?), she revealed her frustration with the lack of discussion about men:

> They say women should be like this, women should be like that, women should work hard, women should be resourceful, women should be virtuous, etc. OK, we understand! But what about men? Do these things not apply to them as well? Are the things that are essential for women not necessary for men? ... They say women are messy, imprudent, and without manners ... Men, on the other hand, are pure gold ... They say our women are ignorant. Are our men educated? They say our women are in the dark. Are our men enlightened? Do our men fulfill their duty so well that they [are entitled to] complain about women?[55]

According to Aziz Haydar, women needed to know how to manage their house, form a family, raise children, and behave; but so did men. She said that if you asked a man what he did to solve the problem he was complaining about, you would only hear him say, "I am a man," which apparently was enough to enjoy all the privileges in the world.

Aziz Haydar felt that it was necessary to make a correct diagnosis in order to cure the disease that inflicted the society:

In our society, we blame everything on women, but we don't consider subjecting our men to careful scrutiny. We don't want to accept that both our women and men are suffering from the same problems ... Our priority should be the things that we desperately need today. First of all, our women should reclaim their lives. They should make their presence known. They should gain enough power to demand their rights. They should be able to show that it is not only women who are responsible for the bad things we observe in our social life, and that our men, as well, have many shortcomings. After that, let men discuss whether women are scientifically inferior or superior to men![56]

In another article, Aziz Haydar said that Turkish men thought that they could elevate themselves by stepping on women, not realizing that by doing so they were harming themselves. She wrote, "Men, who always want to see women as weak, helpless, and insignificant, don't realize that by paying no attention to women, they are ruining their own homes, their own lives, and their own welfare."[57] She believed that women were left so weak and helpless that they could not even loosen the chain that was suffocating them. Conversely, the men who were suffocating the women were not aware that with this chain they were holding themselves captive as well. Aziz Haydar concluded by saying: "God has given people two arms, which work together to nourish the body. Women and men are the two arms of society. If they work together, they will succeed. If there is only one of them left, they will be miserable just like the person who is left with only one arm. One hand is only good for begging, nothing else!"[58]

Fatma Neşide also thought that the country was in misery because of the indifference of men and the weakness of women.[59] She claimed that women were not only crying, but shedding "tears of blood," while men, "indifferent and insensitive," were enjoying themselves instead of feeling sorry. Fatma Neşide blamed men for being bad husbands and fathers, who neither appreciated their wives nor children: "Men should have a strong mind and sound heart to be able to show respect to their mother, compassion to their sister, loyalty to their wife and love to their children. How could you find these in our indifferent men? ... Raki and brandy have destroyed their mental power while syphilis left the bodies of most of them ruined."[60] Fatma Neşide's conclusion was the same as Aziz Haydar's. Because men were not thinking straight and their indifference was causing both themselves and women misery, it was time for women to take control.

In an article she wrote for the journal *Mehasin,* Evliyazade Naciye responded to one of the criticisms raised by male intellectuals of the time, namely that women were indifferent to the journals published for them. "If women do not feel the need for a newspaper and do not understand, unlike the women of other nations, that the press is a tool for progress and advancement, it is because of men," she wrote.[61] She did not understand how men could blame women for their lack of interest in these publications, while they had done nothing to raise women's intellectual level. Sarcastically she asked, "You established women's schools, took an interest in women's education, and encouraged women to engage in literature, but women did not want to learn?"[62] Evliyazade Naciye further complained about the conservative fathers, brothers, and husbands whom she considered obstacles to women's progress. She wrote from Izmir: "We would get upset, thinking that ignorance and bigotry dominated the men in our milieu. Then we realized that the men in our capital are not any different ... Although they live in the capital of education that disseminates knowledge and science to all the provinces, the men of Istanbul don't pay the necessary attention to women's education either."[63] According to Evliyazade Naciye, if women did not feel the need to read newspapers, the fault belonged to men, because it was they who had controlled women's lives for centuries. She concluded her article by asking: "When did women ever receive support, help or protection from men? And what can be expected, but weakness and laziness, from women who are still subjected to men's continuous insults and chains of oppression?"[64]

These examples, which revealed women's frustration with the dominant discourse on women and gender as well as their attempt to change it, indicate a broader change in the style, content, language, and mission of women's journals. Although the Ottoman women's press dated back to the Tanzimat period, it was only after the Revolution of 1908 that women were able to push for their own agenda and to challenge the male monopoly on public discourse in a significant way. This change was mainly due to the dramatic increase in the number of women working in the press, following the 1908 Revolution. As the number of women who owned and edited journals increased, so did their control over the materials they published. Their positions differed on certain issues, but they all wanted to take control of their own lives and play a more influential role in national affairs. Shifting the blame for the problems of the society to men was a way to justify this demand. Fed up with being criticized for everything that went wrong in the empire as well as constantly being told what to do and how to do it, women of the Constitutional Era started drawing people's attention to men's shortcomings. They argued that it was men, not women, who had to be held responsible for the situation the country was in, for women had not been allowed to get involved in national affairs until the recent revolution, leaving men in charge for centuries. If men were genuinely interested in solving the problems of the society, they had to recognize their own incapability and accept women's demands for increased rights and responsibilities, which would help advance both the nation and Ottoman womanhood.

Concrete Changes Women's Journals Helped to Implement

The third issue is the social change that women's journals helped to implement. Not being content with writing and publishing articles, Ottoman women after 1908 also used the press to initiate concrete changes in women's status, which ultimately transformed the society as well. One of the many changes that took place due to the efforts of a women's journal was the employment of Muslim women by a French-led phone company.[65] In a letter addressed to *Kadınlar Dünyası* on 24 April 1913, Bedra Osman—a member of the *Osmanlı Müdafaa-i Hukuk-u Nisvan Cemiyeti* (Ottoman Society for the Defense of Women's Rights)—wrote that she applied, with four of her friends, to work at the phone company. However, a company employee rejected their application because they could not speak French or Greek.[66]

After having received Bedra Osman's letter, *Kadınlar Dünyası* published a series of articles in which they demanded an explanation from the government and an apology from the phone company, because they believed that the company was discriminating against these women for being Muslims.[67] Within ten days, *Kadınlar Dünyası* received a letter from the government, stating that the government had received a memorandum from the phone company in which they denied the allegations and assured that they welcomed all applicants. The letter also informed the editors of *Kadınlar Dünyası* that even though the phone company had not yet reached a decision about Bedra Osman and her friends, they were probably going to be employed in positions that did not require them to speak French.[68] Following this correspondence, the phone company decided to hire the seven women, all of whom were Muslims. Although the editors of *Kadınlar Dünyası* could not force the phone company to change its policy regarding the language requirement, they managed to get them to hire several Muslim women by successfully calling attention to the problem and employing the government's help.[69]

In order to demonstrate Ottoman women's patriotism and interest in science and progress, *Kadınlar Dünyası* also initiated a campaign to purchase a plane for the army, which simultaneously promoted their feminist agenda as well. While the administrative committee was discussing the details of the campaign, Belkıs Şevket, an educator and a regular columnist for the journal, volunteered to fly in a plane and to drop fliers from the air to invite women to make donations to the Ottoman army. After getting the authorities' permission, Belkıs Şevket flew over Istanbul for approximately fifteen minutes, becoming in November 1913 the first Turkish and Muslim woman to fly in an airplane. Belkıs Şevket's flight was a significant act, which not only challenged the traditional gender norms by claiming the public space for women in a new way, but also became a symbol of women's courage, determination, and confidence that served both the nationalist and the feminist movements in the Ottoman Empire.[70]

Women's journals also took important steps to advance women's education and participation in social and economic life. Working in cooperation with women's organizations of the Constitutional Era, they organized lectures, opened schools, assisted women in finding jobs, supported businesses established by women, and in these ways helped move women out of their homes into public places. Furthermore, women's periodicals pressured the government to put certain policies into action, such as opening the doors of the university to women in 1914, sending female students to Europe for education, and introducing the Family Law of 1917.[71] The concrete changes the women's press helped to generate in women's lives are noteworthy, because they prove that Ottoman women, far from being the passive bystanders scholars made them out to be, were active agents of change in the Constitutional Era. In other words, these changes—that we only know of because of women's journals—demonstrate that the revolutionary process, which affected women in various ways, was in turn significantly affected by them.

Conclusion

This article demonstrates that after the Revolution of 1908 the press became a powerful vehicle for women to make their voices heard, to become visible in public, and to initiate social and cultural change. Acting as a forum to criticize patriarchal values, putting women's demands on the public and political agenda, promoting women's rights, and encouraging women to take control of their own lives, women's journals helped plant the seeds of feminism in the Ottoman Empire. Not only did they change women's lives in several ways, but they also changed the power dynamics among the state, the society, and the individual. The fact that both private and state institutions seriously considered women's demands—expressed in women's periodicals—and felt the need to meet them proves how empowering the press could be for organized women and society in the Constitutional Era.

Unfortunately, despite their importance, women's periodicals have been overlooked and have not been utilized to study the Constitutional Era. Accordingly, historical studies about this period have remained limited to the male elite and parliamentary politics, failing to fully explain how revolutionary the 1908 Revolution actually was. Women's periodicals of the postrevolutionary era change our perception of the late Ottoman Empire and even modern Turkey by demonstrating that the transformation was not restricted to the political arena and that social norms, values, relations, and institutions were changing as well. They also illustrate that the 1908 Revolution created an environment in which women, who complained that the changes following the revolution were not satisfactory, could claim equal citizenship, challenge the male monopoly on public discourse, defy established gender roles, and implement change from the bottom up.

In 1913 Feriha Kamran, the head columnist for *Kadınlar Alemi,* wrote: "Women have played important roles in civilized and advanced countries, especially in the history of their constitutions. The struggles that women have gone through, particularly in the social revolutions that follow the political revolutions, are as significant as those of men. … A civilization and a constitutional regime without women are impossible."[72]

Accordingly, we should both value and benefit from the documents that women from the Ottoman Constitutional Era have left us. These documents not only enable us to integrate the story of women into the history of the Constitutional Era, but also require us to reconsider and possibly change some of the premises we have been using while studying this period.

◊ About the Author

Serpil Atamaz-Hazar is an assistant professor of history at SUNY New Paltz. Her research interests focus on nineteenth- and early twentieth-century Turkey and Iran, and women in the modern Middle East. This article is based on a six-month-long archival research on women's journals published in the late Ottoman Empire that was conducted at the National Library in Ankara and at the Women's Library and Information Center in Istanbul. Email address: atamazhs@newpaltz.edu.

◇ Notes

1. These problems included military defeats, loss of territories, weakening of the state authority, increasing dependence on Western powers, and foreign intervention in internal affairs.

2. The Committee of Union and Progress was founded as a secret organization under the name İttihad-ı Osmani (Ottoman Union) in 1889 by Military Medical College students, who opposed Abdülhamid's absolute monarchy. Although the CUP originally consisted of different groups, ranging from extreme Westernists to Islamists and non-Muslims to conservative nationalists, it gained a centralist, Westernist, and nationalist character after it came to power under the leadership of revolutionary figures like Talat Bey, Enver Bey, Cemal Bey, Dr. Nazım, Hüseyin Cahit, and Ziya Gökalp. Having achieved their goal under the umbrella of the CUP by establishing a constitutional regime, other groups that were not pleased with the policies of the Unionists started to leave this party and formed their own political organizations. For a detailed study of the CUP and its policies, see Ernest Ramsaur, *The Young Turks: Prelude to the Revolution of 1908* (Princeton: Princeton University Press, 1957); Feroz Ahmad, *The Young Turks: The Committee of Union and Progress in Turkish Politics, 1908–1914* (Oxford: Clarendon, 1969); Erik J. Zürcher, *The Unionist Factor: The Role of the Committee of Union and Progress in the Turkish National Movement, 1905–1926* (Leiden: E.J. Brill, 1984); Sina Akşin, *Jön Türkler ve İttihat ve Terakki* (The Young Turks and union and progress) (Istanbul: Remzi Kitabevi, 1987); Şükrü Hanioğlu, *The Young Turks in Opposition* (New York: Oxford University Press, 1995); Hasan Kayalı, *Arabs and Young Turks: Ottomanism, Arabism, and Islamism in the Ottoman Empire, 1908–1918* (Berkeley: University of California Press, 1997); and Şükrü Hanioğlu, *Preparation for a Revolution: The Young Turks, 1902–1908* (Oxford and New York: Oxford University Press, 2001).

3. The Unionist leadership made changes in school curricula to encourage the Turkification of the population and established Turkish as the official language. It supported the development of a national economy and a national bourgeoisie by organizing boycotts of foreign products, encouraging the consumption of local manufactures, helping to establish commercial companies, and fostering a Turkish entrepreneurial class. Moreover, it founded national institutions, such as the National Bank, National Library, the National Archive, the National Musical Society, and the National Geographic Society. In addition, it took significant steps to secularize the judicial system by giving the authority to administer and to supervise the religious courts and their employees to the Ministry of Justice (1913), and introducing a secular and uniform family code (1917).

4. Ahmad, *The Young Turks*; Zürcher, *The Unionist Factor*; Sina Akşin, *Turkey from Empire to Revolutionary Republic: The Emergence of the Turkish Nation from 1789 to the Present* (New York: New York University Press, 2007); Şükrü Hanioğlu, *A Brief History of the Late Ottoman Empire* (Princeton: Princeton University Press, 2008).

5. It did not last long and the CUP "never had any intentions of broadening the power structure to include the urban workers or the peasants." But this atmosphere of freedom made it possible for different ethnic, religious, and ideological groups to freely discuss both the future and the identity of the Ottoman state, and to put their ideas and plans into practice (Zürcher, *The Unionist Factor*, 162). The popular demonstrations preceding the Revolution of 1908, such as the uprisings in Kastamonu (1905), Diyarbekir (1905), Trabzon (1906), Erzurum (1906–7), and Van (1907); the demonstrations of joy after the establishment of the Constitution; and the strikes of factory and railway workers in various regions reflected this new sense of involvement of the man and woman in the street and ushered in the emergence of mass politics in the Ottoman Empire (Ahmad, *The Young Turks*).

6. The bureaucratic, military, and intellectual cadre around which the Kemalist state was organized consisted of the key players of the Constitutional Era. It was this era that created the

circumstances that allowed most of the political actors of the early republican period—such as the unionists, the feminists, and the socialists—to organize and to become involved in political affairs. Moreover, policies toward creating a centralized secular state based on a national economy, a priority of the Kemalist regime, were already on the way during the constitutional period.

7. The Balkan Wars, which demonstrated the separatist tendencies among the non-Turkish and non-Muslim population, played an important role in shaping the policies of the CUP after 1913. The loss of almost all the remaining territories in the region, the expulsion of the Turkish population from the area, and the atrocities committed against the Turkish population during the wars caused the CUP to realize the futility of Ottomanism in preserving the empire. The national awakening, emerged among the Turkish population following the Balkan Wars, led many intellectuals and military officers, along with the CUP, to envision a new state centered in Anatolia, where the majority of the population was Turkish. For a discussion of the effect of the Balkan Wars on the psyche of Turkish elite, see Halide Edip, *Turkey Faces West* (New York: Arno Press, 1973); Fuat Dündar, *İttihat ve Terakki'nin Müslümanları İskân Politikası, 1913–1918* (The settlement policy of the Union and Progress for Muslims, 1913–1918) (Istanbul: İletişim, 2001); Zafer Toprak, *Milli İktisat Milli Burjuvazi* (National economy national bourgeoisie) (Istanbul: Tarih Vakfı Yurt Yayınları, 1995); and Erol Köroğlu, *Ottoman Propaganda and Turkish Identity: Literature in Turkey During World War I* (London: I.B. Tauris, 2007).

8. For a more detailed discussion of the continuities between the late Ottoman Empire and early Turkish Republic, see Niyazi Berkes, *The Development of Secularism in Turkey* (Montreal: McGill University Press, 1964); Roderic Davison, *Essays in Ottoman and Turkish History, 1774–1923: The Impact of the West* (Austin: University of Texas Press, 1990); Feroz Ahmad, *The Making of Modern Turkey* (New York: Routledge, 1993); Erik J. Zürcher, *Turkey: A Modern History* (New York: I.B. Tauris, 1997); Akşin, *Turkey from Empire to Revolutionary Republic*; Hanioğlu, *A Brief History of the Late Ottoman Empire*.

9. See, for example, Ahmad, *The Young Turks*; Zürcher, *Turkey: A Modern History*; Akşin, *Turkey from Empire to Revolutionary Republic*; Hanioğlu, *A Brief History of the Late Ottoman Empire*.

10. The Unionists were the first "champions" of women's political rights, at least in principle. Not only did they accept female members to their party, but they also put them on an equal footing with men. As the first political organization that opened its doors to women, the CUP included the following statements in its program: "Ottomans, both women and men can become members of the organization" and "female members have the same rights and responsibilities as male members." Tarık Zafer Tunaya, *Türkiye'de Siyasi Partiler* (Political parties in Turkey) (Istanbul: Hürriyet Vakfı Yayınları, 1984), 44–45.

11. The CUP made primary education compulsory for girls (1913); considerably increased the number of girls' schools in the country; provided them with the chance to receive college-level education by opening to women the doors of the university (1914); promoted vocational education; and sent several female students to Europe for education (1916).

12. The government raised the number of female teachers to thousands; provided training for professional nurses, nurses' aides, female teachers and school administrators; appointed many women as mistresses and inspectors to the newly established girls' schools; started hiring female employees at government offices (1913); facilitated the establishment of factories that allowed women to join the industrial labor force; and founded an organization called the *İslam Kadınlarını Çalıştırma Cemiyeti* (Society for the Employment of Muslim Women) to find employment for Muslim women.

13. The CUP introduced the Ottoman Law of Family Rights (1917), which promoted a family model based on monogamy and mutual consent, recognized women's right to initiate divorce, transferred the conclusion of the marriage contract to the authority of the state, increased the age of marriage to 17 for women and 18 for men; recognized the right of female property

owners to act as guarantors (1911); allowed women to travel without permission from their husbands (1911); introduced equal punishment for female and male adulterers (1911); and secured women's right to receive equal shares from their husbands', fathers', and mothers' inheritance (1912).

14. Ramsaur, *The Young Turks: Prelude to the Revolution of 1908;* Ahmad, *The Young Turks;* Zürcher, *The Unionist Factor;* Sina Akşin, *Jön Türkler ve İttihat ve Terakki* (The Young Turks and Union and Progress) (Istanbul: Remzi Kitabevi, 1987); Hanioğlu, *The Young Turks in Opposition;* Aykut Kansu, *The Revolution of 1908 in Turkey* (Leiden and New York: E.J. Brill, 1997); idem, *Politics in Postrevolutionary Turkey 1908–1913* (Leiden and Boston: E.J. Brill, 2000); Şükrü Hanioğlu, *Preparation for a Revolution: The Young Turks, 1902–1908* (Oxford and New York: Oxford University Press, 2001).

15. These works include Berkes, *The Development of Secularism;* Deniz Kandiyoti, "Slave Girls, Temptresses, and Comrades: Images of Women in the Turkish Novel," *Gender Issues 8,* no.1 (March 1988): 35–50; idem, "Women and The Turkish State: Political Actors or Symbolic Pawns?," in *Woman-Nation-State,* ed. Nira Yuval-Davis and Floya Anthias (London: Macmillan, 1989), 126–149; idem, "End of Empire: Islam, Nationalism, and Women in Turkey," in *Women, Islam, and the State,* ed. Deniz Kandiyoti (Philadelphia: Temple University Press, 1991), 22–47; idem, "Some Awkward Questions on Women and Modernity in Turkey," in *Remaking Women: Feminism and Modernity in the Middle East,* ed. Lila Abu-Lughod (Princeton: Princeton University Press, 1998), 270–287; Faik Bulut, *İttihat ve Terakki'de Milliyetçilik, Din ve Kadın Tartışmaları* (The debates about nationalism, religion, and women within the Union and Progress) (Istanbul: Su Yayınları, 1999); and Palmira Brummett, *Image and Imperialism in the Ottoman Revolutionary Press, 1908–1911* (Albany: State University of New York Press, 2000).

16. Aynur Demirdirek, *Osmanlı Kadınlarının Hayat Hakkı Arayışının Bir Hikayesi* (A story of the Ottoman women's quest for the right to live) (Ankara: İmge Kitabevi, 1993); Serpil Çakır, *Osmanlı Kadın Hareketi* (The Ottoman women's movement) (Istanbul: Metis Press, 1994); Şefika Kurnaz, *II.Meşrutiyet Döneminde Türk Kadını* (Turkish women during the second constitutional era) (Ankara: M.E.B. Press, 1996); Aynur Demirdirek, "In Pursuit of the Ottoman Women's Movement," in *Deconstructing Images of 'The Turkish Woman,* ed. Zehra Arat (New York: St. Martin's Press, 1998), 65–81; Serpil Çakır, "Feminism and Feminist History-Writing in Turkey: The Discovery of Ottoman Feminism," *Aspasia* 1 (2007): 61–83. For a discussion of the women's press in late Ottoman Empire, see also Elizabeth Frierson, "Unimagined Communities: State, Press, and Gender in the Hamidian Era" (PhD diss., Princeton University, 1996); Ayfer Karakaya-Stump, "Debating Progress in a 'Serious Newspaper for Muslim Women': The Periodical *Kadın* of the Postrevolutionary Salonica, 1908–1909," *British Journal of Middle Eastern Studies* 30, no. 2 (November 2003): 155–181; Tülay Keskin, "Feminist/Nationalist Discourse in the First Year of the Ottoman Revolutionary Press: Readings from the Magazines of Demet, Mehasin, and Kadın-Salonica" (MA Thesis, Bilkent University, 2003); and Victoria Rowe, "Armenian Writers and Women's-Rights Discourse in Turn-of-the-Twentieth-Century Constantinople," *Aspasia* 2 (2008): 44–69.

17. There are three published women's memoirs from this era: Halide Edip, *Memoirs of Halide Edip* (New York, London: Century, 1926); Nigar bint-i Osman, *Hayatımın Hikayesi* (The story of my life) (Istanbul: Ekin Basımevi, 1959); and Halide Edip, *House with Wisteria: Memoirs of Halide Edib* (Charlottesville, VA: Leopolis Press, 2003).

18. These translations are mine and may differ from those provided by Serpil Çakır in her article in *Aspasia* 1 (2007).

19. Keskin, "Feminist/Nationalist Discourse." Copies of most of these journals can be found at the National Library Archives in Ankara and at the Women's Library and Information Center in Istanbul.

20. Keskin, "Feminist/Nationalist Discourse," 41.

21. Ibid. *Parça Bohçası* was co-owned by Hatice Semiha and Rabia Kamile, and the editor of *Alem-i Nisvan*—which was located in Bahçesaray—was Şefika Y., the daughter of İsmail Gasprinski, who owned *Tercüman*.

22. Frierson, "Unimagined Communities," 17.

23. The main contributors to the first women's journal of the constitutional period were prominent male intellectuals of the time such as Cenab Şehabeddin, Enis Avni, Selim Sırrı, Mehmet Rauf, Mithat Cemal, and Hakkı Behiç.

24. Halide Edip (1884–1964): novelist, activist, educator, Turkish nationalist, and founder of Teali-i Nisvan (The Society of Elevation of Women). For a discussion of Halide Edip's activities and writings, see Emel Sönmez, "The Novelist Halide Edip Adıvar and Turkish Feminism," *Die Welt des Islams* 14 (1973): 81–115; P. Kucera, "Westernization, Narration, and Gender: Halide Edib and Turkish Modernism," *Archiv Orientální/Oriental Archive* 73, no. 2 (2005): 235–244; Ayşe Durakbaşa, "Edip Adıvar, Halide (1884–1964)," in *A Biographical Dictionary of Women's Movements and Feminisms. Central, Eastern, and South Eastern Europe, 19th and 20th Centuries*, ed. Francisca de Haan, Krassimira Daskalova and Anna Loutfi (Budapest: Central European University Press, 2006), 120–122; and Mushirul Hasan, *Between Modernity and Nationalism: Halide Edip's Encounter with Gandhi's India* (New Delhi: Oxford University Press, 2010).

25. Nigar bint-i Osman (1856–1918): poet and columnist.

26. Hadiye Ebüzziya (1884–1913): playwright, columnist, and cofounder of Teali-i Nisvan (The Society of Elevation of Women), used the pen name "Ruhsan Nevvare." Although our knowledge about most of the women who were involved in publishing, editing, and writing for Ottoman women's journals is based on their own writings in the press and therefore limited, the signatures under their columns do make it possible to identify them by their names.

27. Emine Semiye (1866–1944): daughter of the historian and politician Ahmet Cevdet Paşa, sister of Fatma Aliye Hanım, member of the CUP, newspaper columnist, novelist, and cofounder of the organization, *Şefkat-i Nisvan* (The Compassion of Women, 1898).

28. Şükufe Nihal (1896–1973): poet, novelist, and newspaper columnist.

29. Fatma Aliye: daughter of the historian and politician Ahmet Cevdet Paşa, sister of Emine Semiye, and founder of *Cemiyet-i İmdadiye* (The Society of Help, 1898). For more information, see Serpil Çakır, "Aliye Fatma (1862–1936)," in De Haan et al., *A Biographical Dictionary of Women's Movements and Feminisms*, 21–24.

30. Nakiye (later Elgün) (1882–?): women's rights advocate, columnist, educator, and founder of *Şehit Ailelerine Yardım Birliği* (The Union for Aid to Martyred Soldiers' Families, 1915).

31. The publication of *Kadınlar Dünyası* had to be suspended three times. First, for three months after the publication of the 153rd issue, due to paper shortage; second for four years between 1914 and 1918 because of World War I: and finally for three years between late 1918 and 1921 after it resumed publication on 2 March 1918 (Çakır, *Osmanlı Kadın Hareketi*, 80).

32. Ulviye Mevlan (Civelek) (1893–1964): journalist, women's rights activist, founder of *Osmanlı Müdafaa-i Hukuk-u Nisvan Cemiyeti* (the Ottoman Society for the Defense of Women's Rights), and owner and publisher of *Kadınlar Dünyası*. For more information, see Serpil Çakır, "Mevlan Civelek, Ulviye (1893–1964)," in De Haan et al, *A Biographical Dictionary of Women's Movements and Feminisms*, 336–339.

33. [Editorial Board of *Kadınlar Dünyası*], "Hukuk-u Nisvan" (Women's rights), *Kadınlar Dünyası* (4 April 1914): 1.

34. Çakır, *Osmanlı Kadın Hareketi*, 82.

35. [Editorial Board of *Kadınlar Dünyası*],"Hukuk-u Nisvan," 1.

36. Çakır, *Osmanlı Kadın Hareketi*.

37. Aziz Haydar (1881–?): educator, newspaper columnist, lecturer, fundraiser, and member of *Donanma Cemiyeti* (The Navy Society) and *Hilal-i Ahmer Cemiyeti* (The Red Crescent Society).

38. Belkıs Şevket: the first Turkish and Muslim woman to fly on an airplane in 1913, member of the Ottoman Society for the Defense of Women's Rights, and a regular columnist at *Kadınlar Dünyası*.

39. Yaşar Nezihe (1882–1971): poet.

40. These two names appear often in the pages of *Kadınlar Dünyası*.

41. These pictures were published several times by *Kadınlar Dünyası* in 1913.

42. Demirdirek, "In Pursuit of the Ottoman Women's Movement," 66.

43. For a discussion of the principles of the 1908 Revolution, see Zürcher, *The Unionist Factor*; Kansu, *The Revolution of 1908*; Hanioğlu, *Preparation for a Revolution*; Brummett, *Image and Imperialism*.

44. Çakır, *Osmanlı Kadın Hareketi*; Demirdirek, *Osmanlı Kadınlarının*; Keskin, "Feminist/Nationalist Discourse."

45. *Kadınlar Dünyası* often received and published readers' letters from provincial cities such as Trabzon, Izmir, Konya, Edirne, and Diyarbakır.

46. Naciye, "Erkekler Hakikaten Hürriyetperver Midirler?, Kadınlar Ne İstiyor?" (Are men really advocates of freedom? What do women want?), *Kadınlar Dünyası* (10 April 1913): 3.

47. Ibid.

48. [Editorial Board of *Kadınlar Dünyası*], "Ahlak, Hürriyet-i Mesai" (Morality and the freedom to work), *Kadınlar Dünyası* (13 June 1914): 2.

49. Ibid.

50. Emine Semiye, "Terakkiyat-i Nisvaniyeyi Kimden Bekleyelim?" (From whom should we demand women's progress?), *Mehasin* (September 1909): 733–737.

51. Ibid.

52. Ibid.

53. Mükerrem Belkıs, "İnkılab-ı İctimai Esasları" (The principles of social revolution), *Kadınlar Dünyası* (14 December 1913): 6–7.

54. Ulviye Mevlan, "Kadınlık-Maarif Nazırı" (Womanhood—Minister of Education), *Kadınlar Dünyası* (15 February 1914): 2.

55. Aziz Haydar, "Yalnız Kadınlar Mı Islaha Muhtaç?" (Is it only women who need to be reformed?), *Kadınlık* (17 April 1914): 4–5.

56. Ibid.

57. Aziz Haydar, "Ailede Kadınların Mevkii" (Women's status within the family), *Kadınlar Dünyası* (7 December 1913): 3–4.

58. Ibid.

59. Fatma Neşide, "Lakayıd Erkekler ve Zavallı Kadınlar" (Indifferent men and helpless women), *Hanımlar Alemi* (27 March 1914): 3–4.

60. Ibid.

61. Evliyazade Naciye, "İtiraf-ı Elim, Cevab-ı Elim" (Heartrending confession and heartrending response), *Mehasin* (18 July 1913): 659–661.

62. Ibid.

63. Ibid.

64. Ibid.

65. The exact name of the telephone company I have not been able to find.

66. Kadınlar Dünyası, "Birinci Sene-i Devriye Münasebetiyle" (On the occasion of our first anniversary), *Kadınlar Dünyası* (4 April 1914): 2–7.

67. [Editorial Board of *Kadınlar Dünyası*], "İctima: Telefon İdaresi" (The phone administration), *Kadınlar Dünyası* (26 April 1913); "İzahat Bekleriz, Tarziye İsteriz" (We expect an explanation and demand an apology), *Kadınlar Dünyası* (29 April 1913); "İctihad Kafi Değil, İzahat İsteriz, Tarziye Bekleriz" (Opinion is not enough, we expect an explanation and demand an apology), *Kadınlar Dünyası* (6 May 1913).

68. [Editorial Board of *Kadınlar Dünyası*], "Birinci Sene-i Devriye Münasebetiyle," 2–7.

69. Ibid.

70. Ibid.

71. Ibid.

72. Feriha Kamuran, "Meşrutiyet ve Kadınlar- On Temmuz Münasebetiyle" (Constitution and women—On the occasion of the tenth of July), *Kadınlar Alemi* (10 July 1913): 2–3.

Intimate Passports

The Subversive Performances of Tanja Ostojić

Jehanne-Marie Gavarini

ABSTRACT

The article explores the artwork of Tanja Ostojić, an interdisciplinary artist from Serbia who uses performance art to examine social and political issues. Ostojić in particular expresses the migrant woman's perspective when facing today's world of political and economic inequities. With caustic humor, the artist examines who occupies center positions and who remains in the margins. Ostojić's subversive performances blur the boundaries between art and life. Her use of her own body, personal history, and identity reflects a feminist perspective. Placing Ostojić's work in the longer history of performance art, this article analyzes how this provocative artist pushes the boundaries of art and culture by denouncing the power dynamics that rule exclusive systems such as the Western-dominated art world and the European Union.

KEYWORDS: art world, belonging/exclusion, Eastern Europe, European Union, female migration, feminist artist, identity, performance art, Tanja Ostojić

Passport to the European Union

In 2000, with the provocative title *Looking for a Husband with a EU Passport,* Tanja Ostojić, an interdisciplinary artist from Serbia,[1] started a series of powerful artworks that raise important questions about identity in postsocialist, postwar Yugoslavia, female migration, the East/West divide in European politics, and the position of female artists on the international art scene. *Looking for a Husband with a EU Passport* was an Internet site where the artist posted a nude photograph of herself and provided the address hottanja@hotmail.com for potential suitors to communicate with her. With this evocative email address, Ostojić received over five hundred responses to her Internet ad. After corresponding with German artist Klemens Golf for six months, she organized a public meeting/performance with him outside the Museum of Contemporary Art in Belgrade. Subsequently, she married him, applied for a visa, and moved to Berlin.

aspasia Volume 5, 2011: 112–127
doi:10.3167/asp.2011.050108

After three-and-a-half years she divorced Golf and turned their separation into an-other public event.

Ostojić is not the only artist who has investigated the institution of marriage in her work. With *Wedding Project*, Alix Lambert, a US performance and conceptual art-ist, married and divorced three men and one woman in the space of six months.[2] Both Lambert's and Ostojić's projects conflate the artist's real self with the fiction of a perfor-mance. However, although both artworks can be read and understood as critiques of marriage, in *Looking for a Husband with a EU Passport*, the husband/passport equation gives the work a political dimension that takes it to another level.

Ostojić's artwork departs from a general investigation of the marriage institution; and it differs from usual personal and sexual advertisements. Rather than containing images of women clad in lingerie and erotic costumes, Ostojić presents herself fron-tally, naked, and entirely shaved. The only trace of feminine adornment in Ostojić's photograph is fingernail polish that can be spotted by the more inquisitive viewers. The photo offers Ostojić's body to the public for visual consumption. Although some art critics, and men who responded to the ad, described the photograph as attractive, the combination of nudity and a shaved head creates discomfort in the viewer. This image is rather reminiscent of photographs—as recorded for instance by Hungarian-American photographer Robert Capa—of French women whose heads were shaven because they were accused of being collaborators during World War II.[3] Unlike Capa's images, Ostojić's photograph does not suggest shame; nevertheless, the subject repre-sented is vulnerable and exposed. More than an erotic image, Ostojić's self-portrait is a reminder that women's bodies have been bought and sold to alleviate economic hard-ships throughout history; they have been used for punishment and war retribution, traded and trafficked across borders.

The photograph of the artist is devoid of the conventional signs traditionally used as signifiers of desire, love, and romance. By creating such an image, Ostojić reveals that although romance is used as a tool to construct and perpetuate the heterosexual norm, it is often absent from marriage, in particular marriages of convenience.[4] In *Look-ing for a Husband with a EU Passport*, the nudity of the female subject is an act of erasure. The woman that Ostojić represents is stripped of all visible identity markers. It is as if she must become a blank slate and erase her self to be accepted as a wife. Without physical or symbolic baggage, she has made room for—and is ready to take on—West-ern European customs. Here marriage is the key to living and working legally in the European Union (EU) but it also implies an ability to transform oneself and conform. This work is a reminder that through the marriage contract, women have traditionally become dependent on their husbands. Danish art historian Rune Gade explains that in this work, "the stressing of the quasi-prostitution involved in the exchange occur-ring in the marriage ritual can be seen as a criticism of marriage as a heteronormative, patriarchal institution, maintaining and securing male ownership over women, the 'mail-order wife' being an emblematic example of this."[5]

Ostojić's public display of the wish to obtain a European Union passport signals the importance of such a document. Alexandrovna and Lyon state that for migrant women, "the passport is a major object of concern and care … a guarantee of pos-sible return in both directions."[6] *Looking for a Husband with a EU Passport* reflects the disappointment that followed the 1989 moment when amazing political, social, and economic transformations allowed Eastern Europeans to dream of open borders and a united Europe. However, as of today, the EU has not yet accepted the former Yugoslav republics, except for Slovenia, as equal members. Rosi Braidotti equates this division—and the hierarchy it creates between Europeans—with racism. She argues that "for peoples from the Balkans, or the South-Western regions of Europe, in so far as they are not yet 'good Europeans', they are also not quite as 'white' as others."[7]

What is subversive about *Looking for a Husband with a EU Passport* is that it is both a parody and reality at the same time. When Ostojić makes her search for a husband and her subsequent marriage public, she underscores the legal privileges that are conferred through marriage. With this work, she denounces, but also uses to her advantage, the privileges granted by compulsory heterosexual rule. This work of art is not simply a representation or performance; it has ramifications for her career and also her own life and identity as it allows her to legally move to the West. The work constitutes a legal transgression that alters her status and becomes a way to shun immigration laws. In this case, the artist could be simply one of the many immigrants who arrange marriages of convenience to gain the right to live and work in countries of their choice. But she also clearly works in the tradition of performance art, a form that artists have used to set their work apart from traditional art practice throughout the twentieth century.

Performance Art/Body Art

Art historian RoseLee Goldberg explains that in both Europe and the United States, performance art "came to be accepted by artists as a viable medium" in the late 1950s. She states: "Only ten years after a debilitating major war, many artists felt that they could not accept the essentially apolitical content of the then overwhelmingly popular Abstract Expressionism. It came to be considered socially irresponsible for artists to paint in secluded studios, when so many real political issues were at stake. This politically aware mood encouraged Dada-like manifestations and gestures as a means to attack establishment art values."[8] Subsequently, since the 1960s, artists worldwide have been using performance art, body art, video, happenings, and living sculpture as strategies to explore political ideas and bridge the gap between art and life.

These tendencies also appeared in Eastern European art. Curator and art critic Zdenka Badovinac states that "Body art in Eastern Europe started to emerge back in the time of McLuhan's global village, and it certainly belongs to the developmental line of the broader European-American art tradition."[9] In the early 1970s, Serbian performance artist Marina Abramović created powerful artworks in Yugoslavia that tested the limits of the physical body, the mind, and the self. In *Rhythm 5* (1974, Studenski Kulturni Centar, Belgrade), she built a large five-pointed star, symbol of communism, which she set on fire. After performing ritualistic gestures such as clipping her fingernails and cutting her hair, she lay down in the center of the star. She lost consciousness due to lack of oxygen and was rescued by two audience members.[10] In another famous performance *Thomas Lips* (1975, Krinsinger Gallery, Innsbruck), Abramović explored the weight of politics and culture in relation to one's physical and mental being. This was another endurance performance that involved ritualistic gestures such as slowly eating one kilogram (2.2 pounds) of honey with a silver spoon, drinking one liter (approximately 34 ounces) of red wine from a crystal glass, violently whipping herself, cutting a five-pointed star in the skin of her stomach with a razor blade, and lying on blocks of ice until the audience rescued her.[11] As pointed out by Goldberg, after 1968, "the art object came to be considered entirely superfluous" and "above all, audiences were provoked into asking just what were the boundaries of art: where, for instance, did scientific or philosophical enquiry end and art begin, or what distinguished the fine line between art and life?"[12]

Art historian Amelia Jones explains, "body art asks us to interrogate not only the politics of visuality but also the very structures through which the subject takes place" and further: "Body art in all its permutations (performance, photograph, film, video, text), insists upon subjectivities and identities (gendered, raced, classed, sexed, and otherwise) as absolutely central components of any cultural practice."[13] According to Jones, in the West, the body "faded from view" in the 1980s art world, a period that corresponded with "the disembodied politics of the Reagan/Thatcher era, characterized by political retrenchment and reactionary, exclusionary economic and social policies and by the scrupulous avoidance of addressing the effects of such policies on the increasingly large numbers of bodies/selves living below the poverty line."[14] However, by the 1990s, art practice returned to the body and the works produced at the time began "to acknowledge the deep implications of the politics of representation in relation to the embodied subject (as opposed to the abstracted–conceptual rather than engaged–subject implied by dominant theories of the 1980s)."[15]

Art and National Identity

Ostojić's work, which appeared in the mid-1990s, belongs in this artistic tradition. Ostojić is part of a continuum of feminist artists who have pushed the boundaries of art to question broader topics including economic, cultural, political, and social issues.[16] For the 1997 Venice Biennial, Marina Abramović created a disturbing allegory that commented on the war in Yugoslavia. She combined videos that narrate how "We in the Balkans kill rats" with a performance where she washed 1,500 fresh beef bones while singing folksongs from her childhood.[17] Abramović has influenced a new generation of Eastern European artists including Ostojić. With the dissolution of the "Eastern Bloc" and the aftermath of the war in the former Yugoslavia, these younger artists frequently explore national identity in their work. Historian Maria Todorova explains the ethnic complexity of the Balkans and the origins of "Balkanist discourse." In her words, "Unlike Western Europe where nations lived in more or less homogeneous blocks, in the East they were jumbled in a way that added the word *macédoine* to the vocabulary of menu writers."[18]

The refusal of a specific identity is the subject of Serbian artist Milica Tomić's video *I am Milica Tomić* (1998/1999) in which she appears as a talking head that repeats her name and identity in different languages. In each statement, she claims that she is from one of thirty different countries in thirty different languages.[19] As argued by art curator Yvonne Volkart, "to state, to pronounce one's identity makes one's identity" and here the artist "has rejected any ethnic feeling and explores the whole issue as a rhetorical formation."[20] By changing her national identity in each declaration and turning it into arbitrary statements, Tomić nullifies the power of her own speech acts, and raises questions about the general consensus on the importance of nationhood.[21] With each sentence and each new identity, bloody gouges appear and accumulate on the artist's face and torso. These wounds are inflicted off screen by an invisible weapon or presence but eventually the artist's body inexplicably goes back to its undamaged appearance. As noted by curator Charlotta Kotik: "The way the violence sometimes accompanies identity politics could hardly be more overtly stated."[22] It is important to understand this video in the political context of the Milošević regime. Art historian Jovana Stokić explains that at the time "just being Serbian was often perceived by the international community as a villainous act." With this work, Milica Tomić resisted "being found guilty only for belonging to a certain nation."[23]

In the West, the discrimination against, and bad treatment of, foreigners—particularly Eastern Europeans—is well documented. Ukrainian scholar Mykola Riabchuk writes: "Just try to enter any West European consulate with a Ukrainian passport and you will feel the superiority of the pettiest official who knows in advance that you are not a writer, not a scholar, not a journalist but just human trash, like everybody else is in this land, just one more trickster striving to blunt his/her vigilance and bypass the fence."[24] In *Waiting for a Visa* (2000, Belgrade, Serbia), Ostojić stood in line for six hours in front of the Austrian consulate with no result. Following this performance, Ostojić entered Austria illegally, crossing the Slovenian/Austrian border (2000, *Illegal Border Crossings*). She documented this action, which she described as exciting and less stressful than the legal process that one had to follow to get a proper visa before Slovenia was integrated into the EU.[25] With these works Ostojić drew attention to the plight of citizens of former Yugoslavia who were trying to cross borders and leave their country, which had been devastated by war

Ostojić repeatedly raises her audience's awareness about the situation of Europeans who are considered outsiders to the EU and analyzes the power dynamics that result from the Union's exclusionary rules. Following the traditions of performance art and feminism, the artist blurs the boundaries between art and life, the public and private, and the personal and political. On her website, she comments on the importance of having gone through these experiences in person: "I see myself trying to open the minds of other people, to show them something they might not think of, something they did not experience. For me it was very important to go personally through certain experiences, so that I could tell about this experience. So that I don't use only other people's stories. Then the personal becomes political. Then it's a statement, and it's something that I can, hopefully, use to initiate or to mobilize people."[26]

Art as Mimicry

In 2005–2006, Ostojić was invited to participate in EuroPart, an exhibition mounted on 400 rolling billboards in Vienna and Salzburg to celebrate the Austrian presidency of the EU Council. Ostojić created a photograph, *Untitled/After Courbet (L'Origine du monde)* (Figure 2), a reference to Gustave Courbet's painting *The Origin of the World* (Figure 1), which focuses on a woman's genitals by cropping out the woman's face and pointing the viewer's gaze directly at her pubis. This painting—once owned by Jacques Lacan—was originally commissioned in 1866 by Khalil Bey, a Turkish diplomat and ambassador of the Ottoman empire, for his personal collection of erotic paintings.[27] French historian and psychoanalyst Elisabeth Roudinesco explains that Courbet's scandalous painting was originally covered by a wood panel representing a landscape. When Lacan acquired it, his wife Sylvia Bataille Lacan thought that the painting should remain covered; so she asked surrealist painter André Masson to create a replacement for the original panel. The artist painted an abstract piece that could slide across to hide or to reveal *The Origin of the World*. Roudinesco adds that Courbet's work remained hidden most of the time through this secret system.[28]

Courbet's painting has become an important signifier in Western art history. Both art historians and artists have used it to analyze representations of gender and sexual

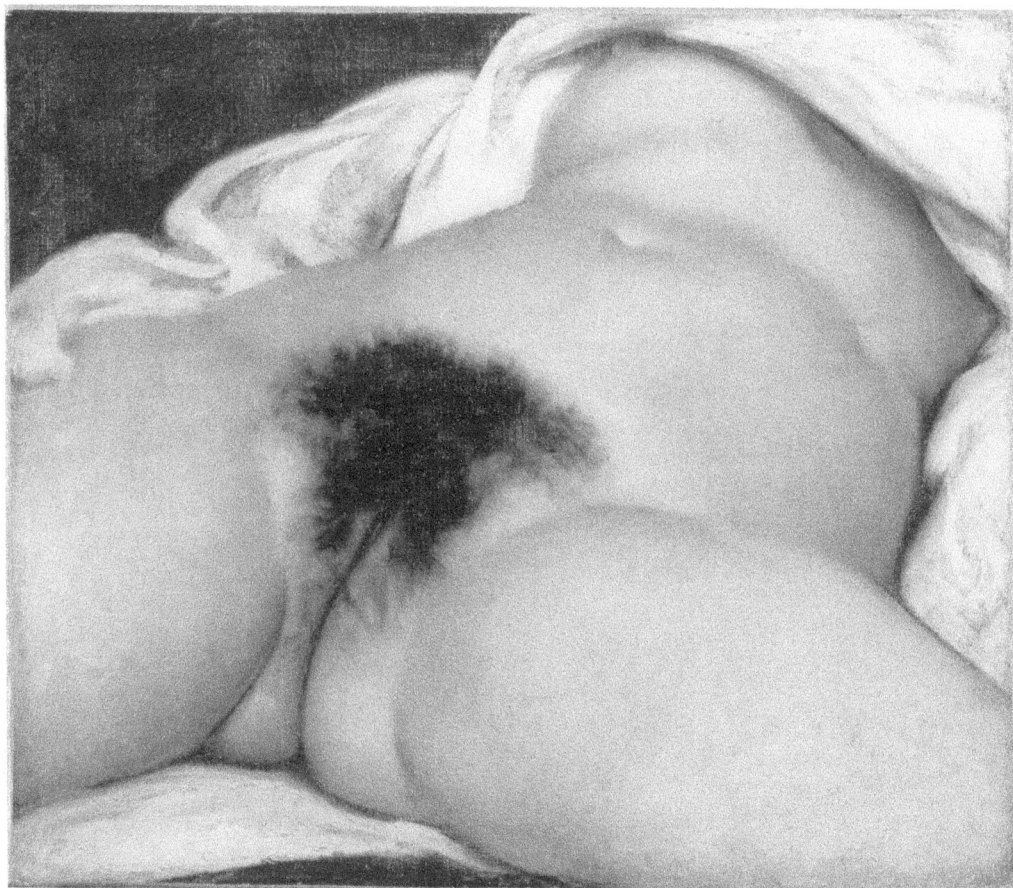

Figure 1. Gustave Courbet, *The Origin of the World*, 1866, Oil on canvas, 46 X 55 cm, Paris, Musée d'Orsay, Photo credit: Réunion des Musées Nationaux/Art Resource, NY, © RMN (Musée d'Orsay)/Hervé Lewandowski

difference.[29] Art theorist Linda Nochlin, describing the art historian's need to "penetrate to *the* ultimate meaning of a work of art," writes that "in the case of Courbet's *Origin*, this ultimate-meaning-to-be-penetrated might be considered the 'reality' of woman herself, the truth of the ultimate Other. The subject represented in *The Origin* is the female sex organ—the cunt—forbidden site of specularity and ultimate object of male desire; repressed or displaced in the classical scene of castration anxiety, it has also been constructed as the very source of artistic creation itself."[30]

In *Untitled/After Courbet (L'Origine du monde)* (Figure 2), what is offered to the viewer's gaze is not a model but the image of Ostojić's own body. She lies in the same position as Courbet's model and wears bright royal blue underpants that feature the EU flag. Art theorist Hilary Robinson revisits French philosopher Luce Irigaray's ideas on masquerade, mimesis, and mimicry to analyze the strategies available to female artists who have to use the visual language of patriarchy, a language that is not their own. Mimicry relates to mime; it is "an excess to strategies of survival, a product of a dislocation of identity."[31] Ostojić's *Untitled/After Courbet (L'Origine du monde)*, can be understood as an act of mimicry where "the subject compromises, represses or adapts

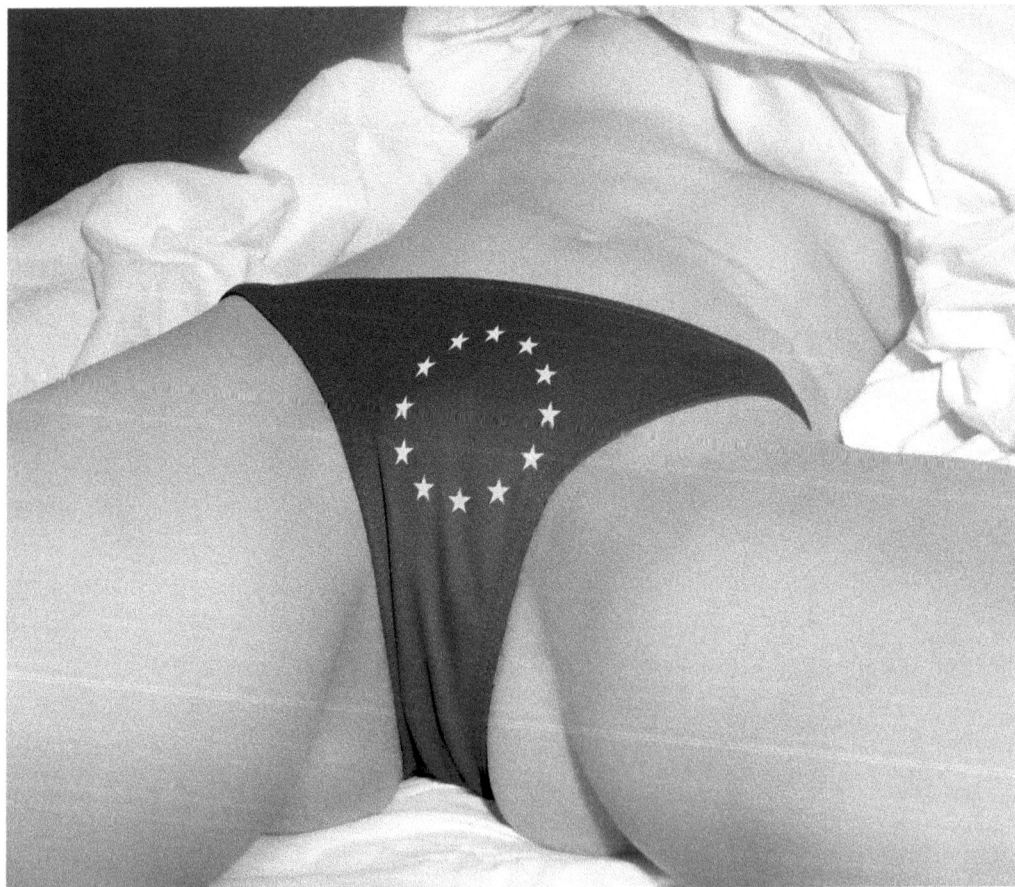

Figure 2. Tanja Ostojić, *Untitled/After Courbet (L'origine du monde, 46 x 55 cm)*, 2004. Color photo, 46 x 55 cm, Photo credit: David Rych, © Ostojić/ Rych

its perspective on its own identity in favour of an *apparent* (literally, in its appearance, visible, perceivable) assimilation to its environment—in this case, assimilation to the virtual world of 'femininity' as erected by the structures of patriarchy."[32] Here Ostojić mimics the traditional objectification of women with an apparent gesture of assimilation. However, she also creates a bold and empowering statement by reappropriating an exploitative image and broadcasting it on billboards.

Ostojić's *Untitled/After Courbet (L'Origine du monde)* provoked uproar. The subtleties and multiple possible interpretations of the work were ignored. It was called pornographic by a tabloid newspaper and removed from public view by the Austrian government after two days of scandalous media coverage. Considering the plethora of sexually explicit images produced by global media, it is difficult to understand how this image is considered more pornographic than so many fashion advertisements displayed all over European cities.[33] Although it may be understandable that the Austrian government would not desire to be associated with this kind of advertising imagery, the fact that women are still represented by such images is rarely questioned.

In this case, the concern about pornography disguises the political aspect of the work. On Ostojić's billboards the EU flag has now replaced the wooden panels that used to conceal the genitals in Courbet's painting. With this work, the artist appropriates the EU flag—a symbol of Europe's alleged unity—despite EU immigration policies that exclude her as a Serb. These billboards refer to the wish of Eastern Europeans for one Europe—a dream that is taking several decades to come true due to the EU's political and economic regulations and slow enlargement process. In *Untitled/After Courbet (L'Origine du monde)*, the flag hides what Nochlin describes as "the ultimate Other."[34] It now covers up the pubic area of a Serbian woman in the aftermath of the war in former Yugoslavia. The sum of all the signs present in *Untitled/After Courbet (L'Origine du monde)*—Courbet's *The Origin of the World* as a reference, the EU flag, and Ostojić's nationality—makes for an interesting equation between a Serbian woman and the origin of the EU. As explained previously, Serbia under the Milošević regime was not perceived very positively. In his essay *Vampires Like Us*, Balkan expert Tomislav Longinović states that the region functions "as the representation of the European unconscious, as a locus where the secrets of violence and sexuality are openly displayed, and as the reminder of the 'old centuries' before civilization."[35] In light of such words, it becomes possible that the cause of censorship is Ostojić's national origin as much as the image she disseminates.[36]

Untitled/After Courbet (L'Origine du monde) is an image that exposes the politics of the gaze, a subject that has been addressed and analyzed by feminist art historians. According to Griselda Pollock, "the sexual politics of looking function around a regime which divides into binary positions, activity/passivity, looking/being seen, voyeur/exhibitionist, subject/object."[37] *Untitled/After Courbet (L'Origine du monde)* imitates images that have been developed for the visual pleasure and consumption of men. The faceless women represented in these images are objectified. Ostojić's photograph comments on the situation of women of Eastern European origin who are still excluded from the EU and have to resort to desperate and degrading means, including pornography and prostitution, to enter and survive in the EU.[38] Yet the artist uses her own body and her international renown as counter elements to the usual anonymity of female models working in the pornographic industry. The half-nude model is offered to the gaze, but the subject is far from anonymous; her agency is undeniable. This image contains a double entendre: the Eastern European female artist appropriates the work of a Western male artist to hijack the symbol of Europe's apparent unity, revealing instead its exclusionist practices.

Passport to the Western Art World

The EU is not the only institution that falls under Ostojić's scrutiny. She has also created a large body of work that examines the Western-dominated art world as a system of privilege and exclusion. In her series *Strategies of Success,* she addresses the relationship between female artists and their curators. In *I'll Be Your Angel* (2001, "Plateau of Humankind," Venice, Italy), a four-day performance, Ostojić transforms herself into the public escort of Harald Szeemann, the curator of the 2001 Venice Biennial. This is another provocative piece where Ostojić addresses the artist's dream of making it to the top and exposes the common means to get there. Dressed in Christian Lacroix haute couture outfits for the duration of the biennial's opening ceremonies, Ostojić carefully plays the effaced female counterpart to the powerful male curator. She becomes the curator's gentle angel who follows him, smiling and pausing for the press. In this piece, Ostojić reveals that traditional gender roles remain accepted and for the most part unquestioned in today's art world and society: men can still hold positions of power while women serve as decorative accessories that visually enhance important male projects.[39] Adherence to a very traditional heterosexual role is presented as a means to become a player in the game and achieve success. Ostojić writes: "It was amusing to see how the ambassadors and ministers of culture we met thought of me as his mistress. They smiled in a special way and with admiration. Some of the artists later approached me on the boat and asked me to give their catalogues to my husband? Crazy."[40] What is interesting here is that the situation Ostojić sets up is almost like a trap into which both the curator and the public throw themselves; and the performance of this heterosexual fantasy goes on as it usually does. According to Ostojić's *Venice Diary,*[41] Harald Szeemann's wife walked into Szeemann's office, kissed him several times and turned to Ostojić to tell her: "You, you are not an angel—angels have to be invisible! You want publicity!"[42] Here the jealous behavior of Szeemann's wife contributes to the conflation of art and life. Ostojić uses the wife's feelings and assigns her a role in her performance and its documentation. The irate jealous wife underlines the performative aspect of social life and reinforces the idea that social behavior is scripted long before being acted out.[43]

Strategies of Success involves several other works in which Ostojić further exposes the sexual contracts of the art world's business. *Be My Guest* (2001, "Gravita 0," Palazzo delle Esposizioni, Rome, Italy), is a performance with the curator Bartolomeo Pietromarchi. The invitation cards for the show announce a "refined civilized dinner for two." After the dinner, Ostojić immerses herself in a Jacuzzi set up in the gallery space, where she is soon joined by both Pietromarchi and art critic and audience member Ludovico Pratesi. The three of them drinking champagne and French kissing in the

tub are recorded on video. Critics perceived this piece as another act of empowerment on the part of the artist. Artist and art historian Bojana Videkanic comments on
the role reversal in this work. In her words, Ostojić's "Eastern European background
further complicates the curator-artist relationship by exposing the often patronizing
attitudes that the West has towards the East."[44] Artist/curator Marina Gržinić argues
that *Be My Guest* performs "the obscene games conducted in the background of the Art
Edifice, now in an overtly visible, public, way." She adds: "What we get is the effect of
coming face-to-face of the everyday of art with its phantasmatic supplement. What we
see cinematically speaking is the concentration of the field (civilized talk and proper
behaviour) and its counter-field (the obscene retro scene of the art world) within the
same framework."[45]

In *Vacation with a Curator* (2003, Tirana, Albania), Ostojić documents a vacation with
Albanian curator Edi Muka with photographs shot in the style of paparazzi. These images are sent to various art magazines. They are also published online and in the book
Strategies of Success that documents the series. Here again, the artist plays with parody
and blurs the line between fiction and her life. She exposes not only the art world's love
for glamor and drama but also the general public's thirst for gossip. Just as with any
tabloid or popular media, what is interesting in the photographs is not so much what
they reveal as what is not there. The photographs trigger the viewers' imagination and
compel them to fantasize about what is implied in the erotic imagery. Both curator and
artist easily fall into the traditional roles assigned by compulsory heteronormativity.
However, Ostojić also reveals different concerns.[46] She writes that these bizarre photographs hide the fact that in that area, small boats carry beaten females who will be
sold as slaves to the West.[47]

Vacation with a Curator evokes ideas that have been explored by other female artists.
In this work, Ostojić addresses the hidden sexual games that rule the art world. She
investigates the fine line between prostitution and an artist's desire for visibility and
success. Likewise, in 1975, Marina Abramović traded roles with a prostitute for the
opening of her show at De Appel Gallery in Amsterdam (*Role Exchange,* Amsterdam,
Netherlands).[48] Also, soon after the completion of Ostojić's *Vacation with a Curator,* as
Guy Trebay noted in a *New York Times* article on 13 June 2004, American artist Andrea
Fraser received twenty thousand dollars through her New York gallery to have sex
with an anonymous collector; the untitled performance was recorded on video. In
1975–1976, Sanja Iveković, an artist from the former Yugoslavia, created *Double Life*—
an artwork that explored the public construction of the self. In this piece, Iveković
juxtaposed magazine advertisements with photographs from her private photo albums underlining similarities in appearance and gestures between the two. *Double
Life* addresses the constant demand on women to compare themselves with models
and celebrities. All of these works interrogate the sexual aspect involved in performing
traditional female roles.[49]

Ostojić belongs to a long continuum of artists who draw on art history and appropriate the work of other artists. Quoting the work of artists who have come before
her is an integral part of her oeuvre. For instance, one of the components of *I'll Be
Your Angel* was *Black Square on White* (2001, "Plateau of Humankind," Venice, Italy).
For this, Ostojić shaved her pubic hair in the shape of a square. Harald Szeemann, the

curator of the Venice Biennial, was the only person who was supposed to see *Black Square on White* before declaring it an official part of the exhibition. This action was in reference to Kasimir Malevich's famous *Black Square,* an abstract painting that has revolutionized twentieth-century art. For Ostojić, choosing Malevich was not incidental: his *Black Square* is a symbol of twentieth-century Russian avant-garde art. The Suprematist movement that Malevich initiated is a pillar of both Eastern and Western modernism. Yet Ostojić's quoting of Malevich is a complete *détournement.*[50] Here, the female artist creates secrecy and drama around the image of her own body to signify the stigma of sex and gender. The representation of the body with its excess of flesh and hair contrasts with the Suprematist and modernist idea that purity of form can be attained through abstraction. The secrecy around the work emphasizes the contradiction between the obsolete—yet still operating—social persuasion that bodies, particularly female bodies, are impure and should be concealed. Simultaneously, it highlights the fact that sexualized images of women are omnipresent in today's society.

With *Black Square on White,* the female artist finds ways to develop her own syntax while using the visual language of patriarchy. This process illustrates Irigaray's concept of mimesis. As explained by Hilary Robinson, for Irigaray, mimesis—unlike hysteria—can be productive: women can use productive mimesis as a strategy without falling into the paralyzing and self-defeating trap of hysteria. Robinson cites Irigaray: "To play with mimesis is thus, for a woman, to try to recover the place of her exploitation by discourse, without allowing herself to be simply reduced to it. It means to resubmit herself—inasmuch as she is on the side of the 'perceptible', of 'matter'—to 'ideas', in particular to ideas about herself, that are elaborated in/by a masculine logic, but so as to make 'visible', by an effect of playful repetition, what was supposed to remain invisible: the cover-up of a possible operation of the feminine in language."[51] In *Black Square on White,* Ostojić's humorous *détournement* is another act of resistance through which she denounces the narrow role socially assigned to women. By making the work accessible to the curator only, Ostojić exposes the power dynamics that bind artists and curators. She also underlines the private exchanges often involved in such relationships.

The Artist as Double Agent

Ostojić successfully blurs the boundaries between her art and her personal life that she repeatedly puts on display. The artist's insistence on using her own body in her work creates confusion between her own life and her constructed narratives, between what is real and what is a copy of the real. This in itself is not a new concept: gossip, sexualized images, and paparazzi-style photography dominate the lives of public figures and stars. However, what is different in Ostojić's work is that her mimicry is a strategy of resistance. Her transgression comes from the questions she raises regarding a system that requires young female artists to be beautiful and make their bodies—as much as their artwork—available for consumption. In the words of art theorist and curator Suzana Milevska, Ostojić "bravely interrogates the limits of this role playing, inhabiting the mind and body of a non-feminist artist in order to understand the way the system functions."[52] Ostojić uses this strategy judiciously. On the one hand, she denounces the objectification of women and the limited roles that femininity offers. On the other hand, she is a shrewd player who takes advantage of the system that she exposes through this very stratagem. Her role as a double agent who uses and twists symbols of female objectification to advance her career is at times uncomfortable; it can even appear as an ambivalent position on her part. Nevertheless, the questions that Ostojić raises about building a career on denouncing exploitation are all the more relevant. Behind the facade of her performance of traditional female passivity is the strategy of a sophisticated intellectual whose astute political awareness infuses the work.

Conclusion

Ostojić uses self-representation to make visible the predicament of migrant women when they face today's world of political, social, and economic inequities. Human trafficking and the trade of women's bodies for money constitute the subtext of this artist's oeuvre. She denounces how women still have to sell their bodies and minds in order to achieve their goals. This, along with the use of her own body, personal history, and identity in her work reflects a feminist perspective. With caustic humor, Ostojić examines who occupies center positions and who remains in the margins. This artist is not afraid of exposing societal taboos and performing embarrassing roles. She is a risk-taker who uses traditional and oppressive notions of femininity to redefine female subjectivity.

Ostojić's work is critical of the parallel forces that rule the EU and the art world, two fabricated entities that are rooted in privilege and exclusion. Ostojić relentlessly comments on the similarities between the desire to belong of the non-EU citizen and that of the female artist. Ostojić repeatedly travels in and out of these worlds, shifting from insider to outsider positions. By doing so, the artist reveals how these positions are interdependent. She underscores how the desire of the underprivileged to belong is born out of carefully constructed systems and power dynamics that are founded on exclusion.

Tanja Ostojić's work does not fit in a single category. Crossing over disciplines, it constantly references art history but it is also located in activism, feminism, postcolonial analysis, and the desire for social justice. The questions that Ostojić raises about the unspoken rules of the art world, as well as the parallels she draws between art and international politics are of critical importance in the process of conceptualizing and constructing a new Europe.

◊ About the Author

Jehanne-Marie Gavarini is a visual artist whose work has been exhibited internationally. In addition to her artistic work, she writes about art, cinema, and visual culture. She is the co-translator of *Tomboy* (University of Nebraska Press, 2007), an autobiographical novel by acclaimed Franco-Algerian writer Nina Bouraoui. Gavarini is the author of "Permeable Borders in *Notre Musique*" in *Zoom in, Zoom out: Crossing Borders in Contemporary European Cinema* (Cambridge Scholars Press, 2007). Gavarini is currently Professor of Art at the University of Massachusetts Lowell and Resident Scholar at the Women's Studies Research Center at Brandeis University. Email: jehannemarie _gavarini@uml.edu.

◊ Notes

1. As cited by Rune Gade, Ostojić says that she is from Belgrade but does not claim Serbia as her country because she has been traveling and living in several European countries since 1998. She says that her nationality is not her choice, but a label that was put on her. Rune Gade, "Making Real: Strategies of Performing Performativity in Tanja Ostojic's *Looking for a Husband with a EU Passport*," in *Performative Realism: Interdisciplinary Studies in Art and Media*, ed. Rune Gade and Anne Jerslev (Copenhagen, Denmark: Museum Tusculanum Press, 2005), 206.

2. Nicolas Bourriaud, *Esthétique relationnelle* (Relational Aesthetics) (Dijon: Les presses du réel, 2001), 35.

3. Not all female collaborators received the same treatment in 1944 France. The women whose heads were shaved were those "whose primary sedition was to have slept with the enemy." See Kristine Stiles, "Shaved Heads and Marked Bodies: Representations from Cultures of Trauma," *Stratégie II: Peuples Méditerranéens* (July–December 1993): 64–65, http://www.duke .edu/web/art/stiles/shaved_heads.html#11 (accessed 1 June 2010).

4. Conradsen and Kronborg explain that in Denmark's recent immigration reform, "love as opposed to marriage has become the object of legal regulation." They add, "concerning family reunion with a partner, the EU countries until now have agreed on marriage as the only relevant criterion. This is an agreement of doubtful content since marriage has lost its traditional meaning as known in family law." Inger Marie Conradsen and Annette Kronborg, "Changing Matrimonial Law in the Image of Immigration Law," in *Women Migrants from East to West: Gender Mobility and Belonging in Contemporary Europe*, ed. Luisa Passerini et al. (New York and Oxford: Berghahn Books, 2007), 240.

5. Gade, "Making Real," 196.

6. Nadejda Alexandrovna and Dawn Lyon, "Imaginary Geographies: Border-places and 'Home' in the Narratives of Migrant Women," in Passerini et al., *Women Migrants from East to West*, 97.

7. Rosi Braidotti, "On Becoming Europeans," in Passerini et al., *Women Migrants from East to West*, 23–44, esp. 34.

8. RoseLee Goldberg, *Performance Art: From Futurism to the Present*, rev. and enl. ed. (New York: H.N. Abrams, 1988), 144.

9. Zdenka Badovinac, *Body and the East: From the 1960s to the Present* (Cambridge and London: MIT Press, 1999), 14.

10. Marina Abramović, *Artist Body* (Milano: Edizioni Charta, 1998), 62–69.

11. Ibid., 98–105.

12. Goldberg, *Performance Art*, 152, 154.

13. Amelia Jones, *Body Art/Performing the Subject* (Minneapolis: University of Minnesota Press, 1998), 24, 31.

14. Ibid., 198.

15. Ibid.

16. Ample literature is available on feminist artists; for instance, Peggy Phelan and Helena Reckitt, *Art and Feminism: Themes and Movements* (London and New York: Phaidon, 2001). There have been several comprehensive feminist exhibitions in the last few years including *WACK!*

Art and the Feminist Revolution (2007–2008) and *Global Feminisms: New Directions in Contemporary Art*, which featured Ostojić's work and whose catalog is available. Maura Reilly and Linda Nochlin, eds., *Global Feminisms: New Directions in Contemporary Art* (London: Merrell, 2007).

17. Abramović was chosen as the artist representing Yugoslavia for this exhibition. Curator Bojana Pejić unveils the complicated historical and political context behind this controversial curatorial choice. Bojana Pejić, "Balkan for Beginners," in *Primary Documents: A Sourcebook for Eastern and Central European Art since the 1950s*, ed. Laura Hoptman and Tomáš Pospiszyl (Cambridge and London: MIT Press, 2002), 325–339.

18. Maria Todorova, *Imagining the Balkans* (New York and Oxford: Oxford University Press, 1997), 128.

19. In an interview Tomić told Jovana Stokić that these were the languages in which she could get instructions on pronunciation. Jovana Stokić, "Un-Doing Monoculture: Women Artists from the 'Blind Spot of Europe'—the Former Yugoslavia," *Artmargins Contemporary Central and East European Visual Culture*, New Critical Approaches Series, 10 March 2006, http://www.artmargins.com/index.php/archive/532-un-doing-monoculture-women-artists-from-the-blind-spot-of-europe-the-former-yugoslavia (accessed 1 June 2010).

20. Yvonne Volkart, "Imag(in)ing Identities," http://www.treibsand.ch/assets/files/Compiler/MTomic.pdf (accessed 1 June 2010).

21. For more information on new subjectivities in the postcolonial era, see Ioanna Laliotou, "'I want to see the world': Mobility and Subjectivity in the European Context," in Passerini et al., *Women Migrants from East to West*, 45–67.

22. Charlotta Kotik, "Eastern Europe and Post-Totalitarian Art," in Reilly and Nochlin, *Global Feminisms*, 30.

23. Jovana Stokić, "Un-Doing Monoculture," http://www.artmargins.com/index.php/archive/532-un-doing-monoculture-women-artists-from-the-blind-spot-of-europe-the-former-yugoslavia (accessed 1 June 2010).

24. Mykola Riabchuk, "Diversity or Divergence?" *Third Text* 21, no. 5 (2007): 538.

25. "Tanja Ostojić," http://www.van.at/see/tanja/index.htm (accessed 1 June 2010).

26. "Integration Impossible? The Politics of Migration in the Artwork of Tanja Ostojić," http://www.van.at/see/tanja/book02/20-21-dd.pdf (accessed 1 June 2010).

27. As investigated by Linda Nochlin in "Courbet's '*L'origine du monde*': The Origin without an Original," *October* 37 (1986): 76–86.

28. Elisabeth Roudinesco, *Jacques Lacan: Esquisse d'une vie, histoire d'un système de pensée* (Outline of a life, history of a system of thought) (Paris: Fayard, 1993), 248–249. The Musée d'Orsay that currently owns the painting explains on its web site: "Until it joined the collections of the Musée d'Orsay in 1995—by which time it belonged to the psychoanalyst Jacques Lacan—*The Origin of the World* epitomised the paradox of a famous painting that is seldom actually seen . . . *The Origin of the World*, now openly displayed, has taken its proper place in the history of modern painting. But it still raises the troubling question of voyeurism." "Gustave Courbet, *The Origin of the World*," at http://www.musee-orsay.fr/index.php?id=851&L=1&tx_commentaire_pi1%5BshowUid%5D=125&no_cache=1 (accessed 1 June 2010).

29. One appropriation of Courbet's painting by Jenny Saville represents Del LaGrace Volcano, a female to male transgender photographer. Peggy Phelan explains that: "As Saville makes her bid to enter the pantheon of great figurative painters, she returns to Courbet's *Origin* and paints it again. But Saville, unlike Courbet, includes her model's face, and the catalogue of the show contains an interview about the model's experience of her performance. Saville's model, in short, is not art history's nude woman, but contemporary art's former-woman." Phelan and Reckitt, *Art and Feminism*, 46.

30. Nochlin, "Courbet's '*L'origine du monde*'," 76.

31. Hilary Robinson, *Reading Art, Reading Irigaray: The Politics of Art by Women* (London: I.B. Tauris, 2006), 30.

32. Ibid., 31.

33. The Musée d'Orsay resorts to formal analysis to dodge the subject of pornography in relation to the original Courbet painting. The anonymous writer confirms that the painting belongs in the Western canon of high art by comparing it with Italian Renaissance painting. "The almost anatomical description of female sex organs is not attenuated by any historical or literary device. Yet thanks to Courbet's great virtuosity and the refinement of his amber colour scheme, the painting escapes pornographic status. This audacious, forthright new language had nonetheless not severed all links with tradition: the ample, sensual brushstrokes and the use of colour recall Venetian painting and Courbet himself claimed descent from Titian and Veronese, Correggio and the tradition of carnal, lyrical painting." "Gustave Courbet: *The Origin of the World*," http://www.musee-orsay.fr/index.php?id=851&L=1&tx_commentaire_pi1%5BshowUid%5D=125&no_cache=1 (accessed 1 June 2010).

34. Nochlin, "Courbet's '*L'origine du monde*'," 76.

35. Tomislav Z. Longinović, "Vampires Like Us: Gothic Imaginary and 'the serbs'," in *Balkan as Metaphor,* ed. Dušan I. Bjelić and Obrad Savić (Cambridge, MA: MIT Press, 2002), 53.

36. With regard to the representation of people from Serbia after the implosion of former Yugoslavia Longinović also writes: "Both the self-glorification of the nation by some Serbian intellectuals and politicians and the 'orientalization' of the entire Serbian population by the West have created this phantasmatic collective construct which I call 'the serbs' ... 'The serbs' therefore emerge as a figure and a reflection of a new form or racism that demands distance and separation between the protagonists of different civilizations within the same symbolic territory of Europe." Tomislav Z. Longinović, "Yugoslavism and Its Discontents," in *Thresholds of Western Culture: Identity, Postcoloniality, Transnationalism,* ed. John Burt Foster and Wayne Jeffrey Froman (New York: Continuum, 2002), 242–243. Longinović explains that in the Western European imaginary the Serbs have been viewed as a bloodthirsty people. Demonizing them represents the return to an older phenomenon. "In the post-Cold War rhetoric, this spectral collective has replaced both Marx and his big Soviet brother as the threatening apparition of human nature gone astray," Longinović, *Vampires Like Us,* 47.

37. Pollock also argues that female artists have challenged traditional power dynamics, in particular that of the gaze. In the work of Morisot and Cassatt, "one of the major means by which femininity is thus reworked is by the rearticulation of traditional space so that it ceases to function primarily as the space of sight for a mastering gaze, but becomes the locus of relationships." Griselda Pollock, *Vision and Difference: Femininity, Feminism and Histories of Art* (London and New York: Routledge, 2003), 87. This applies to *Untitled/After Courbet (L'Origine du monde)*, an image that was meant to be in a private collection and that is boldly delivered to the public.

38. As stated earlier, the former Yugoslav republics except for Slovenia are not integrated in the EU. Bulgaria and Romania were accepted in 2007 but on unequal terms.

39. Some progress has been made since Linda Nochlin's 1971 essay "Why Have There Been No Great Women Artists?" in her *Women, Art, and Power: And Other Essays* (New York: Harper & Row, 1988), 145–176. However, the art world and the media still marginalize female artists. The Guerrilla Girls take deserved credit for the improvement of women's representation in museums and galleries. Despite their activist work that started in the 1980s, they explain that they have not put an end to institutionalized sexism in the art world. Regarding one of their (in)famous posters that was displayed on New York City buses, the Guerrilla Girls state: "In 1989, when the GGs did the first version of this poster, less than 5% of the artists hanging in the Modern and Contemporary Sections of New York's Metropolitan Museum of Art were women, but 85% of the nudes were female. In the fall of 2004 we went back and recounted. SURPRISE.

Not much had changed. In fact, there were a few less women artists than fifteen years before!" See Guerrilla Girls, "Introduction and Conclusion to the Guerrilla Girls' Bedside Companion to the History of Western Art," in *The Feminism and Visual Culture Reader,* ed. Amelia Jones (New York: Routledge, 2003), 349–353. Also see "Guerrilla Girls: Reinventing the 'f' Word: Feminism!" http://www.guerrillagirls.com/posters/venicewallf.shtml (accessed 1 June 2010).

40. Tanja Ostojić, "Venice Diary," in *Strategies of Success: Curators Series 2001–2003,* ed. Tanja Ostojić (Belgrade: SKC Belgrade, 2004), 60.

41. The published biennial diary, although in documentary form, does not reflect Ostojić's usual political analysis and caustic irony. The style is closer to a romance novel and Ostojić explains that it owes its naive and innocent tone to the journal of Galina Vishnevskaya, a Russian diva and opera singer. The journal was read as part of the public performance *Strategies of Success* (2003, Gallery La Box, Bourges, France).

42. Ostojić, "Venice Diary," 54.

43. Analyzing the role of the "social machine" in the production of subjectivities, Félix Guattari writes: "Rather, it's a question of being aware of the existence of machines of subjectivation which don't simply work within the 'faculties of the soul,' inter-personal relations or intra-familial complexes. Subjectivity does not only produce itself through the psychogenetic stages of psychoanalysis or the 'mathemes' of the Unconscious, but also in the large-scale social machines of language and the mass media–which cannot be described as human." Félix Guattari, *Chaosmosis: an ethico-aesthetic paradigm,* trans. Paul Bains and Julian Pefanis (Bloomington: Indiana University Press, 1995), 9.

44. Bojana Videkanic, "Tanja Ostojić's Aesthetics of Affect and PostIdentity," *Artmargins Contemporary Central and East European Visual Culture,* New Critical Approaches Series, September 10, 2009, http://www.artmargins.com/index.php/featured-articles/414-tanja-ostojis-aesthetics-of-affect-and-postidentity-series-qnew-critical-approachesq (accessed 1 June 2010).

45. Marina Gržinić, "Tanja Ostojić: 'Yes It's Fucking Political'—Skunk Anansie," in Ostojić, *Strategies of Success,* 24.

46. Tanja Ostojić, "Remarks," in Ostojić, *Strategies of Success,* 68.

47. For information on Albanian female migrants working in the sex industry, see Iordanis Psimmenos, "The Making of Periphractic Spaces: The Case of Albanian Undocumented Female Migrants in the Sex Industry of Athens," in *Gender and Migration in Southern Europe: Women on the Move,* ed. Floya Anthias and Gabriella Lazaridis (Oxford and New York: Berg, 2000), 81–101.

48. Abramović, *Artist Body,* 112–117.

49. Nada Beroš, "Sanja Iveković: Double Life: Documents for Autobiography, 1959–75," in Hoptman and Pospiszyl, *Primary Documents,* 203–219.

50. As explained by Tony Godfrey, *détournement* was a strategy used by the Situationist movement. The word "can be roughly translated as 'diversion' or 're-routing'. It was an attempt to extend the notions of parody, plagiarism and collage." Two of the main actors of the movement were Guy Debord and Asger Jorn, an artist who believed that "artists must heal society's wounds, not merely design band-aids." Situationists who were inspired by political events in the early 1960s "saw the possibility of a new society and the end of traditional art in favour of a 'unitary urbanism', in which art and life co-existed." Tony Godfrey, *Conceptual Art* (London: Phaidon, 1998), 75–77. Unsurprisingly, Ostojić's performances have been labeled Situationist by art critics: http://www.medienkunstnetz.de/artist/ostojic/biography/ (accessed 1 June 2010).

51. Robinson, *Reading Art, Reading Irigaray,* 41.

52. Suzana Milevska, "The Portrait of an Artist as a Young 'Strategic Essentialist'," in Ostojić, *Strategies of Success,* 35.

forum

◈ ————————————————————————————

The Birth of a Field

Women's and Gender Studies in Central, Eastern
and Southeastern Europe, Part II

Edited by Krassimira Daskalova

———————————————————————— ◈

Women's Movements and Gender Studies
in Bosnia and Herzegovina

Gorana Mlinarević and Lamija Kosović

In Bosnia and Herzegovina (BiH), histories of women's movements, feminism, and gender studies are an underexplored field of study. Given the unstable historical situation in this country this is understandable, but that situation is not a sufficient excuse for the lack of initiatives in this direction.[1] This article is a starting point in fixing this omission.

As elsewhere, the disregard—or even negation—of the existence of women's movements, feminists and academics dealing with gender issues allows present-day opportunists to contest their relevance in the society. More important, whatever the political direction of the regime, this attitude is beneficial for those in power. The former socialist regime negated the influence of historical feminism of the late nineteenth and early twentieth century on women's emancipation, because it considered feminism to be a "bourgeois" movement. For socialist ideology in power the only "real road" to equality of all people was through socialist revolution and class liberation. Some attention was given to women's issues through formal instruments—such as the right to vote and political representation, equal access to (school and university) education, equal access to the labor market and equal pay for equal work, access to abortion, right to divorce, and access to some benefits with regard to maternity and childcare. However, the debates within the socialist governing circles remained mainly gender-blind. Women were still marginalized, experiencing domestic violence, political and public disempowerment and carrying a double or even triple burden in respect to providing income for the household, child caring, and house work. Similarly, during the war of the 1990s, women were among the most affected parts of the population and were subjected to the ethnonationalistic conquest campaigns. This situation continued into the postwar time.

The current regime, although refraining from military actions, fuels the ethnoreligious conflict in order to remain in power and it tries to keep women—and every-

aspasia Volume 5, 2011: 128–203
doi:10.3167/asp.2011.050109

one who does not conform to the current religious, ethnonationalistic, and neoliberal dominance—in subordinated positions. Feminist and gender studies, which promote human emancipation in many aspects of contemporary life, are very often seen as threatening to ethnonationalistic political elites in power in many parts of the world, and BiH is no exception. Although the elites in power try to appear as disinterested in women's, feminist, and gender issues, they are aware of their progressive potential (and antinationalistic and anti-neoliberal sentiments), which they witnessed during the turmoil of the 1990s in the neighboring countries (for example, the feminist peace activism in Croatia and Serbia).

Although not historians, through working in different academic areas dealing with contemporary women's issues in BiH and the Balkans, we have discovered the importance of exploring the history of women's movements and feminisms in BiH. Unfortunately, this endeavor is only beginning and we can only represent our initial observations and information. Our memory as "participant observers" dates back to the 1980s, but under the current circumstances—when "all ethnic sides" are in constant struggle for fabrication of their historic truths—even this flash-back free from the dominant ethnonationalistic vision can be a small contribution to women's, feminist, and gender studies research in BiH. Given the scarcity of the sources we had the access to for this research and the subjectivity of our memories, we are aware that our contribution to the *Aspasia* Forum can raise questions and criticism and we welcome discussion. Finally, we acknowledge that there are differences among the histories of women's movements, feminism, and gender studies, yet we see these histories as intertwined and treat them as such in this article.

Although the greatest part of the histories of women's movements, feminist, and gender studies in BiH cannot be detached from similar developments within the territories once known as Yugoslavia, some specificities apply only to this country. Within that broader context, we need to stress that, unfortunately, BiH was in a peripheral position in relation to the academic centers of Belgrade and Zagreb. Much of the academic thought was imported from those centers while at the same time "the brains" from BiH were "exported" and began their activity in those centers. Consequently, particularities concerning women in BiH were not adequately addressed. Our academia is not mature enough to resist ideological influences, engage in independent research, or to benefit from its capacity to influence social changes.

A History of Women's Organizing in Bosnia and Herzegovina

The existing literature dealing with the history of women's and feminist movements in former Yugoslavia refers to the suffrage movements of the late nineteenth and early twentieth century, as well as to socialist women's movements from the beginning of the twentieth century. However, most of the feminist texts written on this topic mainly refer to the movements in Serbia, Croatia, and Slovenia, and only imply that those movements also influenced the situation in BiH.[2]

The only texts dedicated to the history of women's movements in BiH were written through the lenses of socialist and communist ideologies. They state that prior to the end of World War II, women in BiH lived in a very conservative and backward society with retrograde traditionalist attitudes toward women. To support this, these texts refer to the rural character of BiH and the fact that in some rural areas the feudal system remained well in place, even until the beginning of the twentieth century.[3] One of the consequences of this socio-political situation was the high illiteracy rate among women. According to the 1910 census, there were only around 40,000 women who could write and read.[4] Nevertheless, those texts acknowledge that women's movements did exist in BiH continuously from the beginning of the twentieth century.

Although some publications recognize the existence of historical feminism and suffrage movements in the majority of the cities in the country,[5] they emphasize the greater significance of women in the workers' and socialist movements.[6] The texts dealing with the history of women's movements written during the socialist era showed the tensions between socialism and feminism. Moreover, the stigmatized "bourgeois" women, "most often intellectuals, conscious members of the bourgeois class, who led the struggle for the recognition of certain women's rights in isolation from the class and political struggle, very often directing this struggle against the male sex."[7] Scholars from the socialist time focused only on the history of women's movements through their participation in the workers', socialist, and communist movements. Hence, within the Socialist Federal Republic of Yugoslavia publications about the history of women's movements in BiH, if nothing else, provided significant insights into the socialist women's movements in the country and acknowledged that women's activism existed outside the socialist and communist ideologies as well. Thanks to such texts, we now know that before World War II women in BiH actively participated in the struggles for gaining economic and political rights. Moreover, they were actors in workers' strikes (e.g., of women workers in the textile and tobacco industries in Sarajevo in 1905). They worked for the education of other women through cultural, educational, and even ethnic and religious societies. In addition, women worked alongside women's feminist associations in other parts of former Yugoslavia; actively participated in both legal and illegal activities of the Communist Party of Yugoslavia. Finally, they collected aid for participants in the Spanish Civil War and for political prisoners; wrote for journals such as *Žena danas* (Woman today); organized International Women's Day celebrations; and were active participants in the trade unions.[8] Recognizing the significance of women's efforts, these publications emphasize that women had to struggle continuously in order to achieve equality within the context of the Socialist Republic of Bosnia and Herzegovina (SRBiH).

World War II marked even greater involvement and active participation by women. It is generally acknowledged that women in Yugoslavia and correspondingly in BiH contributed greatly to the partisan fight against fascism, both as active participants in combat and as illegal activists in the cities. Their involvement is visible in their contributions to journals such as *Žena danas*, their participation in women's county conferences throughout BiH, and—most significantly—in the *Antifašistički front žena* (AFŽ, Anti-Fascist Women's Front). World War II years—notwithstanding the ideological aspects of this coverage—are probably the best documented historical period in respect to the women's movement in BiH. Due to their World War II involvement in the People's Liberation Army and in the Communist Party, women in BiH as well as women in the Socialist Federal Republic of Yugoslavia in general gained significant social and political rights after 1946.[9]

As Svetlana Slapšak notes, the Communist Party distanced itself from the political activism and participation of women in the late 1940s, a process culminating in the disbanding of the AFŽ in 1952.[10] This was a big blow to the women's movement both in Yugoslavia and BiH, and after that, the women's movement was reduced mainly to individual actions. Within BiH, women's issues were assigned and institutionalized only through the Conference for Social Activities of Women of the Socialist Republic of Bosnia, which was operating within the advisory and monitoring body for social and political issues called the Republic Conference of the Socialist Association of the Working Class of BiH. The Conference for Social Activities of Women secured the quotas for women's participation within the one-party system and monitored and advised on social, economic, and political issues related to women. Apart from this political organization, women's issues were not high on the agenda of the Communist Party. Other issues of gender and sexuality were not yet acknowledged; for example, homosexuality was a completely taboo subject in society during state socialism and prohibited by the BiH Criminal Code until 1997. Not a single aspect of feminism or gender issues was recognized at any educational level.[11]

Nevertheless, within the ideological framework overseen by the Communist Party of Yugoslavia, some publications did focus exclusively on women in BiH. An interesting example is the 1977 book prepared by the Institute for History in Sarajevo, *Žene Bosne i Hercegovine u Narodno oslobodilačkoj borbi, 1941–1945* (Bosnian and Herzegovinian women in the struggles for the people's liberation, 1941–1945).[12] In the introduction Rato Dugonjić, the then president of the Socialist Republic of BiH, recognized the neglect of histories of women and women's movements in mainstream scholarship: "... with this book one big gap has been filled in respect to the already rich literature, both chronologies and scholarly publications, dedicated to the history and development of the Communist party, workers' movement, people's liberation army and revolution on our territory."[13] The book also admitted that the Communist Party made a mistake in ignoring the activities of the pre–World War II feminists by acting toward them in a sectarian way. A collection of stories of women participants in the women's movement in BiH during World War II, this book continues to be a valuable resource for gender research.

Finally, some BiH women participated in the activities of the leftist students' movements in 1968 within which feminists academic circles started to develop primarily in Belgrade, Zagreb, and Ljubljana. Probably the best-known feminist from BiH is Nada Ler Sofronić, whose name appears as the only participant from the country during the famous feminist conference *Drug-ca žena* (Comrade woman) held in Belgrade in 1978.[14] Nada Ler Sofronić's doctoral dissertation is considered to be one of the first feminist dissertations defended in socialist Yugoslavia.[15] Unfortunately, she remained an exception and it is important to highlight that, unlike Belgrade and Zagreb, Sarajevo lacked feminist activism at the end of the 1980s. No wonder then that Rada Iveković's observation regarding the revival of the patriarchal discourse in late 1980s Yugoslavia most noticeably applied to BiH. Iveković pointed to the strengthening of religious ethnonationalism and the economic crisis, which together led to women's higher unemployment rates and their increased isolation within the private sphere, even in urban areas.[16]

During the war of the 1990s, women's political consciousness was raised, but only after they experienced great suffering. Targeted by the ethnonationalists and reduced to victim positions, women of BiH courageously spoke up against the atrocities committed against them. They were not afraid to speak publicly about the rapes and other forms of sexual violence they experienced during the war, imprisonment in concentration camps, or the disappearances of family members. It should be mentioned that the rights and self-confidence achieved under the previous socialist regime played a role in women's speaking up about sexual violence. Whereas domestic violence and rape in marriage were not criminalized and were among the big social taboos, rape outside marriage was criminalized and punishable as a severe crime since the 1950s.[17] Thus, unlike the situation of women who survived rape and sexual violence during World War II and remained silent, women in the 1990s knew that rape and sexual violence were crimes and bravely spoke out about them. The contribution of the feminist antiwar activists and academics from the neighboring countries of Serbia and Croatia cannot be ignored in terms of their support for women-refugees from BiH and their participation in the mobilization of the international movement for lobbying for the recognition of rape as war crime. However, this would not have been possible without the courageous women from BiH, who publicly spoke aloud about the crimes committed against them.

The significant efforts of the women's movement in BiH during the war of the 1990s were related to the organization of women to help other women and to peace activism. Probably the best-known organization of women for women established during the war, which still plays a significant role providing assistance for women survivors, is Medica (Therapy center) in Zenica. Although international feminists played a significant role in the establishment of Medica, it was the local women who joined them and who have continued to work there.[18] The second issue was women's antiwar activism. The noteworthy peace movement that was initiated by women in the neighboring Yugoslav countries, primarily in Croatia and Serbia (with Women in Black from Belgrade as the most prominent example) did not leave the women from BiH uninvolved. Inspired by the Croatian and Serbian examples, women in BiH in the late 1990s established their own organizations with similar goals. Women's mobilization in BiH in helping themselves raised their feminist consciousness. This became visible internationally during the UN Women's Conference in Beijing in 1995, where they actively participated in the discussions.[19]

After the end of the war, several gender sensitive developments occurred. Among them were the rights of the Lesbian, Gay, Bisexual, Transgender, Transsexual, Intersexual and Queer (LGBTTIQ) population and discussions regarding the construction of masculinity in BiH. International donors came in huge numbers after the war due to the change in policies of international human rights institutions and their proclaimed dedication to gender mainstreaming. Many women's nongovernmental organizations were formed under their influence. The donor policies were too wide and lacked focus and local knowledge, so many initiatives were lost in the course of the postwar reconstruction of BiH.[20] These interventions clearly played into the hands of religious ethnonationalistic elite in power who easily accepted the formal changes stripped of any meaning. It allowed them to pretend to be "progressive and pro-European integration," while applying retrogressive policies toward women, even taking away rights they had enjoyed under the socialist government. Without the assistance of international donors not even a formal recognition of women's rights would have been possible in BiH. But an approach that would have valued local experiences and had given the initiative to the local women would have resulted in greater and more meaningful progress and would have played less into the strategies of repatriarchalization and retraditionalization of the current regime.

As noted, the joint actions of women's organizations, along with the donors' presence, did achieve progress regarding the institutionalization of women's rights. The most significant examples are the Gender Equality Law (2003), changes in the Criminal Codes concerning the criminalization of domestic violence, and the securing of pensions (reparations) for women who survived wartime sexual violence. However, given the really complex structure of the state of BiH and the resistance of the religious ethnonationalistic elite toward women's and gender issues, the implementation of these formally secured rights is still far from satisfactory. As Nejra Nuna Čengić has noted, although we have more mechanisms and procedures to register violence and more shelters for women, those are just institutionalized measures that treat the consequences of an unjust world, but do not really affect social relationships.[21] Given the fact that women's groups spent a great deal of energy on lobbying for those laws, their disappointment with the lack of implementation resulted in the weakening of the women's networks/movement, rather than in the revival of their activism. Another problem is that women's groups, out of fear of losing their funding, did not try to shift the donors' agenda towards project more relevant to the local social needs or to regain the rights enjoyed by women during state socialism. It is not surprising that women's NGOs, which became dependent on foreign donors, are spending more time on defining what to do and what to live off once the foreign funding will have ended and the donors will have left the country than on women's rights-related issues.

The current situation with respect to the women's movement is bleak. Nada Ler Sofronić more than eight years ago warned about the serious threat women in the region were facing with regard to the neutralization of their social and political subjectivity, the wiping out of their connection to the global progressive women's movement, and the dramatic worsening of their social status, which now seems to have become a reality.[22]

The Institutionalization of Gender Studies in Bosnia and Herzegovina

Keeping this historical background in mind, we now turn to the institutionalization of gender studies in BiH, which started in the context of postwar BiH in the early 2000s. We are well aware of the danger of gender studies losing their critical edge after having entered academia. Yet we idealistically believe that in the context of BiH, the introduction of gender studies in the country can open a space for contextualized and situated discussions and create potential for actions aimed at changing the repressive institutions and systems, and gender relations in general. Given the historical background of the feminist movement of former Yugoslavia where antiwar and antinationalist progressive actions were carried out by feminist activists from academia, we believe that students of gender studies can become creative and innovative forces in the struggle for the emancipation of BiH society.

Gender studies in BiH are still mainly perceived as "women's studies," but have slowly started to address and include such issues as heteronormativity, body politics, masculinity, and queer studies. Gender studies in BiH have been established as an innovative interdisciplinary field of knowledge and research, aiming to participate in the democratization of academia and society of BiH.[23] So far, the publications have mainly dealt with the current issues affecting women in BiH, connected with the consequences of the war and the period of transition. Most texts appear in the local language in the journals published in BiH. However, some gender specialists have started to publish in English in prestigious international journals, challenging mainstream perceptions about the contemporary gender contract in the country. During the late 1990s and the beginning of the 2000s, the most prominent authors were Nirman Moranjak-Bamburać, Marina Katnić-Bakaršić, Nada Ler Sofronić, Jasminka-Babić Avdispahić, and Jasna Bakšić-Muftić, later joined by younger authors like Jasmina Husanović, Damir Arsenijević, and Zilka Šiljak Spahić.

The introduction of feminist and gender studies in academia started at the end of the 1990s and in the early 2000s with the individual efforts of feminist scholars— Nirman Moranjak-Bamburać, Jasminka Babić-Avdispahić, and Jasna Bakšić-Muftić— working primarily in the fields of literature, philosophy, and law, respectively. Initially their influence consisted of introducing feminist topics in the university courses they taught in Sarajevo, but later they also introduced entire feminist courses. Most recently these scholars have been joined by Jasmina Husanović and Damir Arsenijević in Tuzla, and Danijela Majstorović in Banja Luka.

Parallel to this, some of them started to organize alternative spaces for women's and gender studies. Already during the late 1990s and early 2000s they organized workshops, which gathered academic activists dealing with various women's issues. One such workshop was held under the auspices of the International Forum Bosnae, the NGO that mainly focuses on the organization of academic debates.[24] As a result of these discussions—and thanks to the dedicated work of Nirman Moranjak-Bamburać— a Center for Gender Research was established within the International Forum Bosnae in 2002. In 2004 the International Forum Bosnae also supported the feminist studies in BiH by publishing a special issue of its own journal—*Izazovi feminizma* (The challenges of feminism). There for the first time appeared texts by feminists scholars working in Sarajevo, namely Jasminka Babić-Avdispahić, Jasna Bakšić-Muftić, Marina Katnić-Bakaršić, and Nirman Moranjak-Bamburać. Their essays covered feminist approaches to law, language, and (de)construction of gender, values, and models in feminist literary criticism, as well as the connections among feminism, ethics, and politics.[25]

Similarly, in 2002 feminists in BiH got together and started an alternative (outside the official academia) women's studies school, established under the auspices of the nongovernmental organization *Žene ženama* (Women to Women).[26] This project was a cornerstone for further endeavors of the feminists from BiH's academia in developing gender studies. Within this alternative women's studies school, professors from different universities merged their individual intellectual capacities and efforts into a high quality program in women studies for nearly five consecutive years.[27] Parallel to the realization of women's studies and under the umbrella of the same project, Nirman Moranjak-Bamburać, together with the journalist Angelina Šimić, created *Patchwork*, the first feminist magazine in BiH.

Activities were also undertaken for the opening of postgraduate programs in gender studies. The idea to enter into the formal academic world with such a program matured, but various obstacles postponed its launching. Jasna Bakšić-Muftić, Nirman Moranjak-Bamburać, and Jasminka Babić-Avdispahić spearheaded this initiative and were joined and supported by Zdravko Grebo (professor of law and at that time director of the Center for Interdisciplinary Studies of the University of Sarajevo [CIPS]) and Nejra Nuna Čengić (then academic tutor and coordinator of the European MA Program in Human Rights and Democracy in Southeastern Europe of the CIPS). In the 2006–2007 academic year they finally succeeded in creating a postgraduate program in gender studies. The Gender Studies MA program was implemented within the Center for Interdisciplinary Postgraduate Studies of the University of Sarajevo.

The program tried to engage a number of prominent academics interested in the different aspects of gender relations in BiH, as well as the rest of the Balkan region. In addition to the three initiators, an active role in the realization of the program was played by Marina Katnić-Bakaršić (linguistics and stylistics), Jasmina Husanović (political science and cultural studies), Ugo Vlaisavljević (philosophy), Asim Mujkić (political science), Damir Arsenijević (literature, queer studies and cultural studies), Nebojša Jovanović (media studies), Zilka Šiljak-Spahić (religious studies), Lejla Somun-Krupalija (political studies and nationalism), Ajla Demiragić (literature), Svjetlana Đurković (queer studies), and Olivera Simić (peace studies) from BiH; Marina Blagojević (sociology and methodology of feminist research), and Daša Duhaček (political studies) from Serbia; Biljana Kašić (history) from Croatia; Katerina Kolozova (philosophy) from Macedonia, and Svetlana Slapšak (anthropology) from Slovenia. Some of them are also members of the Scientific/Advisory Board of the Program.

The idea behind the establishment of the MA program in gender studies[28] was to start an epistemological project that critically revises mainstream academic knowledge, while spreading legitimate fields of scholarly research and introducing new knowledge and epistemology.[29] This interdisciplinary gender studies program aimed to cover topics in the humanities and the social sciences in general by offering courses in gender theories, women's human rights, gender and politics, gender and development, and gender, ideology, and culture. It focused on Southeastern Europe through gender-sensitive courses dealing with political, cultural, economic, and social aspects of the Balkans. Initially the list of courses was rather limited but this has been changing thanks to the increased number of teaching staff especially from BiH as well as the considerable interest shown by students. Other courses offered focus on gender, peace studies, and transitional justice, gender and economy, gender and economic and social rights, gender and transition, and gender and EU integration. Although formally within academia, the program still has not secured sufficient recognition on the same level as the graduate programs of other long-established academic fields. It is interesting to note that gender studies is not included in undergraduate teaching, which is considered to be more fundamental by the elite in power because it affects more people. The postgraduate program in Gender Studies has been formally supported by the Agency for Gender Equality BiH, thanks to the Agency's director, Samra Filipović-Hadžiabdić. Formal support was also provided by the University of Sarajevo, but the program would not have been able to survive without funding by the Norwegian and Swedish governments.

The existence of a gender program at the postgraduate level facilitated, both directly and indirectly, the opening of more gender-sensitive courses not only within different faculties of Sarajevo University, but also in other institutions of higher learning in BiH.[30] Some of the lecturers and former students have already initiated gender courses at their institutions, thus influencing other colleagues to start considering gender studies as an important scholarly field.

Conclusions

We have provided a personal perspective on the history of women's activism and its relations with the establishment of gender studies in BiH. The biggest problem with the histories and actions of local women's organizations is their invisibility. Numerous factors have played and continue to play a role in sustaining that state of affairs. These histories were erased or reshaped depending on the ideological regimes that subsequently ruled over the territory of BiH. These regimes, patriarchal in their nature, were only slightly different in their policy toward women. The "concessions" that were "given" to women from various modernizing elites were superficial and did not affect deeply gender relations in the society.

In contemporary BiH, individual activism is more prominent than the organized women's movement. That is probably why there is a reversal of the usual order of things (the top down approach) in the country and the postgraduate program in Gender Studies was established in Sarajevo without the existence of strong women's or feminist movements.

◊ Acknowledgments

We thank our collegues Melina Sadiković and Nejra Nuna Čengić for their valuable comments and contributions to this article. This text reflects our personal positions and does not reflect the view of our employer—the Master's program in gender studies of the Center for Interdisciplinary Postgraduate Studies, University of Sarajevo.

◊ Notes

1. It is true that different political systems that changed over time prevented historical retrospectives without ideological connotations on events in past regimes. However, the lack of texts about the histories of feminisms and women's movements and gender studies in BiH—even of the texts that come from within the ideologies in power—is striking.

2. See, for example, Lydia Sklevicky *Konji, žene, ratovi* (Horses, women, wars) (Zagreb: Ženska infoteka, 1996); Sabrina P. Ramet, ed., *Gender Politics in the Western Balkans* (University Park: Pennsylvania State University Press, 1999); Neda Božinović, *Žensko pitanje u Srbiji u XIX i XX veku* (The woman question in Serbia in the nineteenth and twentieth centuries) (Belgrade: Žene u Crnom, 1996).

3. Uglješa Danilović, "Pitanje ravnopravnosti žena u Politici KPJ" (The question of equality of women in the politics of the Communist Party of Yugoslavia), in *Žene Bosne i Hercegovine u Narodno oslobodilačkoj borbi* (Bosnian and Herzegovinian women and the struggle for the people's liberation), ed. Zdravko Antonić, Nevenka Bajić, Drago Borovčanin, Rasim Hurem, and Dubravka Škarica (Sarajevo: Svjetlost, 1977), 11–20.

4. Dana Begić, "Osvrt na napredni pokret žena u Bosni i Hercegovini između dva svjetska rata" (Review of the advanced movement of women in Bosnia and Herzegovina between two world wars), in *Žene Bosne i Hercegovine u Narodno oslobodilačkoj borbi*, ed. Antonić et al., 21–33.

5. Danilović, "Pitanje ravnopravnosti žena u Politici KPJ."

6. Begić, "Osvrt na napredni pokret žena u Bosni i Hercegovini između dva svjetska rata"; see also Zlata Grebo, "Ravnopravnost žene—dio borbe za socijalističko društvo" (Equality of women – a part of the struggle for socialist society), *Patchwork* 1, nos. 1–2 (2003): 135–144.

7. Danilović, "Pitanje ravnopravnosti žena u Politici KPJ," 13.

8. Begić, "Osvrt na napredni pokret žena u Bosni i Hercegovini između dva svjetska rata;" see also Grebo, "Ravnopravnost žene—dio borbe za socijalističko društvo."

9. For Yugoslavia, see for example, Svetlana Slapšak, "Nationalist and Women's Discourse in Post-Yugoslavia," in *Transitions, Environments, Translations: Feminisms in International Politics*, ed. Joan W. Scott, Cora Kaplan, and Debra Keates (New York and London: Routledge, 1997), 72–78.

10. Svetlana Slapšak, "Identities under Threat on the Eastern Borders," in *Thinking Differently: A Reader in European Women's Studies*, ed. Gabriele Griffin and Rosi Braidotti (London and New York: Zed Books, 2002), 145–157, esp. 149.

11. Jasminka Babić-Avdispahić, "Feminizam, etika i politika" (Feminism, ethics and politics), in *Izazovi feminizma*, ed. Jasminka Babić-Avdispahić, Jasna Bakšić-Muftić, Marina Katnić-Bakaršić, and Nirman Moranjak-Bamburać, vol. 26, no. 4 (Sarajevo: Forum Bosnae, 2004): 193–247, esp. 195.

12. Antonić et al., eds., *Žene Bosne i Hercegovine u Narodno oslobodilačkoj borbi*.

13. Rato Dugonjić, "Uvod" (Introduction), in Antonić et al., *Žene Bosne i Hercegovine u Narodno oslobodilačkoj borbi*, v–vi, esp. v.

14. See, for example, Lepa Mlađenovic, *Počeci feminizma: ženski pokret u Beogradu, Zagrebu, Ljubljani* (Beginnings of feminism: The women's movement in Belgrade, Zagreb and Ljubljana), Belgrade: Autonomni zenski centar, http://www.womenngo.org.rs/content/blogcategory/28/61/#per2 (accessed 25 July 2010).

15. See Nada Ler Sofronić's biography, published on the website of the Novi Sad Association "Women's studies and research," http://www.zenskestudije.org.rs/01_o_nama/biografije.html (accessed 25 July 2010).

16. Rada Iveković, "Tela na frontu: (Ne)predstavljivost ženskog u simboličkoj ekonomiji: Žene, Nacija i rat nakon 1989. godine" (Bodies on the front: The (non)representationality of femaleness in the symbolic economy: Women, nation and war after 1989), in *Žene, slike, izmišljaji* (Women, images, imaginaries), ed. Branka Arsić (Beograd: Centar za ženske studije, 2000), 9–30.

17. Only during the late 1980s did women's activists in Belgrade, Zagreb, and Ljubljana start to set up rape hotline numbers and raise issues concerning rape in marriage.

18. About the establishment and work of Medica in Zenica, see Cynthia Cockburn, *The Spaces between Us: Negotiating Gender and National Identities in Conflict* (London and New York: Zed Books, 1998), 174–210.

19. Jasna Bakšić-Muftić from Sarajevo was among the participants from BiH.

20. Although an analysis of the donor funding policies in BiH is not the topic of this article, it should be pointed out that the donors' intervention was really a replica of the intervention in most of the developing countries, using a depoliticized and instrumental version of "women's empowerment." As some feminists have argued, the "women's empowerment" agenda was "stolen by the high priests of neo-liberalism only to be foisted onto women in the global south as their putative salvation." Attention was given to the formal acknowledgment of some issues related to women's human rights and to a futile calculation of women and men in different institutions (educational, political, or administrative). The donors applied a top-down approach without encouraging on the ground research or an assessment of real needs. The international donors instrumentalized women and the concept of empowerment and focused on demands for evidence-based results rather than on the implementation of quality programs valued and owned by the local population. Andrea Cornwall and Nana Akua Anyidoho, "Introduction: Women's Empowerment: Contentions and contestations," *Development* 53, no. 2 (June 2010): 145. For a gender and development critique of such an approach by international donors, see Wendy Harcourt, "Editorial: Lady Gaga Meets Ban Ki-Moon," Ibid, 141–143.

21. Nejra Čengić, "Gender and Empowerment: A Conversation with Nejra Cengic." This is a response to Wendy Harcourt's editorial "Lady Gaga meets ban Ki-Moon" in *Development* 53, no. 2 (June 2010), http://www.sidint.net/gender-and-empowerment-a-conversation-with-nejra-cengic/ (accessed 1 August 2010).

22. Nada Ler Sofronić, "Engendered End of Transitional Paradigm," in *Women and Politics: Gender in Political Theory*, ed. Đurđa Knežević and Koraljka Dilić (Zagreb: Ženska infoteka, 2002), 113–126, esp. 117.

23. Jaminka Babić-Avdispahić, "Uvod: Rodni studije i demokratizacija znanosti" (Introduction: Gender studies and the democratization of science), in *Rod i nauka* (Gender and science), ed. Jasminka Babić-Avdispahić, Jasna Bakšić-Muftić, and Ugo Vlaisavljević (Sarajevo: Center for interdisciplinary postgraduate studies at the University of Sarajevo, 2009), 7–16.

24. For more information about the International Forum Bosnae, see its website, http://www.ifbosna.org.ba/_site/indexen.php?lang=en&sel=1 (accessed 1 August 2010).

25. See Jasminka Babić-Avdispahić, Jasna Bakšić-Muftić, Marina Katnić-Bakaršić, and Nirman Moranjak-Bamburać, eds., *Izazovi feminizma*, vol. 26, no. 4 (Sarajevo: Forum Bosnae, 2004).

26. For information about Žene ženama, see http://www.zenezenama.org/ (accessed 1 August 2010).

27. Babić-Avdispahić, "Uvod: Rodni studije i demokratizacija znanosti," 8.

28. For additional information about the MA in Gender Studies see www.cps.edu.ba (accessed 1 August 2010).

29. Babić-Avdispahić, "Uvod: Rodni studije i demokratizacija znanosti."

30. We are aware that there are young feminists such as Jasmina Husanović and Damir Arsenijević who were educated abroad and have been actively engaged—independently from the Master's program in gender studies and even prior to its establishment—in the work on emancipation of the society, once they returned to BiH.

The Beginnings of Gender Studies in Bulgarian Academia

Kornelia Slavova

This article is a reflection on the process of introducing gender studies in Bulgarian institutions of higher learning, both in terms of curriculum transformation and teaching. It traces the history of the discipline from its initial stages in the early 1990s to the end of the first decade of the twenty-first century. In addition, it discusses the two major strategies that have been adopted in the Bulgarian context, more precisely, the pros and cons of teaching gender studies as a separate (stand-alone) program as well as integrating it into already established traditional disciplines. The analysis looks both backward and forward: to some of the successful practices of gendering the academia in the past twenty years, but also to the stumbling blocks in the development of gender studies as a fully fledged discipline in the future.

Gender studies as a separate field of research and educational practice emerged in Bulgarian science and academia in the early 1990s after the collapse of the communist regime.[1] This delay can be explained by a series of cultural and political voids in Bulgarian society. These include the absence of a politically significant independent women's movement backing up intellectual efforts to generate a greater gender awareness, the lack of critical social research on gender issues before 1989, the absence of gender theory to articulate the concerns of postcommunist women, the lack of corresponding gender-oriented discourses in Bulgarian culture, as well as lack of support at the institutional and political levels.

The gaps between practice and theory, and between local experience and Western feminist knowledge have further complicated the situation, leading to an eclectic borrowing of already existing and proven Western models of women's and gender studies, and adapting them to the specificities of the postcommunist situation.[2] In addition to the resistance of academic institutions to opening their doors to this innovative discipline—seen as antifoundational and antidisciplinary as it destabilizes the foundations of traditional disciplines—the teaching of gender studies at the university level has been delayed because of at least three other important gaps: a lack of trained scholars to carry out the teaching process; a lack of sufficient knowledge/interest in the many aspects of gender relations on behalf of potential students; and a lack of teaching materials and resources in the Bulgarian language. Without these three basic pillars, no university education can exist; let alone the process of gendering university education.

The unavailability of these conditions has predetermined the winding road to institutionalizing gender studies in Bulgaria. The ideal option would have been to introduce the subject as an autonomous stand-alone program, and then work on its implementation and growth, further extending it into other areas. Instead, the Bulgarian strategy has involved processes pulling in different (at times contradictory) directions: that is, pushing from within the established traditional disciplines to make changes, revisions, and infuse gender components in their content and methodology, while sporadically trying to teach the discipline in self-contained academic units. This is why in the last twenty years of postcommunist transition we can trace a trajectory that combines both approaches—the autonomous and the integrated ones—simultaneously.

Gendering the Traditional Disciplines

The process of infusing gender as an important social and cultural category of analysis into the overall educational process began in the mid-1990s at several Bulgarian universities. Scholars from diverse fields in the humanities started integrating gender perspectives into traditional disciplines by gendering as many educational programs/courses/departments as possible. This process has been the work of a feminist ad hoc group belonging to the Bulgarian Association of University Women (BAUW)—originally founded in 1924, suspended by the communist regime in 1950, and re-established in 1991—still the only openly pronounced feminist nongovernmental organization in Bulgaria.[3] Around this relatively small organization a circle of devoted academics and nonacademics was created, who were actively engaged in various public activities such as promoting new gender politics, fighting against pornography and homophobia, and addressing other serious gender-related social and political issues.

At the same time, the members of the BAUW—mostly academics from the St. Kliment Ohridski University of Sofia—have paved the way for introducing gender studies as an academic subject through the implementation of many educational projects. For example, they produced the first books of translated gender and feminist theory[4] as well as the first collections of gender research by Bulgarian scholars.[5] These publications have turned out to be useful resources for the general reading audience, students, and professors involved in the teaching of gender studies. The very first issues of the Newsletter of the BAUW (edited by the philosopher Maria Dimitrova, published from 1992 to 1994), offered practical information and theory/criticism in translation, and further expanded the circle of people interested in gender issues in the early 1990s.

Furthermore, the BAUW organized five annual national writing contests on women's issues in high schools and universities, as well as many national and international conferences,[6] public talks, lectures and hearings, exhibitions of women artists, and many other events. These activities have drawn public attention to the significance of women's issues and have created a more gender-sensitive climate in academia. As early as 1991, the BAUW organized the first extra-curricular courses on gender issues (such as "English cum Feminism" and "Women Writers"), as well as interdisciplinary courses on gender, culture, and representation for students from several faculties of Sofia University, focusing on the intersection of gender with media and cultural studies, literary and psychoanalytic theory, history, linguistics, and sociology. This ad hoc gender studies group has had a strong impact on gender sensitization in Bulgarian society and on putting gender issues on the educational agenda. Thus the connection between the BAUW and the academic world has turned out to be a natural (though not institutionalized) entry point for feminism in academia. It bridged the gap between feminist theory and practice, emphasized the importance of feminist approaches to history, sociology, literature, cultural studies, philosophy, political science, and other disciplines that are at the heart of any culture.

Connections with Western academic institutions have been another important factor in the initial stages of building gender studies as a legitimate academic discipline. Many visiting scholars from American, British, and Canadian universities have put gender on the map of Bulgarian university education by acting as mediators between Western academic knowledge and the Bulgarian tradition. For example, the very first women's studies course—"Introduction to Women's Studies in the United States"—was taught at the University of Sofia in 1993 by Professor Francine Frank from the State University of New York (SUNY) at Albany. The next year she taught "Gender and Language." She was followed by a number of visiting American scholars teaching various courses about women's literature and culture, such as Judith Barlow ("Women Playwrights" and "Feminist Literary Theory") and Iliana Semmler ("Growing Up in America")—both from SUNY. Gradually, under the influence of the revised and more inclusive Western canons of literature, culture, and history, women's issues became an indispensable part of area studies—British, American, and European—taught at Sofia University

Despite the fact that this more scattered and less organized strategy of infusing gender into the overall curriculum transformation was not attached to a specific program/degree in gender studies, it has turned out to be quite efficient because it made gender analysis more visible in academia and deemed worthy of attention. Working at the intersection between gender and other significant categories of human difference such as class, ethnicity, religion, and nationality, more and more Bulgarian lecturers introduced gendered approaches in their research and teaching of history, culture, literature, film, popular culture, theory, and many other educational fields (predominantly in the humanities). Thus the number of gender-oriented courses offered at Bulgarian universities has constantly grown, as seen from the following selected list of available courses at the St. Kliment Ohridski University of Sofia around the turn of the twenty-first century. These included "European Women's History" and "The History of Women in the Balkans" (taught by Krassimira Daskalova), "Biblical Femininity" (taught by Milena Kirova), "Gender Representation in American Culture" (taught by Kornelia Slavova), "Gender, Power and Communication" (taught by Ralitsa Muharska), "Identity and Culture" (taught by Tatyana Stoicheva), "The British Female Tradition: From Realism to Postmodernism," and "Contemporary British Women Writers" (taught by Vessela Katsarova). The University of Veliko Turnovo St. Cyril and Methodius also offered a series of gender-related courses, including "Introduction to Gender Studies" and "Images of Childhood in British Writing from Mary Wollstonecraft to Angela Carter" (taught by Ludmilla Kostova), "Feminist Transformations in Contemporary British Literature and Drama" (taught by Elena Zlatanova), "Gender in American Literature" (taught by Yonka Krasteva), "Gender in Austrian Literature" (taught by Penka Angelova), and "Gender and Politics" (taught by Teodora Kaleinska). Although fewer in number, similar courses were introduced at other Bulgarian institutions such as the University of Plovdiv ("Introduction to Gender Studies" by Zlatka Chervenkova), at the New Bulgarian University ("Constructions of Gender in Everyday Communication" by Petar Vodenicharov), and at the South Western University in

Blagoevgrad ("Masculinity and Representation" by Stephan Dechev). This list is not exhaustive; it is growing constantly and signals the gradual process of gendering more and more areas of teaching and research. Although not part of an institutionalized gender studies program, the vicarious strategy has acted as a catalyst for familiarizing a greater number of students (both men and women) with the significance of gender issues and methodology in all areas of human knowledge.

Gender Studies as a Stand-Alone Discipline

This gradual infusion of a gender perspective at different Bulgarian universities during the 1990s has paved the way for the second strategy of introducing gender studies as a separate and self-sufficient program. In fact, the very first attempt at launching a gender studies MA program in Bulgaria, at the New Bulgarian University in 2000 (Director Tanya Kotzeva), was a collective effort of academics from several universities but it failed to materialize due to insufficient student enrolment and lack of institutional support. This option was successfully put into practice in 2001 at the Faculty of Philosophy at the St. Kliment Ohridski University of Sofia, where the first MA gender studies program was set up (Director Nedyalka Videva). The program (taught today at both the MA and PhD levels) offers over thirty courses in gender by bringing together diverse perspectives in the humanities, but it still needs to fully develop its potential in terms of attracting a greater number of students. The existence of the program is a significant contribution to the development of the discipline of gender studies on a national level.

The program also illustrates most of the risks involved in similar academic practices, evolving in a context of insufficient gender awareness and institutional support. The greatest pitfall in developing such self-contained gender studies programs concerns the missing rationale for the professional development of students graduating from such programs: What will be their future professional career? In the transitional period, often marked by financial instability and employment crises, the job prospects for such graduates are scarce. This is why the idea of combining the MA degree in gender studies with a BA in another discipline (law, political sciences, and literature) has turned out to be a good solution. Such arrangement offers better job prospects for gender studies graduates, and has a multiplier effect on changing the traditional perceptions of gender in a wider section of the academia and culture as a whole. Also, this combination will positively affect the future of the discipline by eliminating the danger of marginalizing or ghettoizing gender studies by turning it into a kind of elitist education for a very small circle of interested students and faculty.

Internationalizing Gender Studies

An important strategy for gendering higher education in Bulgaria is related to the very process of internationalizing the content and methodologies of teaching. Gender studies can act as a mediator to bring together the study of women and men from various regions, locations, and periods, which is especially topical nowadays in view of the current "decentering" of Europe. What is more, in a global age of instant communication and shifting boundaries, where all factors of experience coexist in a state of active and continual interplay, gender studies cannot be developed serially any longer. The recent accession of Bulgaria to the European Union has served as a strong catalyst for improving gender-related legislation as well as creating a much more tolerant environment for the development of gender-related educational and cultural projects.[7] European integration processes have helped legitimize gender studies as a significant and necessary discipline before Bulgarian official institutions.

In addition, European gender and women's studies departments and research centers have proved to be most open to international components and experiences, global in nature, and more inclusive. Some of the first fruitful practices of European academic collaboration in the field of gender research were initiated by the Program on Gender and Culture at the Central European University (CEU) in Budapest. As early as 1995 it started organizing regional conferences, summer schools, and seminars (sponsored by the Higher Education Support Program [HESP]) to promote the teaching and research in gender studies in Central, Eastern, and Southeastern Europe. Most notable among these initiatives was the series of regional seminars on gender issues, held four times a year at the CEU in Budapest and Warsaw, which brought together scholars from former Soviet bloc countries where women's studies and gender studies had not existed before. The seminars were thematically arranged around important topics such as "Women and Citizenship," "Multiple Feminisms," "Feminist Methodologies," "Women and Politics," "Gender Representations in Visual Arts." They stimulated a broader public debate on the importance of gender analysis during the transitional period, encouraged comparative perspectives, and created a Southeast European network of gender scholars/organizations,[8] which enriched the dialogue among universities and NGOs in the region, and increased the mobility of scholars throughout the region.

HESP research grants have further stimulated the development of gender-oriented research as well as the creation of new gender-related courses at institutions of higher learning in Central and Eastern Europe. In addition, the Open Society Institute–Budapest funded a number of translation projects throughout the region. These included the translation into Bulgarian of ground-breaking books such as *Our Bodies, Ourselves* (by the Boston Women's Health Book Collective) as well as publications in feminist theory and methodology by renowned scholars such as Seyla Benhabib, Judith Butler, Joan Scott, and Gisela Bock. In 2000, financial support from the Open Society Foundation in Bulgaria made possible the publication of the first Bulgarian journal for gender issues, *P.S.* (*Post Scriptum;* Virga Publishing House, chief editor Virginia Zaharieva). In 2005 the avant-garde journal *Altera* was launched (Altera Publishers, chief editor Miglena Nikolchina). It focused on gender, language and culture, and some of its thematic issues broke new grounds by introducing discussions on queer theory and practice; lesbian, gay, bisexual, and transgender studies, and masculinity studies. For the first time in Bulgaria, this journal was produced by an independent Bulgarian publisher, which further extended its interest in the field of gender studies and culture by launching a second, more specialized journal, *Altera Academica,* in 2008.

Another positive example of the fruitful cooperation among European universities through the network of gender studies programs is the recent creation of the first European MA Program in women's and gender history. Called MATILDA, this program is a partnership of five European universities: Vienna University (Austria), Central European University (Budapest, Hungary), Nottingham University (Great Britain), University of Lyon 2 (France), and Sofia University St. Kliment Ohridski (Bulgaria).[9] It was set up in the 2008–2009 academic year to promote the exchange of students, faculty, teaching models, successful practices, and ideas among partner institutions. All students from the program are expected to spend at least a semester at one of the partner institutions abroad, and they are offered the possibility to work with a supervisor from another European university. This is a truly "feminist nomadic project" in the sense of not simply creating a greater mobility of students and faculty, but also of opening up new frontiers of comparative exploration.[10] Under the Erasmus scheme additional opportunities for collaborative teaching, research, and summer schools have developed; for example, the European Summer University on Gender and Genre, held in June 2009 at the University Charles de Gaulle, Lille-3, co-organized by Lille-3 University and St. Kliment Ohridski University of Sofia, with participants from five other European universities.

Similar joint European academic projects around gender studies can have an invigorating effect on both emerging gender studies departments/centers in Central and Eastern Europe and Western feminist knowledge. As Rosi Braidotti observes, women's studies is "a critical project not only in terms of its revisions of how science perpetuates forms of women's discrimination and exclusion," but it is also "far more critical of ethnocentrism and nationalism than the official guidelines from the European Commission."[11] This new European dialogical space has already opened up wider and more critical vistas on Eurocentric myths and gendered identities, has generated debates and pushed for revisions in the global agenda of women—thus proving the significance of feminist location politics.

In conclusion, the integrationist and autonomous approaches of teaching gender studies, combined with the internationalizing processes in a context of globalization and European Union enlargement, have produced broader and more solid gender knowledge in Bulgarian academia. Both the autonomous and integrated forms of gender studies have shown potential for enriching and improving educational standards through the prism of gender by invigorating existing theories and methodologies. On a larger scale, the increasing academic visibility of gender will further develop the critical apparatus and thinking in public spaces and institutions to restrict the very social and psychic mechanisms that construct and perpetuate gender inequality and insensitivity in Bulgarian society and culture.

◊ Notes

1. Unlike North America and Western Europe, in most Eastern and Central European countries the label "gender studies" has been preferred to women's studies or feminist studies—not simply in terms of labels but also because of the politics involved in them. "Women's studies" is seen as related with earlier standpoint feminism, whereas "feminist studies" is associated

with radical feminism stemming from postmodernist and poststructuralist identity politics. Because of the overall hostility toward feminist ideas and practices in the postcommunist world (accompanied by a great dosage of misinformation and ignorance), "gender studies" seems the most acceptable term, sounding more academic and inclusive, and less discriminatory to men.

2. For reflections by Bulgarian feminist scholars, see Ralitsa Muharska, "Silences and Parodies in the East–West Feminist Dialogue," *L'HOMME. Europäische Zeitschrift für Feministische Geschichtswissenschaft*, Special Issue: *Übergänge. Ost-West-Feminismen*, ed. Ute Gerhard and Krassimira Daskalova 16, no. 1 (2005): 36–47; Kornelia Slavova, "Looking at Western Feminisms through the Double Lens of Eastern Europe and the Third World," in *Women and Citizenship in Central and Eastern Europe*, ed. Jasmina Lukić, Joanna Regulska, and Darja Zavirsek (Burlington, VT: Ashgate, 2006), 259–277.

3. On the early history of the BAUW, see Georgeta Nazarska, "The Bulgarian Association of University Women, 1924–1950," *Aspasia* 1 (2007): 153–175.

4. Among the most significant translated books are Miglena Nikolchina et al., eds., *Vremeto na zhenite* (Women's time: Feminist critical theory) (Sofia: Sofia University Press, 1997); Sneja Gunew, ed., *Feministkoto znanie* (Feminist knowledge: Critique and construct) (Sofia: Polis, 2002); Krassimira Daskalova and Kornelia Slavova, eds., *Zhenski identichnosti na Balkanite* (Women's identities in the Balkans) (Sofia: Polis, 2004).

5. These include Krassimira Daskalova, ed., *Ot syankata na istoriyata: Zhenite v bulgarskoto obshtestvo i kultura* (From the shadow of history: Women in Bulgarian society and culture) (Sofia: Dom na naukite za choveka i obshtestvoto, 1998); Ralitsa Muharska, ed., *Maiki i dushteri. Pokoleniya v bulgarskiya feminism* (Mothers and daughters. Generations and themes in Bulgarian feminism) (Sofia: Polis, 1999); Kornelia Slavova and Milena Kirova, eds., *Teoria prez granitsite: Vuvedenie v izsledvaniyata na roda* (Theory across boundaries: Introduction into gender studies) (Sofia: Polis, 2001); Krassimira Daskalova et al., eds., *Tehnite sobstveni glasove* (Voices of their own: Oral history interviews of women) (Sofia: Polis, 2003; English edition, 2004); Milena Kirova and Kornelia Slavova, eds., *Rod i red v bulgarskata kultura* (Gender and power in Bulgarian culture) (Sofia: Tip Top Press, 2005); Milena Kirova and Kornelia Slavova, eds., *Identichnosti v prehod: rod, medii i populyarna kultura* (Identities in transition: Gender, media and popular culture) (Sofia: Polis, 2010).

6. Among the most influential conferences and workshops were "Women in Politics" (1991); "Feminism in Literary Theory and Cultural Studies" (1992); "Women and the Media" (1995); "Transition and Transcendence: Women East/West" (1995); a five-week seminar on the works of Julia Kristeva (1996); as well as the international conferences "Limits of Citizenship: European Women between Tradition and Modernity" (1998); and "Women, Gender and the Cultural Production of Knowledge" (2007).

7. Despite the overall positive effect of EU enlargement on the gender legislation in former communist countries, some critics cautiously warm of the dangers of such practices coming from above instead of below, or what Mihaela Miroiu calls "room-service feminism" (see her article "State Men, Market Women," *Feminismos*, no. 3 [June 2004]). See also Elaine Weiner's article "Dirigism and Déjà-vu Logic: The Gender Politics and Perils of EU Enlargement," *European Journal of Women's Studies* 16 (2009): 211–228.

8. The Open Society Institute, more precisely its Network Women's Program, published the first and so far only *Gender Studies and Women's Studies Directory*, which includes information about research centers, teaching institutions, and women's NGOs in the countries of Central and Eastern Europe, the former Soviet Union, and Mongolia. Published in 1999, this is still a valuable resource for gender-oriented scholars and activists in the region. Debra Schultz, ed., *Gender Studies and Women's Studies Directory* (Budapest: Open Society Institute, 1999).

9. For more detailed information on the Matilda Program see its website: http://www.ned
.univie.ac.at/matilda/, accessed 12/12/2010.

10. I use the term nomadic project in Rosi Braidotti's understanding of "feminist nomadism" as a movement not only of "critical opposition of the false universality of the subject, but
also as a positive affirmation of women's desire to affirm and enact different forms of subjectivity." See her *Nomadic Subjects: Embodiment and Sexual Difference in Contemporary Feminist Theory*
(New York: Columbia University Press, 1994), 158.

11. For a more detailed understanding of her argument, see Rosi Braidotti's article "Uneasy
Transitions: Women's Studies in the European Union," in *Transitions, Environments, Translations,*
ed. Joan Scott, Cora Kaplan, and Debra Keates (New York and London: Routledge, 1998), 363.

◆

Establishing Gender Studies in Czech Society
Hana Hašková

Before 1989, feminism in the countries of Central and Eastern Europe was defined in
the framework of communist propaganda as an ideology that undermined the unity
of the working class and the effort to achieve communism.[1] Consequently, the second
wave of the women's movement did not penetrate the borders of the Central and Eastern European region before 1989. Mass women's organizations established during the
state-socialist period in all the countries of the former Eastern bloc were fully under
the control of the respective Communist Party. Although some issues about the status
of women were publicly discussed, in such a context it was impossible for a critical
analysis of gender relations to develop.[2]

In the Czech Republic, like elsewhere in Central, Eastern, and Southeastern Europe, the beginning of gender studies and feminist research dates from the 1990s.
The women who founded gender studies departments at universities and research
institutions in the Czech Republic concur that they had never encountered women's
studies or feminist theories before 1989. Women's and gender studies and feminist
epistemologies were not known (or amid the lack of information were rejected) before
1989, not just in official scholarly circles, but even among male and female members
of the dissident movement.[3] For the founders of gender studies in the Czech Republic,
the early 1990s were a time of fascination with and rapid immersion into the concepts
of gender and feminist theories. Coming in contact with Western feminist researchers
and literature represented an important impetus for these scholars to introduce feminist reflection into their fields.

Marie Čermáková, the founder of the research team and later of the Department of
Gender and Sociology at the Institute of Sociology of the Academy of Sciences of the
Czech Republic, describes this personal experience:

I long had a growing personal interest in Czech women's literature and the
issues and problems affecting the fate of women. But I didn't like the way this
was approached. So until 1989 I took up the idea of how I would do it if I had

the opportunity for independent work ... [Only after 1989] for the first time I was able to choose my field. I had no idea how much gender and women's studies had expanded. ... I really didn't know how elaborated feminist sociology was, it fascinated me. In the first years I had an opportunity to take part in numerous conferences and exchanges that opened up opportunities for me.[4]

Hana Havelková, the founder of the Gender Studies Department at the Faculty of Humanities at Charles University in Prague recalls the beginning of the 1990s as a time when there was an absence of any expert discussion in the field of gender studies and feminist theories in Czech society: "I needed urging, and even then, sometime in 1990, I, like many others, believed that there is not much to say on the theme of men and women in this country and there aren't that many problems. I soon realized how wrong I was."[5]

However, this new opportunity to learn about developments in the field of gender studies and feminist epistemologies and even any external impulses to do so were not themselves enough for these scholars to become as engaged as they did in the subsequent development of gender studies and feminist research in the Czech Republic. As with the younger generation of gender studies scholars, equally important for promoting an interest in this field for both generations was their personal experience in everyday life. Gender studies and feminist epistemologies provided them with a framework for new ways of interpreting these experiences. The only difference was that the younger generation of scholars usually had their first encounter with gender-sensitive research and feminist theories while studying on student exchanges at Western universities or in the first gender studies courses offered at Czech universities.[6]

Hana Havelková described her situation in the early 1990s as a bit "schizophrenic." She educated others in the Czech Republic on the usefulness of feminist theories, but she had to explain to Western colleagues "why we can't automatically adopt some of their concepts without critically confronting them with the differences in reality in this country."[7] Jiřina Šiklová experienced a similar situation. She describes her early involvement in the field of gender studies in the Czech Republic as follows:

Since the early 1960s I had been interested in the problems of the status of women, without having any idea about feminism as a developed worldview or even philosophy. ... During the so-called normalization period, in the 1970s, it was possible to write about family issues, the relationship between men and women, and criticism about these issues, including the burden on women and their double shift, was tolerated. And so I ... wrote several popular educational books on women and their position in society. But I knew almost nothing about feminism at that time ... I first encountered theories of feminism after November 1989, when journalists from the West, including some feminists, began coming to Czechoslovakia, and they brought with them books and periodicals. ... I was very resistant to any kind of ideology, so I tended to argue with them, and I didn't just immediately become an enthusiast for feminism.[8]

Despite this initially ambivalent view of feminism, Šiklová agreed with Ann Snitow, founder of the Network of East-West Women (NEWW), on informal support in founding a nongovernmental organization (NGO), which was intended to serve as a special gender studies library and a curriculum center for developing teaching programs in the field of gender studies.[9] Then, with the help of Jana Hradílková and Laura Busheikin, a collection of feminist literature was assembled in Šiklová's apartment, initially comprised mainly of donations from the United States, Germany, and Scandinavia. Lectures, discussions, and public meetings with foreign feminists were also held there. Finally, a group began to meet there that went on to officially register an NGO in 1992—the Gender Studies Foundation. The foundation took on all the activities that had been discussed as necessary for supporting gender studies, including the specialized gender studies library. With the assistance of Saša M. Lienau, the Foundation began to receive long-term support from the German foundation *FrauenAnstiftung;* among other things it was able to set up its first office.[10]

In 1992 the Czech *Filosofický časopis* (Journal of philosophy) in its first thematic issue on feminist philosophy published several original papers by Czech scholars and some foreign feminist philosophers focusing on feminist philosophy. This was followed in 1995 by the publication of the first thematic issue of the bilingual—Czech-English—*Sociologický časopis/Czech Sociological Review* on gender in the social sciences.[11] All the articles in the first issue were written by Czech authors.

In the editorial introduction to this special issue, "Gender in the Social Sciences and Feminist Issues," the editors outlined the current discussion on adapting the concepts of gender and feminism to the Czech environment. The discussion revealed the negative connotations associated with the concept of feminism, not just by the general public, but also by Czech social scientists. "In the general awareness of the Czech public feminism is usually ascribed a very narrow meaning, most often articulated as women trying to imitate men, erasing the differences between the sexes, or the emancipation of women through work, and it is perceived as a threat to femininity."[12]

Other scholars similarly recalled the Czech public's resistance to feminism, though not primarily in terms of content, because even in the early 1990s most Czech women and even some men fully identified with some of the ideas of leading feminists. Overwhelmingly, they rejected the term feminism, which had been the target of substantial ridicule during the state- socialist period.[13] Alena Heitlinger has explained the opposition to feminism in the Czech environment in terms of its lack of resonance with Czech women's framings of their own situation, and with the distorted way in which Western feminism was framed locally:

> because most Czechs mistrust utopian and emancipatory ideologies, associate concepts such as "women's emancipation" and "women's movement" with the policies of the discredited communist regime, are disinclined to engage in collective action, regard themselves as strong women rather than as victims, assign highly positive meaning to motherhood and the family, and perceive feminism to be anti-male.[14]

In the late 1990s, however, Šiklová noted an increase in the strength with which feminism was rejected in Czech society after 1989, due to the hegemonic behavior of some Western feminists. She noted in particular the distorted interpretation of gender relations and social realities with which some Western researchers came to Eastern and Central Europe after 1989.[15] In this context, a form of discourse dismissive of feminism emerged in Czech society, while the concept of gender or gender studies was viewed more positively.[16] Unlike feminism, gender was an unknown term in Czech society. Free from any connotations, negative or positive, it was perceived more favorably than feminism. In interviews with university professors in the field of gender studies, Lenka Václavíková-Helšusová has observed a tendency to avoid the term feminism in describing their lectures: "when we asked students whether they would have enrolled in our course 'An Introduction to Gender Studies' if it had been called 'An Introduction to Feminist Theory' or 'Feminism,' the majority of them said no."[17] For some the concept of gender was and still is more agreeable because they regard the application of the feminist approach as ideological interference in the construed neutrality of science.[18]

During the 1990s the activists who established the Prague Gender Studies Foundation discussed whether to use the unknown foreign term "gender" in Czech or to use a Czech translation of the word instead. A consensus has been reached for leaving the term "gender" untranslated.[19] The Czech translation of the term "gender" —rod— primairly refers to the grammatical category of gender or to concepts such as kinship and family. So there were fears that the use of rod instead of "gender" would be misleading. For this reason, all the teaching and research institutions founded in the field of gender studies in the Czech Republic use the name "gender (studies)" and not the Czech translation rodová (studia) or (studia) rodu.[20]

Unlike the situation in the US or the UK, where discussions took place over whether the introduction of gender studies at universities would weaken the feminist charge that had driven the founding of women's studies programs at universities in these countries in the 1970s, no such discussion appeared in the Czech Republic. The reason for this is that women's studies had never existed, so a de-politicization of the field via the politics of naming was not a topic of controversy among Czech academics. Czech gender studies developed within the context of feminist scholarship and the institutionalization of women's and gender studies worldwide at a time when theoretical debates had already moved away from the task of including women in the academic realm as physical bodies and subjects of inquiry to the task of using gender as an analytical tool for understanding inequalities and other social orders.[21]

Amid the discussions (or absence of them), the first series of lectures in the field of gender studies began at Charles University in Prague during the first half of the 1990s. Some of the founders of the Gender Studies Foundation were involved in the lectures. Among the founders and the first participants, there were Jiřina Šiklová, a historian and sociologist; Jana Hradílková, a scholar, actor, writer and women's activist; Laura Busheikin, a Canadian journalist living in Prague; Marie Čermáková, a sociologist; Hana Havelková, also a sociologist; Jiřina Šmejkalová, a linguist, psychologist, and sociologist; Libora Oates-Indruchová, an English and American Studies scholar; Pavla Frýdlová, a film scholar and filmmaker; Pavla Jonssonová, a rock musician, English translator, and cultural activist; Mirek Vodrážka, a former member of the dissident underground and publisher of the underground magazine Vokno; Hana Maříková, a sociologist; Šárka Gjuričová, a clinical psychologist; Eva Hauserová, a science fiction writer; and Eva Věšínová-Kalivodová, a Czech and American Studies scholar.[22]

In the academic year 1993–1994, lectures on gender studies and feminist theory were offered at several faculties of Charles University, specifically at the Arts, Social Sciences, and Pedagogical faculties, as well as the Institute for Liberal Education.[23] Alongside lectures for university students, the Gender Studies Foundation also organized lectures for the wider public, many of them given by foreign experts. During the academic year 1994–1995, the Gender Studies Foundation launched a series of gender studies lectures at Masaryk University in Brno and at Palacký University in Olomouc. However, none of these courses were integrated in the internal structure of any of the universities. Instead, they were entirely organized and financed by the Gender Studies Foundation.

The Institutionalization of Gender Studies in the Czech Republic

The Establishment of the First Czech MA Gender Studies Program in Prague

Toward the end of the 1990s it was clear that the current situation, in which the Gender Studies Foundation served as a substitute for an academic institution, was untenable. In order to start teaching gender studies at the universities it was necessary to move beyond isolated lectures and courses and to establish a separate center for gender studies that would be part of the university; in other words, to establish gender studies as an independent field of study at the universities. To this end, in 1998, with the financial support of the American Ford Foundation, an independent "Gender Studies Center" was founded within the Department of Social Work at the Faculty of Arts of Charles University. The Gender Studies Foundation abandoned its initial academic focus and turned its attention instead to media and project activities, while at the same time continuing to build its public gender studies library.[24]

Starting with the academic year 2000–2001, the Gender Studies Center at the Faculty of Arts at Charles University offered a master's degree in "social work and gender studies" as part of the field of social work, developed there by Šiklová.[25] This program was the first study program related to gender issues in the Czech Republic. The Gender Studies Center opened its courses to all students at the Faculty of Arts, and also served as a center for interdisciplinary research in the field of gender studies. Despite the evident success of the Gender Studies Center in fundraising, the Academic Senate of the Faculty of Arts in 2003 decided to shut it down. The reason given for this decision was that gender studies was not regarded as a field that the Faculty of Arts wanted to cultivate as part of its educational agenda.

Because of this situation, in 2003 a large part of the Gender Studies Center moved to the Faculty of Humanities of Charles University,[26] where the management was supportive of advancing this field. This move led to the foundation of an independent Gender Studies Department at the Faculty of Humanities at Charles University,[27] the first and so far only gender studies department at a Czech university. In 2005 the department began to offer a master's degree in gender studies, the first and still only MA program in gender studies in the Czech Republic. The professors in the gender studies MA program of the Department of Gender Studies are experts in various fields of social sciences and humanities: sociology, anthropology, cultural studies, history, communications, philosophy, and British and American studies. In addition to the master's program, the Department of Gender Studies also offers a number of bachelor level courses for students. However, rather than pursuing a BA program in gender studies, the department's principal objective is to develop a PhD program in the field.

The Establishment of the First Czech BA Gender Studies Program in Brno

The first and thus far only baccalaureate program in gender studies was developed in 2004 at the Department of Sociology, Faculty of Social Studies at Masaryk University in Brno.[28] As in Prague, here the introduction of courses on gender also came about as a result of the personal interest of some lecturers. In 2000 a circle of female and male students with a strong interest and very involved in this field set up a Gender Center students association with financial support from the Open Society Fund.[29] The students from this association, who were gradually involved in the teaching, were also invited by the faculty's management to develop a curriculum for gender studies. It should be added that around this time both the Faculty of Social Studies of Masaryk University in Brno and the Faculty of the Humanities of Charles University in Prague were expanding the selection of study programs they were offering. Thus, the conditions were favorable for the introduction of gender studies at both these institutions, and the gender studies lecturers in Prague and Brno managed to take advantage of this fact. In Brno, the institutional grounding benefited substantially when the sociology professor Gerlinda Šmausová began teaching at Masaryk University and joined the gender studies team of lecturers.

Unlike in Prague, in Brno no separate gender studies department was established, and that has been the source of some of the difficulties faced by the Brno program. Although the latter is part of the Department of Sociology, around one-quarter of its courses are interdisciplinary by nature, and the baccalaureate program in gender studies is designed as part of a double major. Thus, students have to choose whether to study gender studies in combination with sociology, social anthropology, environmental studies, European studies, media studies and journalism, political sciences, psychology, or social policy and social work. In the future the faculty plans to further expand the range of options of second major students studying gender studies in the baccalaureate program. Then they will be able to choose not only those fields offered by the Faculty of Social Studies of Masaryk University, but can also look for options among programs offered by the Faculty of Arts and the Faculty of Education. None of the other Czech universities offer any gender studies program. Some of them offer individual courses on gender issues when there are lecturers interested in teaching them.

Gender Studies within the Czech Academy of Sciences

It was not until the end of the 1990s that the first program in gender studies was established at a Czech university, but the establishment of the first and thus far only research gender department at the Academy of Sciences happened relatively early. The Gender and Sociology Department at the Institute of Sociology of the Academy of

Sciences of the Czech Republic was established in 1990 based on a research proposal aiming to investigate the position of women in Czech society.[30] With the support of a national research grant received a year later, the research team "Gender in Sociology" was formed around the sociologist Marie Čermáková and the foundations were laid for developing feminist gender-oriented sociology in the Czech Republic. This research team was the forerunner of the Gender and Sociology Department that emerged out of the transformation of the Institute of Sociology, when its research teams were turned into research departments. Over the course of the 1990s the team gradually expanded to fifteen researchers. The large size of the team was due to the number of successful research grants obtained from national and foreign agencies and from the Fifth and Sixth Framework Programs of the European Union (EU). It has become one of the key Czech scholarly institutions focusing on gender relations and feminist research. Although the Academy of Sciences is purely a research institution consisting mainly of senior researchers and PhD students, most of the researchers in the Gender and Sociology Department also teach courses in gender studies and sociology at Czech universities.[31]

Since 2000, the Gender and Sociology Department has been publishing the journal *Gender, rovné příležitosti, výzkum* (Gender, equal opportunities, research).[32] Since 2005, the abstracts of articles published in the journal have been published in the e-journal *Central European Journal of Social Sciences and Humanities* (http://cejsh.icm.edu.pl). In 2006, the biannual *Gender, rovné příležitosti, výzkum* became the first and so far only Czech peer-reviewed journal in gender studies and feminist theories. The intention of its international editorial board is to increase the number of articles by feminist scholars coming from Central and Eastern Europe published in English, as well as by other feminist scholars working on this region. *Gender, rovné příležitosti, výzkum* is very well known in the feminist community and among social scientists in the Czech Republic, and it has become an important teaching resource in Czech university programs. In recent years the number of foreign contributors has been growing. However, the publication of the journal is entirely dependent on grant support and is thus very insecure in a long-term perspective. So far the Gender and Sociology Department has managed to attract financial support for the publication of the journal. However, the transformation of the system of financing science and research has resulted in a significant reduction in financial support: the Grant Agency of the Academy of Sciences no longer exists; the EU does not favor funding to support the publication of national periodicals; and since the Czech Republic joined the EU, sponsors have for the most part left the region and moved farther east, or are themselves cutting back on the volume of grant funding because of the economic crisis.

Other Initiatives and Actions

In 2001, the "National Contact Center—Women in Science" was established as part of the Gender and Sociology Department in a project funded by the Czech Ministry of Education, Youth, and Sports.[33] This National Contact Center has two foci. On the one hand, it develops activities to promote gender equality, and to shape science and human resources policy in the Czech Republic, especially with respect to the position of women in science. On the other hand, it studies the gender aspects of knowledge production and develops feminist epistemologies. On an international level, it has established close ties with Unit C-5: Women in Science at the Directorate for Science, Technology, and Innovation of the European Union, and it is involved in the activities of the European Platform for Women Scientists. Although the National Contact Center is well known not just in Czech circles of science policy but also at the European level, its existence is wholly dependent on grant support. Thus far the National Contact Center has been successful at gaining and coordinating national as well as EU research grants, but its day-to-day operations have been supported by grant funding from the Ministry of Education, Youth, and Sports of the Czech Republic, which is not a reliable source for long-term support. The ministry is one of the most important actors in Czech science policy to which the National Contact Center proposes changes in science policy designed to attain gender equality, and in this way it helps the government satisfy some of the requirements of the European Union. Without the EU requirements, the activities in support of women and gender equality in science would most certainly not have achieved the results they already have in the Czech Republic. The National Contact Center also critically assesses science policy and helps publicize cases of discrimination (especially gender-based) in higher education, science, and research, whereby it negates loyalty to the ministry that has thus far been sponsoring it.

Finally, a milestone testifying to the institutionalization of gender studies in the Czech Republic, which created an opportunity for gender studies scholars and institutions to work together, was the first conference on Czech and Slovak feminist studies, held in Prague in September 2005. The conference was co-organized by the Department of Gender Studies of Charles University in Prague, the Gender Center of Masaryk University in Brno, the Department of Gender and Sociology at the Institute of Sociology of the Academy of Sciences of the Czech Republic, and the Center for Gender Studies at the Faculty of Arts of Komenský University in Bratislava (Slovakia). The conference showcased the current state of feminist discourse and research in the Czech Republic and Slovakia and created space for discussing and bringing together feminist knowledge and activities in the academic and nonacademic sphere. There were more than eighty papers submitted, mainly from the scholars in the Czech Republic, demonstrating a very broad range of research topics.

It should be noted that cooperation takes place not just between institutions and individuals within the academic sphere of gender studies and feminist research. Both individuals and the academic institutions also establish contacts and cooperation with feminist organizations outside academic circles. But unlike the situation in the early 1990s, individual positions and spheres of activity have come to be staked out. Most (though not all) gender studies scholars understand their scientific and research work as engaged, activist, and political work, and in contrast to the early 1990s they are avowedly feminist.[34]

As already mentioned, without significant and long-term assistance from Western gender studies scholars, universities, research institutes, and funding bodies, the institutionalization of gender studies in the Czech Republic would not have come about. Personal contacts, joint projects, funding and material support aimed at the advancement of the field, research exchanges, and discussions were all of fundamental importance for the development of gender studies in the Czech Republic. Also, the importance the EU has placed on promoting gender equality played a decisive positive role in the advancement of gender studies at Czech universities and in research in the field. Czech gender studies scholars are integrated into international research networks of gender studies and feminist scholars; they participate in and/or coordinate work on international research projects, publish, and organize international conferences and summer schools together with their colleagues abroad.

However, experience with Western feminist scholars has not always been just positive. Jiřina Šmejkalová has described the inequality of the interaction between local and some Western researchers who behaved toward local researchers as if they were undergraduate students whose function was to provide the research data and interpretations the Western researchers needed. The latter then published the results in international journals or in books produced by major publishing houses. These articles were full of misspelled (local) names, misunderstood points, unconfirmed information, and even unreferenced quotations from underpaid unpublished research reports written by Czech researchers.[35] It should be added that such research conduct by no means applied to all Western researchers writing on Central and Eastern Europe, nor was limited to the early 1990s. Such attitude persisted through the period when these countries were negotiating membership in the EU, when international research on gender aspects of social reality in these countries began to be funded by the European Commission.

Concluding Remarks

To sum up, gender studies and feminist research have become conceptually and institutionally established in the Czech Republic. Existing research centers and even individuals working at Czech universities and research institutions where gender studies has not yet been institutionalized are exploring new areas of research and new issues that were not discussed under the pre-1989 political regime. They are significantly contributing to the description and analysis of gender orders, and the development of feminist research in various areas, such as the labor market, family, parenthood, partnerships, childbirth, and childcare, sexuality, science and research, media, political participation, civil society, citizenship, the penal system, art, domestic violence, and education. They also contribute to the development of new scholarly fields in the Czech context (e.g., queer studies, men's studies) or gender specializations within established scientific fields in the country (e.g., Czech literature, literary and film theory, history, psychology, sociology, political science).

Nevertheless, in the academic milieu, among the wider public, or in state policy, even today it is still possible to find many individuals who refuse to accept gender studies as a scientific field. The growing insecurity produced by the economic crisis and the related slashes in funding for science, research, and education resulting from the current budget cuts, driven by rhetoric on the need to reduce the state debt, also contribute to the negative potential to bring about stagnation or regression in the field of gender studies and feminist research in the Czech Republic.

◊ Notes

1. This text was written with the support of grant project of the Grant Agency of the Czech Republic, no. P404/10/0021 "Changes in partnership and family forms and arrangements from the life course perspective."

2. Martina Kampichler, "Reflexe k tématu genderových studií na pozadí feministického rozvoje" (Reflection on gender studies against the background of feminist development), *Gender, rovné příležitosti, výzkum* 7, no. 2 (2006): 1–6; Hana Hašková, "Czech Women's Civic Organising under the State Socialist Regime, Socio-economic Transformation and the EU Accession Period," *Sociologický časopis/Czech Sociological Review* 41 (2005): 1077–1110.

3. Jiřina Šmejkalová, "Feminist Sociology in the Czech Republic after 1989," *European Societies* 6 (2004): 169–180, esp. 169.

4. Barbora Tupá et al., *Vlastní pokoj: 10 pohledů* (Own room: 10 views) (Prague: Sociologický ústav Akademie věd České republiky, 2004), 12–13.

5. Hana Havelková, "Affidamento," in *Nové čtení světa—feminismus devadesátých let českýma očima* (New reading of the world—feminism of the nineties with Czech lenses), ed. Marie Chřipková, Josef Chuchma, and Eva Klimentová (Prague: Marie Chřipková Press, 1999), 46.

6. Lenka Václavíková-Helšusová, "Genderová vědecká komunita vlastníma očima" (The gender community through its own eyes), in *Mnohohlasem. Vyjednávání ženských prostorů po roce 1989* (Multiple voices. Negotiating women's spaces after the 1989), ed. Hana Hašková, Alena Křížková, and Marcela Linková (Prague: Sociologický ústav Akademie věd České republiky, 2006), 174–175.

7. Havelková, "Affidamento," 58–59.

8. Šiklová, "Únava z vysvětlování," 131–133.

9. This happened at a European meeting of women from the East and the West in Dubrovnik in the summer of 1991 and at the conference "Women in a Changing Europe" in Denmark.

10. After *FrauenAnstiftung* merged with other foundations, the Czech NGO received significant funds from the Heinrich Böll Foundation, which continues to support its activities today.

11. Several more thematic issues of *Sociologický časopis/Czech Sociological Review* on gender studies were published over the years. In 1999 Marie Čermáková guest-edited an English-language issue of the journal. In 2005 Hana Hašková guest-edited an English-language issue on the civic and political representation of women in Central and Eastern Europe. In 2009 Alena Křížková guest-edited a Czech-language issue titled "Soukromé je veřejné: gender, péče a veřejnost" (The private is public: Gender, care and the public). Also in 2009 Iva Šmídová guest-edited a Czech-language issue titled "Instituce genderu a gender institucí" (The institution of gender and the gender of institutions).

12. Marie Čermáková and Hana Havelková, "Úvodem k monotematickému číslu Gender v sociálních vědách a otázky feminismu" (Introduction to the thematic issue "Gender in Social Sciences and the Questions of Feminism), *Sociologický časopis/Czech Sociological Review* 31 (1995): 3–4.

13. See, e.g., Sharon Wolchik, "Women and the Politics of Transition in the Czech and Slovak Republic," in *Women in the Politics of Postcommunist Eastern Europe*, ed. Marilyn Rueschemeyer (Armonk, NY: M.E. Sharpe, 1994), 3–27.

14. Alena Heitlinger, "Framing Feminism in Post-Communist Czech Republic," *Communist and Post-Communist Studies* 29 (1996): 77–93, esp. 77.

15. Šiklová, "Únava z vysvětlování," 128–140.

16. Kampichler, "Reflexe k tématu genderových studií na pozadí feministického rozvoje," 1–6.

17. Václavíková-Helšusová, "Genderová vědecká komunita vlastníma očima," 166.

18. Ibid., 166–167; see also Marcela Linková, "Co si neuděláš, to nemáš, aneb ženský aktivismus, produkce znalosti a gender" (What you don't get done alone, you don't get at all, or women's activism, the production of knowledge and gender), in *Mnohohlasem. Vyjednávání ženských prostorů po roce 1989* (Multiple voices. Negotiating women's spaces after 1989), ed. Hana Hašková, Alena Křížková and Marcela Linková (Prague: Sociologický ústav Akademie věd České republiky, 2006), 145–146.

19. Čermáková and Havelková, "Úvodem k monotematickému číslu Gender v sociálních vědách a otázky feminismu," 3.

20. The Gender Studies Center at the Faculty of Arts of Charles University was the only exception to the rule of using the term gender in the name of an institution, as it is the only institution that has used both "Gender studies center" and "Centrum studií rodu."

21. Karen Marie Kapusta-Pofahl, "Legitimating Czech Gender Studies: Articulating Transnational Feminist Expertise in the New Europe" (PhD diss., University of Minnesota, 2008).

22. Ibid., 94–95.

23. The Institute for Liberal Education of Charles University became the Faculty of the Humanities of Charles University in 2000.

24. Owing to a change in the law, in 1998 the Gender Studies Foundation changed its legal status and name and became a benevolent association called Gender Studies, o.p.s. In the late 1990s, Gender Studies, o.p.s. began to be more involved in work for the media, through which it appealed to the wider public. It founded the first Czech website of its kind on feminism, women, men, and gender (http://www.feminismus.cz). Since 2002, it has also functioned as a feminist bookstore. Moreover, it operates an electronic mailing list that distributes information about feminist activities (in arts, sciences and humanities, political lobbying, NGO networking) in the Czech Republic to the Czech feminist community beyond the academic circles. More information on Gender Studies, o.p.s. is available at http://www.genderstudies.cz (accessed 24 August 2010).

25. For more information on the Gender Studies Center at the Faculty of Arts, Charles University, see http://gender.ff.cuni.cz (accessed 24 August 2010).

26. The part of the Gender Studies Center that remained within the Faculty of Arts of Charles University became a research center under the Institute of Czech Literature and Literary Theory, where, alongside its research work, it offers courses on gender, literary criticism, and film studies.

27. For information on the Department of Gender Studies, Faculty of the Humanities, Charles University in Prague, see http://www.fhs.cuni.cz/gender/o_nas_eng.html (accessed 24 August 2010).

28. For more information on the BA in gender studies at the Department of Sociology, the Faculty of Social Studies of Masaryk University, see http://gender.fss.muni.cz/index.php/home.html (accessed 24 August 2010).

29. For more information on the Gender Center students association, see http://www.gendercentrum.unas.cz/web/index.php?catid=2 (accessed 24 August 2010).

30. For information on the Gender and Sociology Department of the Institute of Sociology of the Academy of Sciences of the Czech Republic, see http://www.soc.cas.cz/departments/en/4/42/Gender-Sociology.html (accessed 24 August 2010).

31. Unlike in Western Europe and in the United States, in the Czech Republic a separation is maintained between scientific research, conducted primarily at the institutes of the Academy of Sciences, and teaching at the universities. However, researchers from the Academy of Sciences often also teach at the universities, and the institutes of the Academy of Sciences provide a base at which PhD students can work. Meanwhile, university professors are under increasing pressure to engage in research work, and many of them successfully do so.

32. The journal appeared thanks to support from the Grant Agency of the Czech Republic and later from the Academy of Sciences of the Czech Republic. For more information on *Gender, rovné příležitosti, výzkum,* see www.genderonline.cz (accessed 24 August 2010).

33. For more information on the National Contact Center—Women in Science, see http://www.zenyaveda.cz/html/ (accessed 24 August 2010).

34. Václavíková-Helšusová, "Genderová vědecká komunita vlastníma očima," 178.

35. Jiřina Šmejkalová-Strickland, "Revival? Gender Studies in the Other Europa," *Signs* 20, no. 4 (1995): 171.

◆

Out of the Room of One's Own?
Gender Studies in Estonia

Raili Põldsaar Marling

Gender studies in Estonia are shaped by the unspoken presence of the forty-year Soviet annexation that removed Estonian society from the international exchange of ideas during the time when gender became, first, a political issue and, second, an object of academic study. According to Soviet ideology, gender was irrelevant in the Soviet Union as the equality of men and women had supposedly been achieved. The fact that the reality men and women lived in was anything but gender-neutral or gender-equal did nothing to alter the ideological dictums. In a situation where even sociology did not effectively exist, it is no surprise that gender was not academically explored in Estonia under the Soviet regime. The total lack of gender awareness is reflected in an entry in the Soviet Estonian Encyclopedia from 1987, which defines feminism as a medical term, "the existence of feminine physical features or characteristics in a male."[1]

Thus, gender became visible and, subsequently, political, only after Estonia regained independence in the early 1990s. Estonia joined gender-related international debates and discussions but the period of separation also shaped the way in which new ideas were received. Local circumstances created an ideological sieve that let some ideas in and left others out. Feminism was one of the notions caught in the net. In the 1990s, Estonia sought to turn its back on all that was assumed to be Soviet, including the Soviet ideology of gender equality. Instead of asking how to make the vacuous equality slogans of the past meaningful, they were thrown out completely, replaced by neoliberalist economic policies and the hedonistic pleasures of Western consumer culture, neither of which was interested in investigating the politics of gender. Estonian women were exhorted to bear more children to guarantee the survival of the nation and make themselves beautiful to please their husbands. The rediscovery of gender was channeled into sexualization and Estonian women were congratulated on their ability to ignore the supposedly misguided feminist ideologies of their Western and, especially, Nordic sisters.[2] Awareness of gender equality as a serious social issue emerged only gradually as the income gap between men and women increased and as the toll of hegemonic masculinity on Estonian men became more and more evident.[3]

Gender studies as an academic field has grown out of these historical and social constraints. By the time Estonian society started to ask whether gender was in any way different from biological sex, Western academic discourse had developed very complex theories of gender. Estonia had to join the dialogue in mid-conversation. While the gender studies theorists in the West were debating Judith Butler, Estonians had yet to have a public discussion on whether gender was a political issue. This is not to suggest that gender studies have to follow the same evolutionary stages in all countries, but rather to explain why the transmission of ideas could not be automatic and direct. Ideas are exchanged and shared, but they are interpreted and domesticated locally. The local filter explains the coexistence of some of the newer theories (for example queer studies) with very traditionalist family ideologies in Estonian discussions of gender. As Estonian mainstream culture is relatively antifeminist, it has had an easier time embracing some aspects of poststructuralist theory rather than the political claims of the second-wave feminist movements. It is less dangerous to the status quo to mull the slipperiness of the category of "woman" than to try to explain the large wage gap. This paradox has to some extent made the more theoretical aspects of academic gender studies easier to domesticate, but it has also created complications in the relations between gender studies and feminist politics.

The ideological nature of naturalized gender notions becomes obvious only with the existence of an alternative viewpoint. Perhaps this is why international influences play such a prominent part in Estonian gender studies. The first people to raise gendered concerns in Estonian public, and gradually also academic discourse, were either educated in the West, especially in the neighboring Nordic countries, or self-educated based on international theoretical literature. Expatriate Estonian academics, most notably the literary scholar Tiina Kirss, helped to combine international ideas with the Estonian context. Her theoretically sophisticated seminars on feminist philosophy showed that something can be gained, not lost, in translation.

The gender studies of the 1990s pioneers were clearly feminist-identified endeavors. It took guts as well as conviction to go against the grain of public opinion to make Estonian society understand that feminism was not a medical term, but a philosophy and social movement the aim of which was to guarantee human rights for all. Barbi Pilvre played an invaluable role in this process. For many years a journalist at the *Eesti Ekspress,* the country's leading weekly newspaper, she was Estonia's "watchdog feminist" despite the deeply antifeminist and misogynist responses her work evoked. Her efforts in mainstreaming feminism were supported by such academic voices as Eve Annuk from the Estonian Literary Museum, Katrin Kivimaa from the Estonian Academy of Art, and Anu Laas from the University of Tartu.

Academic gender studies took root gradually and unevenly in Estonian soil. Individual scholars working in various departments at different universities started to use women's and gender studies ideas in their research or teaching within their traditional disciplines, usually in the form of elective courses. These efforts were primarily located within the humanities and social sciences: in literary scholarship (Leena Kurvet-Käosaar, Eve Annuk, Tiina Kirss), art history (Katrin Kivimaa, Reet Varblane), sociology (Anu Laas, Leeni Hansson), media studies (Barbi Pilvre), and English (Raili Põldsaar). In the context of academic freedom scholars could teach gender if they chose to, but there was—and still is—no encouragement from either the university administrations or the Ministry of Education and Research. It was and still is a matter of personal commitment by the pioneering women's and gender studies researchers. Their enthusiasm gained a student following and schooled new scholars, whose gender awareness was rooted in Estonian reality. One no longer had to go abroad to see a live academic feminist. We can now, with justice, talk of a second generation of Estonian gender studies students currently working on their PhDs, in Estonia and abroad.

The process of establishing Estonian gender studies was encouraged by the supportive international gender studies community. This is not to say that gender studies was an international import, but rather that the enquiring minds from Estonia could develop their local knowledge in a dialogue with their international peers. For example, the gender studies program at the Central European University in Budapest was for many years the academic center where Estonian scholars interested in the field went to, to not only get graduate degrees and use the library resources, but also to imbibe the spirit of feminist inquiry. Many first- and second-generation Estonian gender studies scholars got their degrees there and the program's graduates have also worked for the gender equality unit of the Ministry of Social Affairs as well as for various NGOs. The other important influences were the Christina Institute at the University of Helsinki and the Nordic Gender Institute.

International contacts offered encouragement that the Estonian scholars lacked in their own university communities and in the society at large. Estonia had a backlash against feminism before it could develop feminism. It is telling that the early feminist texts that were published in the mainstream media were mostly dedicated to the question of why Estonia was so afraid of feminism.[4] An atmosphere of interest in the gender approach mixed with suspicion characterized the reception of gender studies courses. Most students who studied at Estonian universities in the 1990s probably never heard of the politics of gender and considered feminism a radical social ideology. Feminist ideas were often stigmatized because they challenged the newly formed status quo of the fledgling nation state and feminist-identified or gender-related research was viewed with, at best, suspicion and, at worst, ridicule. Thus the challenge before the first generation of gender studies researchers and teachers was to normalize the discipline as a legitimate field of study.

Most of the first-generation gender studies scholars were young and only starting to work on their PhDs. This meant that, on the one hand, they had considerable enthusiasm and energy, but, on the other hand, that they lacked influence in institutional discussions. This exclusion from the corridors of power has remained a problem for Estonian gender studies to this day. The field can now boast a considerable number of PhDs, defended within other disciplines, such as comparative or Estonian literature, art history, or English, studied from a gender/feminist angle, but not greater institutional legitimization. There are no degree programs at any level, not even minor, and thus gender studies is a kind of an ancillary discipline. As a result, Estonian gender studies is relatively loose—we can speak of excellent individual scholars doing intriguing work but not a consistent sense of community or a clear identification as gender studies scholars.

The awareness that individual courses were unlikely to gain public attention led to the creation of the first gender studies centers in Estonia. A gender studies unit was founded in 1995 by the Department of Sociology at the University of Tartu by Anu

Laas, focusing on teaching, international research projects, and applied research. Because of its institutional affiliation, its emphasis was on sociology and social sciences. It has continued its work up to this day, in its area of specialization. At the same time, a relatively short-lived Women's Studies Center was established at Tallinn University of Educational Sciences, on the initiative of Voldemar Kolga and Barbi Pilvre. In addition to individual scholars who worked on topics of gender, gender studies research was conducted, in a more concentrated manner, for example, in the family sociology unit of the Institute of International and Social Studies in Tallinn and, in the field of oral history at the Estonian Literary Museum in Tartu. Neither, however, had a formal gender identification.

An explicitly gender-oriented NGO, Eesti Naisuurimus- ja Teabekeskus (ENUT; Estonian Women's Studies and Research Center) was founded in 1997 within Tallinn University of Educational Sciences (now Tallinn University), but financially and institutionally independent from it. Initiated by Eda Sepp and Suzanne Lie, it was envisioned as both an information center and an academic resource center for scholars all over Estonia. The experience of the previous centers had shown that association with one university or department can hinder effective inter-institutional and interdisciplinary cooperation. ENUT's mission, according to its founders, was to provide Estonian women's and gender studies with a room of their own. Everybody, regardless of their institutional background, was to be welcome in a shared space.[5] The center's library collection, covering a wide variety of disciplines, made it an important resource for a generation of gender studies scholars. Thanks to its cooperation with the Open Estonia Foundation, the United Nations Development Program, the EU Phare program and other international organizations, the center was able to coordinate both academic activities and more socially oriented NGO work (e.g., in connection with trafficking in women and violence against women). Its independence from the Tallinn University meant that it could continue its feminist work without having to make compromises. But it depended on grants and other irregular sources of funding, which considerably limited the scope of its activities. This problem has become even more acute now that international funding is more scarce. This has led to the further scaling down of ENUT's activities, especially academic ones.

ENUT started two initiatives in gender studies that have been of lasting importance to the field in Estonia. The first was the establishment of *Ariadne Lõng* (Ariadne's thread), the peer-reviewed Estonian journal of gender studies in 2000. It was to become an Estonian-language forum for academic research on gender and a platform for the development of Estonian terminology for gender studies. This is especially important at a time when most international academic discussions are conducted in English and thus do not impact the local community or contribute to the development of its gender awareness. To assist in domesticating the field, each issue of the journal also publishes a translation of a classic text in women's and gender studies. Authors translated include Sherry Ortner, Julia Kristeva, and Joan Scott. Initially edited by Eve Annuk and Mirjam Hinrikus, currently by Eve Annuk as the editor-in-chief and Raili Põldsaar Marling as the acting editor, it has established itself not only in the Estonian gender studies community but also within the Estonian scholarly community at large, as one of the few reputable local peer-reviewed academic journals. *Ariadne Lõng* is an interdisciplinary periodical. It publishes works from the humanities and social sciences and has thus become probably the most vibrant meeting place of Estonian gender studies scholars from different disciplines. Despite having to seek funding for each issue from different foundations and not having a set publishing home, the journal has retained its quality and stability of publication already for ten years, possibly making it the most successful Estonian project in gender studies.[6]

The other initiative was an academic minor program in gender studies at Tallinn University. Activists working at the Estonian Women's and Resource Center recognized that the scattering of gender studies courses in different departments of different universities did not allow students to get a full interdisciplinary picture of the field. The proposed gender studies minor was met with considerable suspicion from the university community, but became a reality, thanks to the enthusiastic efforts of the organizers, especially Liina Järviste. The program brought together the best gender specialists from three Estonian institutions (Tallinn University, University of Tartu, and Estonian Academy of Art). The first year was a success and the program attracted students from as far as Tartu. Students as well as instructors traveled between institutions and cities to take part in stimulating discussions. However, in a few years the minor fell victim to curriculum reform and the fact that no single department could support the interdisciplinary range of courses. Financial-institutional pressures also contributed to the eventual demise of the minor.

From 2004 to 2006, Raili Põldsaar and Leena Kurvet-Käosaar attempted to create an informal gender studies minor at the University of Tartu to encourage student mobility between faculties and curricula, but their efforts were frustrated by the rigidity of administrative boundaries and a lack of institutional support. At the time of this writing, no coherent gender studies program exists in Estonia. Students can only take individual courses in different universities/departments and combine them, but they have to make the effort themselves. A new Gender Studies Center has been established at the Institute of International and Social Research, Tallinn University, but it is too early to speak of its longevity or outreach into teaching, in view of the lack of funding and administrative barriers that have hampered previous efforts.

Gender studies is thus not really institutionalized in Estonian academia. This can be viewed from both a positive and a negative perspective. The negative side is obvious—courses are scattered among different departments and universities and thus gender studies scholars do not appear as a power bloc in academic discussions nor do they develop stable interdisciplinary links. Moreover, because gender studies scholars work in institutions teaching, for example, sociology or literature, they cannot fully dedicate themselves to gender and this makes them suffer from the pressures of an academic "double day." However, there is a positive aspect to this situation as well. A separate gender studies center or program could become a kind of a ghetto, marginalized in the wider academic world. Moreover, it would increase the still widespread opinion that gender studies is the only field that needs to deal with gender and that nobody else has to worry about it. At present, gender studies scholars who are teaching in a variety of contexts help to mainstream gender-aware and even feminist thought, and students who might not bother or dare to take a full gender studies pro-

3. Estonia has the widest gendered income gap in the EU, 30.3 percent (compared to the 17.6 percent average for the EU), according to 2010 figures, http://www.eurofound.europa.eu/areas/gender/internationalwomensday2010.htm (accessed 20 July 2010). Estonian men have over decades been shown to engage in overwork and other self-destructive health practices, and there is a ten-year gap between female and male life expectancy. These two gaps are contributing factors to why Estonian public opinion has started to see gender equality as an issue that concerns both men and women.

4. Some of the examples are Eve Annuk, "Kas Eestis on feminismi"? (Is there feminism in Estonia), *Postimees*, 10 February 1993; Barbi Pilvre, "Feminismitondi taltsutamine Eestis" (Taming the spectre of feminism in Estonia), originally published in the weekly *Eesti Ekspress* in 1999, reprinted in Pilvre's essay collection *Formaat: valitud tekste klassivõitlusest ja naisküsimusest 1996–2002* (Format: Selected texts on class struggle and the woman question, 1996–2002), (Tallinn: Eesti Ekspressi Kirjastuse AS, 2002), 140–159; Katrin Kivimaa, "Teise tootmine" (Production of the other), *Ariadne Lõng* 1/2 (2000): 41–49.

5. Suzanne Lie, "Milleks naiste teabekeskus Eestis" (Why do we need a women's resource center in Estonia), *ENUTi uudised* 1 (1999), http://www.enut.ee/enut.php?id=29 (accessed 14 August 2010).

6. The tables of contents of all past issues of *Ariadne Lõng* are available on the English-language website of the Estonian Women's and Resource Center, http://www.enut.ee/enut.php?keel=ENG&id=94 (accessed 12 August 2010).

7. Three collections have also been published in English: Tiina Kirss, Ene Kõresaar, and Marju Lauristin, eds., *She Who Remembers Survives. Interpreting Estonian Women's Post-Soviet Life Stories* (Tartu: Tartu University Press, 2004); Suzanne Stiever Lie, Lynda Malik, Ilvi Jõe-Cannon, and Rutt Hinrikus, eds., *Carrying Linda's Stones. An Anthology of Estonian Women's Life Stories* (Tallinn: Tallinn University Press, 2007); Tiina Kirss and Rutt Hinrikus, eds., *Estonian Life Stories* (Budapest: Central European University Press, 2009). There are some notable collections in Estonian as well. For example, Eve Annuk was the first to put together a collection of women's life stories: Eve Annuk, ed., *Naised kõnelevad* (Women speak) (Tartu: Eesti Kirjandusmuuseum, 1997).

8. Katrin Kivimaa, *Rahvuslik ja modernne naiselikkus eesti kunstis, 1850–2000* (National and modern femininity in Estonian art, 1850–2000) (Tallinn and Tartu: Tartu University Press, 2009). This book is especially important because of its critical analysis of the interrelations of gender and the nation.

9. Tiina Kirss has, for example, intriguingly used postcolonial theory in the Estonian context. Eve Annuk has in recent years worked in parallel on the representation of Estonia's first feminist author, Lilli Suburg (1841–1923), and the archival policies of the Soviet period. Leena Kurvet-Käosaar has dedicated herself to the study of women's autobiography and, in recent years, also trauma theory.

10. Kadi Mänd et al., ed., *Tilliga ja tillita: retsepte Eesti feministidelt* (With or without a wiener: Recipes from Estonian feminists) (Tallinn: ENUT, 2003). This collection successfully combined the traditional feminine world of cooking with edgy humor. It was timed to coincide with a number of feminist public interventions and was so popular that it went through a second printing. The second collection was more clearly focused on the body, largely from a personal perspective, with articles on the maternal body and breastfeeding to the HIV-positive body: Redi Koobak and Raili Põldsaar, eds., *Ihakeha ja lihakeha. Valik esseid ja arutlusi kehast* (Desired body, carnal body: Selected essays and musings on the body) (Tallinn: Kirjastuskeskus, 2008). Both collections have used eye-catching illustrations and involved both female and male authors.

11. For example, a collection edited by Raili Marling, Liina Järviste, and Käthlin Sander, *Teel tasakaalustatud ühiskonda II* (On the road to a balanced society II) (Tallinn: Sotsiaalministeerium,

2010), which is the follow-up collection to a similar volume published in 2000 that recorded the state of gender equality in the Estonian society at the time. The fact that the preface of the upcoming volume will be written by the president of Estonia is, to an extent, a sign of the mainstreaming of gender in Estonian politics. This is but one example of important publications of the Ministry of Social Affairs that make productive use of gender studies scholarship.

12. Anu Koivunen and Marianne Liljeström, eds., *Võtmesõnad. 10 sammu feministliku uurimiseni* (Keywords: Ten steps towards feminist studies) (Tallinn: Eesti Keele Sihtasutus, 2003). Its translation was the most systematic effort to systematize Estonian-language terminology for women's and gender studies.

13. Virginia Woolf, *Oma tuba* (A room of one's own) (Tallinn: Perioodika, 1994); Virginia Woolf, *Esseed* (Essays) (Tallinn: Hortus Litterarum, 1997); Simone de Beauvoir, *Teine sugupool* (Le deuxième sexe) (Tallinn: Vagabund, 1997); Evelyn Fox Keller, *Mõtisklusi soost ja teadusest* (Reflections on gender and science) (Tartu: Tartu Ülikooli Kirjastus, 2001); Julia Kristeva, *Jälestuse jõud: essee abjektsioonist* (Pouvoirs de l'horreur. Essai sur l'abjection) (Tallinn: Tänapäev, 2006); Seyla Benhabib, Judith Butler, Nancy Fraser, and Drucilla Cornell, *Feministlikud vaidlused* (Feminist contentions) (Tallinn: Tallinna Ülikooli Kirjastus, 2006).

◆

Gender Studies at Greek Universities
Theodossia-Soula Pavlidou

Within the formal university structure in Greece, there were no Women's or Gender Studies (WGS) until recently. Even today, there are no autonomous departments or official studies programs leading to degrees in WGS per se, although certain changes in the direction of the institutionalization of WGS did take place in the last decade.[1] In the 1980s, there were a few collective attempts to problematize issues of gender in knowledge production. Outside the university, the most prominent effort in this field came from the group (consisting to a large extent of women academics), which since 1986 published the feminist journal *Dini* (Whirlpool) in Athens. Most members of this group were very skeptical about the institutionalization of WGS. Inside academia, there has been only one collective and long-enduring effort, namely the Women's Studies Group at the Aristotle University in Thessaloniki. This group was founded in 1983, in an attempt to bridge the gap between the "feminist" selves outside the university and the "scientific" presence within it; this led to a questioning of scholarly views, which were taken for granted, "trying to examine the foundations of knowledge and seeking new forms for its production that ensure collectivity and collaboration."[2] The activities developed by the Women's Studies Group at the Aristotle University provided a space for colleagues from various Greek universities to meet and exchange ideas.[3]

The first gender-related academic courses at Greek Universities were mainly within the humanities and also date back to the 1980s. Although the number of courses and research activities—including MA and PhD theses—pertaining to gender increased in the 1990s, it remained small and the field stayed invisible within the university structure. This was due to the lack of an institutional frame that would support and enhance the individual activities. Over the years, members of the Women's Studies Group have been submitting proposals for more permanent (and visible) forms of WGS at the Aristotle University—all of which were turned down. So were, for example, the projects for the establishment of a center for documentation and information on women's studies issues in 1995, and an interdepartmental postgraduate studies program on women's studies in 1997.[4]

As none of these attempts bore fruit, how is it that we can talk about gender studies at Greek universities? In contrast to what happened in Northern American and Western European universities in earlier years, in Greece the establishment of WGS was not the outcome of internal processes or requests from within, let alone the free choice of the state, namely the Greek Ministry of Education. The latter's decision to finance measures for the "Support of Women" at an academic level was imposed externally by a European Union (EU) directive in 2000. According to this directive, 10 percent of the budget for education coming from the Third Community Support Framework (2000–2006) was to be spent on measures for the promotion of "Gender and Equality." In other words, this money could be utilized only to the specified end, otherwise it would have to be returned.[5] Thus, educational policymakers in Greece—with scarcely any prior experience on these matters—found themselves designing a number of activities that entailed the first steps towards the institutionalization of WGS in the country. These included a call for postgraduate studies programs on gender and equality in 2002, which resulted in the establishment of four postgraduate programs, followed by a call for "cycles of courses" at the undergraduate level, titled "Interdepartmental Study Programs for Topics in Gender and Equality" (ISPs).[6] Eight such (time-limited) programs at different universities were approved.[7] As one can imagine, the Ministry's reaction to the EU directive could not but be mechanical, given that the Ministry had turned a deaf ear to relevant proposals. Moreover, the response of the academic community to the ministry's measures did not necessarily reflect a pre-existing interest in gender issues and/or relevant experience.

The ISPs were originally designed to run for two years and end in 2005,[8] according to the initial specifications of the Greek Ministry of Education. After several extensions, they expired unequivocally in August 2008. Given the institutional frame of the Greek universities, their designation as "undergraduate study programs" is quite misleading, because there has never been a university department in Greece in the field of "gender and equality" or anything similar; but, according to Greek legislation, only a department can issue study programs. What was actually intended was the modification of existing study programs to include courses dealing with gender and equality.[9] Subsuming all such courses under the so-called "cycle of courses," coordinated by an interdepartmental committee, helped establish connections among the different courses at a university and spotlighted them as a whole. However, the ISPs did not lead to a diploma or degree. Therefore, right from the beginning, these programs were imparted with two decisive features, besides their limited duration: their institutionally up-in-the-air position within Greek universities and their low appeal to students, which is partly connected to the former.

The ISPs were heterogeneous in many respects. Thirteen departments, for example, participated in the program at the Aristotle University of Thessaloniki, with more than fifty gender-related courses during the first year of the program's existence. However, in the respective program at the National Technical University of Athens only the Department of Architecture/Engineering was involved and offered a more limited number of courses. Inevitably, such quantitative differences led to qualitative differences in the overall design and implementation of the programs. Again taking the two programs just mentioned as an example, it is clear that the former was much broader in scope, while the latter focused on topics relevant to the field of knowledge specific to this department. There were further qualitative differences among the various programs, related — at least in part — to the prehistory of each and the basis on which they were organized. For example, in the case of the Panteion University program, most members of the scientific committee running the program had a common feminist background, according to Maria Stratigaki.[10] This enabled their common understanding that promoting feminist thinking and supporting activities coming from the women's movement would be the leading criteria in spending the available funds.

Despite their differences, the ISPs had to overcome similar problems in their implementation. Primarily, at the university level, a department's decision to participate in the interdepartmental program did not guarantee either its unimpeded cooperation in the ISP or its commitment to securing appropriate space, for example, for temporarily hosting the secretariat of the program. Moreover, nowhere were the ISPs received enthusiastically. On the contrary, in some cases, their reception was blatantly negative, which cannot be accounted for by the usual competition for funds, because these programs had their own budgets. It seems, however, that the usual resistance (both institutional and personal) to something innovative, or simply new, in this case combined with mistrust/hostility toward anything related to "gender" (read "women") and "equality" in the official university discourse. In addition, the institutional context (and, accordingly, the rationale that informs both the way courses are designed as well as the criteria for getting selected by the students) undermined one of the constitutive features of these programs, namely their interdepartmentality. The reasons for this are to be traced in the structural inflexibility of the Greek universities and the lack of an interdisciplinary tradition,[11] which are not unique to the Greek situation. As some gender studies specialists in Western countries have stressed,[12] monodisciplinarity and the enormous power of individual disciplines have more generally created great obstacles to the development and integration of WGS at the universities of Continental Europe.[13]

In its turn, the Ministry of Education failed to ensure some of the necessary conditions for the consolidation and smooth running of these programs. Apart from the huge bureaucracy involved in the transactions with the Ministry, the delays on its part in providing the (long approved) funds significantly stunted the implementation of the ISPs. For example, the purchase of technical equipment could not be accomplished even by the end of the second year — a rather bizarre situation, given the original two-year duration of these programs. Even graver was the fact that some new courses were not offered as planned because the (short-termed) employment of outside scholars with the necessary training in WGS was not feasible in time. Such problems, together with the uncalled-for cancellation of relevant university positions, as originally planned and approved of, shaped a situation that was quite distant from the one promised.

Despite their inherent weaknesses and the difficulties with implementation, the first two years of the programs had some positive effects, which are acknowledged in the existing publications.[14] Among them are the enrichment of university libraries with gender-relevant literature; the renewal of the curricula of various departments through the introduction of a gender perspective in new or existing courses; the organization of activities such as lectures, symposia, and conferences on gender issues; and the fact that gender-related initiatives at a university were put on the map through their inclusion into the common framework of the ISP. At the same time, the impact of these programs on the academic community, as well as on society at large, is much more difficult to assess. The numbers of students attending one or more courses of the ISPs or of faculty members participating in program-related activities can be viewed as "big" or "small," depending on the circumstances and the specificities of each program, not to mention the expectations of its representatives. Some accounts or evaluations of female students—although limited in number—reflect the sort of "sensitization" or "awareness" that these programs have led to.[15] In any case, the academic community, from students and colleagues to administrative staff and rectors, *nolens volens* stumbled over "gender" at some point.[16] This fact can gather momentum. The question is: in which direction?

The institutionalization of WGS certainly facilitates a greater awareness of gender-related issues, including the problematization of the role of gender in knowledge production. As Jalna Hanmer argues, based on a survey in eleven European countries, one of the factors in students' selection of WGS courses is "the visibility and prominence of the discipline [of WGS] within the country's educational system" and these "are influenced by the degree of institutionalization in higher education."[17] Taking into account previous attempts, the experience gained through the programs, and the academic reality in Greece, a feasible option would be the foundation of interdepartmental WGS centers at those universities, which have already shown that they have the human potential and the necessary experience for such an institution. The question, however, is not only which form of institutionalization of WGS will be adopted in Greece but also what its content will be. Personally, I am skeptical on this issue because the attempts within the ISPs to intervene in the dominant knowledge production or, at least, to unfold discussions on theoretical issues with which WGS have been traditionally preoccupied, have remained extremely restricted. Let us take "interdisciplinarity" as an example—one of the constituent notions and values of the field of WGS. Despite the almost inflationary use of this word in the framework of the ISPs, in most cases "interdisciplinarity" has been just a buzzword, meaning at best "multidisciplinarity," with no regard to its differences from the latter nor the complications and consequences involved that make it a recurrent topic in debates.[18]

Part of the explanation for this lack of theoretical discussion rests in the immense workload[19] required for the implementation of the ISPs, along with all other obligations that we had as faculty members. This additional workload did not leave room for us to reflect on what we were doing (and how), on how we would position ourselves with respect to the fundamental questions raised by WGS, and on how we would contextualize such issues in the Greek reality. A less optimistic interpretation is that the situation might be indicative of our reliance on the beaten track of the academic establishment, where crucial issues of WGS such as critical thinking, the relation between theory and praxis, between teachers and students, and between subjects and objects of research cannot flourish. In such a context, the perspective of gender in science may be downsized to "equality," while the production of gender-related knowledge may be easily replaced by "equality policies" at the university. Given the position of women at Greek universities, the enactment of equality policies is undoubtedly useful.[20] But if my concerns as to complacency prove right, then the Greek version of WGS might run the risk of losing its potentially subversive power before really entering academia and ending up as a fully integrated component of the academic establishment.

I believe that the challenge for the future of WGS in Greece lies in whether the opportunity opened up by the ISPs will result in more radical changes in Greek universities or whether WGS will exhaust itself in political correctness. In my opinion, this will crucially depend on whether we will be able to:

a) transform our monologues into collective discourse(s);
b) ground gender issues in the Greek context;
c) connect theoretical thinking inside academia with the actions/activities of women (who have started to form their own collectivities again) outside academia, or, to put it in more general terms, connect the university with society, and theory with praxis;
d) inspire and educate a new generation of students who will get involved in gender studies in a meaningful, rather than casual and utilitarian, way.

Then, perhaps, some old values of the women's movement and WGS, such as solidarity and collectivity, will reappear along the way.

Afterword

The ISPs—together with all other measures for the promotion of "Gender and Equality"—ended in 2008, as already mentioned. Undoubtedly, more time (coupled with sufficient funding to ensure the bare essentials) would have been required for the consolidation of the positive outcomes (those extending beyond books and equipment) of the ISPs.

What has remained from the ISPs in the Greek academic landscape? First, some gender-related courses that were introduced in certain departments' curricula are still offered. Overall the number of such courses has shrunk, and in some departments, it was almost impossible to retain at least some of the courses that the respective ISP originally comprised. Second, certain initiatives undertaken and/or accomplished within an ISP have continued, such as the electronic encyclopedia for gender topics developed in the ISP of the National and Kapodistrian University of Athens.[21] Third, a number of other initiatives for ensuring some form of continuation have been attempted. For example, at the Aristotle University of Thessaloniki a Center for Space, Technology, and Gender[22] and a Committee for Gender and Equality[23] were established at the faculties of Engineering and Philosophy, respectively. Moreover, by the end of 2008, the *Elliniki Etaireia Gynaikon Panepistimiakon*, EL.E.GY.P. (Hellenic Society of University Women) was launched and has been since organizing annual conferences.[24] Finally, activities/formations, developed within an ISP but with some institutional footing outside it, have become part of the university infrastructure, such as the Laboratory for Gender Studies and Equality at Panteion University, which has been very active in organizing lectures and seminars on gender issues.[25]

More generally, of all beneficial measures for the promotion of "Gender and Equality," those conceived on the basis of the available institutional infrastructure of the universities generated outcomes that survived the measures' expiring date. This was most prominently the case with those measures pertaining to the postgraduate level and the establishment of gender-related MA (and PhD) programs. Because there are no WGS departments at Greek universities, such programs were carried by various other departments. Three such programs are still running: "Women and Genders: Anthropological and Historical Approaches" (Aegean University at Lesvos: School of Social Anthropology and History);[26] "Training Teachers and School Psychologists in Education on Gender Equality: Promoting Equality Ideologies in the Educational Process" (Aristotle University of Thessaloniki: School of Education & Philosophy and School of Psychology);[27] and "Gender and New Educational & Working Environments in the Society of Information" (Aegean University at Rhodes: School of Elementary Education and Educational Planning).[28]

Finally, it should be mentioned that the measures for the promotion of "Gender and Equality" have triggered the publication of a number of books and articles on original research in Greek[29] as well as the translation of some classical feminist works in WGS into Greek.[30] Although there are still reasons to be skeptical, one could say that a gender perspective is more present in Greek academia and in related practices nowadays than it was a decade ago.

◊ Notes

This forum contribution is a modified version of my introduction "Spoudes Fylou sta Ellinika Panepistimia: Apotimisi kai Prooptikes" (Gender studies at Greek universities: Assessment and perspectives), to the volume that I edited *Spoudes Fylou: Taseis/Entaseis stin Ellada kai se Alles Europaikes Chores* (Gender Studies: Trends/Tensions in Greece and other European countries) (Thessaloniki: Zitis, 2006). All reflections relating to the second ("Gender Studies in SE European Countries") and third ("Crucial Issues for the Future") parts of the book have been omitted. However, at the end of this version a short afterword is included with information about the current situation.

1. For earlier reports on WGS in Greece, see Eleni Varikas, "Women's Studies in Greece," GRACE Report 1988 (reprinted in *The Making of European Women's Studies* 7 (2006): 159–163); Liana-Evangelia Sakelliou, "Women's Studies in Greece," SIGMA National Report Training and Education in Europe (Coimbra, 1995; reprinted in *The Making of European Women's Studies* 7 (2006): 164–177); Sasa Lada, "Women's Studies in Greece," in *Women's Studies—From Institutional Innovations to New Job Qualifications*, Report from ATHENA panel of experts 1.a., ed. Nina Lykke, Christine Michel, and Maria Puig de la Bellacasa (University of Southern Denmark, 2001), I, 91–96; Theodossia-Soula Pavlidou, "Women's Studies in Greece: An Update," *The Making of European Women's Studies* 7 (2006): 178–185.

2. See Theodossia-Soula Pavlidou, Gianna Savvidou, and Zogia Chronaki, "Gynaikeies Spoudes: I Gynaikeia Optiki stin Epistimi" (Women studies: Women's perspective in science), *Archaiologia* (Archeology) 21 (1986): 58–60.

3. The Women's Studies Group organized several conferences on "Objectivity in science" (1989), "Women's Studies in Greece and the European Experience" (1993), and "The role of women in the German and Greek resistance against National-Socialism and the German occupation" (1994). Two of these conferences are documented in the volumes: Women's Studies Group of the Aristotle University of Thessaloniki, ed., *Gynaikeies Spoudes stin Ellada kai i Europaiki Empeiria* (Women's studies in Greece and the European experience), (Thessaloniki: Paratiritis, 1996); Theodossia-Soula Pavlidou and Rüdiger Boltz, eds., *"Min Apaleipheis Pote ta Ichni ...": O Rolos ton Gynaikon sti Germaniki kai Elliniki Antistasi enantia ston Ethnikososialismo kai sti Germaniki Katochi* ("Don't ever erase the traces ...": The role of women in the German and Greek resistance against National-Socialism and the German occupation) (Thessaloniki: Paratiritis, 1999). The group also organized thematic cycles of lectures, open to all members of the academic community and the broader Thessaloniki public, e.g., on "Women creators," and "The artist is a woman."

4. The last one of these attempts, signifying in a sense also the end of the Women's Studies Group at the Aristotle University of Thessaloniki, was in 2000—the year that marked the beginning of a new era.

5. It should be stressed here that by that time the EU had already long developed multifarious activities concerning gender and equality issues. In particular, with respect to WGS, the Advanced Thematic Network in European Women's Studies (ATHENA) has played the most significant role. The basic aim of ATHENA, besides the enhancement/promotion of WGS, has been to develop a European feminist discourse that takes the different "voices" of Europe into account—especially in crucial areas, such as the Balkans and the Baltic countries—and does not cede to the Anglo-Saxon hegemony in feminist theorizing. See, for example, Gabriele Griffin and Rosi Braidotti, "Introduction: Configuring European Women's Studies," in *Thinking Differently: A Reader in European Women's Studies*, ed. Gabriele Griffin and Rosi Braidotti (London: Zed Books, 2002), 1–28; and the series *The Making of European Women's Studies* published by the ATHENA headquarters at Utrecht University. For the significance of this European project, see, among others, Jalna Hanmer, "What Is the Relevance of Women's and Gender Studies?" in *Spoudes Fylou: Taseis/Entaseis stin Ellada kai se Alles Europaikes Chores* (Gender Studies: Trends/Tensions in Greece and other European countries), ed. Theodossia-Soula Pavlidou (Thessaloniki: Zitis, 2006), 141–149; and Rosi Braidotti, "Beyond Interdisciplinarity: The New Transversal Feminist Theories," in ibid., 161–177.

6. In the same time interval, there were two calls for time-limited gender-related research programs (in the framework of PYTHAGORAS I and II) running over a couple of years, as well as for support of gender-relevant doctoral theses (in the framework of HERAKLEITOS).

7. Such ISPs were developed at the following universities: Aristotle University of Thessaloniki (http://web.auth.gr/genderstudies), University of Thessaly (http://www.uth.gr/main/newsevents/ProgFylo.htm), National & Kapodistrian University of Athens (http://www.isotita.uoa.gr/home/index.html), National Technical University of Athens (http://www.arch.ntua.gr), Panteion University (Athens) (http://www.genderpanteion.gr), University of Piraeus (http://www.unipi.gr/ypires/epeaekIsotitaFulon/index.html), University of Crete (http://www.soc.uoc.gr/gender), Aegean University at Rhodes (http://www.rhodes.aegean.gr/genderstudies/graduate).

8. The fact that 2005 was supposed to be their last year motivated the organization to host a conference with the Interdepartmental Undergraduate Program for Topics in Gender and Equality at the Aristotle University of Thessaloniki, titled "Gender Studies at Greek and other

European Universities: Assessment and Perspectives." The Conference comprised three thematic sessions: one with presentations of the individual ISPs by representatives of these programs; the second on the development of WGS in Southeastern European countries and their
specific geopolitical context; and the third about the future of WGS from the perspective of
colleagues with a long service at European universities, where WGS had materialized since
decades. It is this conference that led to the publication of the already-mentioned volume Pavlidou, ed. *Spoudes Fylou.*

9. This applied both to new and old courses.

10. Maria Stratigaki, "Panteio Panepistimio: Spoudes Fylou kai Isotitas stis Politikes kai
Koinonikes Epistimes" (Panteion University: Studies of gender and equality in the political and
social sciences), in Pavlidou, ed. *Spoudes Fylou,* 65–69.

11. See, e.g., Bessie Dendrinos and Stella Vosniadou, "Ethniko kai Kapodistriako Panepis-
timio Athinon: To Programma THE.FY.LIS" (National and Kapodistrian University of Athens:
The program THE.FY.LIS), in Pavlidou, ed. *Spoudes Fylou;* 35–44, and Sasa Lada, "Aristoteleio
Panepistimio Thessalonikis: To DPPS gia Themata Fylou kai Isotitas" (Aristotle University of
Thessaloniki: The DPPS for topics in gender and equality), in ibid., 79–86.

12. Gabriele Griffin and Rosi Braidotti, "Introduction: Configuring European Women's
Studies," 5. Similar problems are also encountered in Turkey.

13. See, for example, Serpil Üsür Sancar, "Women's Studies in Turkish Academic life," in
Pavlidou, ed., *Spoudes Fylou,* 129–138.

14. See notes 10, 11, and 16.

15. For example, "This course changed my life," "I can't talk about anything any more without referring to gender," are some of the comments that my colleagues and I have been hearing
from students in personal communications.

16. See also Benveniste's concluding remarks in Rica Benveniste, "Panepistimio Thessalias:
Oi Spoudes Fylou sto Panepistimio Thessalias—Enas Protos Apologismos" (University of Thes-
saly: Gender studies at the University of Thessaly—an assessment), in Pavlidou, ed. *Spoudes
Fylou,* 71–78.

17. Hanmer, "What is the Relevance of Women's and Gender Studies?" 142.

18. See, for example, Nina Lykke, "Women's/Gender/Feminist Studies — a Post-disciplin-
ary Discipline?" *The Making of European Women's Studies,* ed. Rosi Braidotti, Edyta Just, and
Marlise Mensink, vol. 5 (2004): 91–101; Veronica Vasterling et al., *Practicing Interdisciplinarity in
Gender Studies* (London: Raw Nerve, 2006).

19. On this see Dina Vaiou's comments, "Ethniko Metsovio Polytechneio: Fylo kai Choros"
(National Technical University of Athens: Gender and Space), in Pavlidou, ed. *Spoudes Fylou,*
29–33.

20. See Stella Vosniadou and Lydia Vaiou, "I Thesi ton Gynaikon sto Akadimaiko Prosop-
iko ton Ellinikon Panepistimion" (Women's place in the academic staff of Greek universities),
Pavlidou, ed. *Spoudes Fylou,* 45–49; and Stella Vosniadou and Lydia Vaiou, "Protaseis gia tin
Anaptyksi Politikon Fylou kai Isotitas sta Ellinika Panepstimia" (Proposals for the development
of gender and equality policies at Greek universities), in ibid., 51–58.

21. "Fylopedia," http://www.thefylis.uoa.gr/fylopedia (accessed 31 August 2010).

22. See http://newton.ee.auth.gr/genderIssues (accessed 31 August 2010).

23. See http://www.phil.auth.gr/gender (accessed 31 August 2010).

24. *Elliniki Etaireia Gynaikon Panepistimiakon, EL.E.GY.P.* (Hellenic Society of University
Women), http://www.isotita.uoa.gr/elegyp (accessed 31 August 2010).

25. For more information, see http://www.genderpanteion.gr/gr/ergastirio.php (accessed
31 August 2010).

26. See http://www.aegean.gr/gender-postgraduate) (accessed 31 August 2010).

27. http://web.auth.gr/edlit/html/spoudes/meta/diatmim/central) (accessed 31 August 2010).

28. http://www.rhodes.aegean.gr/genderstudies/postgrad) (accessed 31 August 2010).

29. See, e.g., *Fylo kai Koinonikes Epistimes sti Synchroni Ellada* (Gender and social sciences in contemporary Greece), ed. Venetia Kantsa, Vassiliki Moutafi, and Euthymios Papataxiarchiseds (Athens: Alexandria, 2010); *Glossa-Genos-Fylo* (Language—grammatical gender—social gender), ed. Theodossia-Soula Pavlidou (second edition, Thessaloniki: Institute for Modern Greek Studies, 2006); Maria Stratigaki, *To Fylo tis Koinonikis Politikis* (The gender of social policy) (Athens: Metaichmio, 2007).

30. See, for example, the collection of translated papers *Feministiki Theoria kai Politismiki Kritiki* (Feminist theory and cultural criticism), ed. and with an introduction by Athina Athanasiou (Athens: Nisos, 2006); Judith Butler, *Anatarachi Fylou* (Gender trouble) (Athens: Alexandria, 2009); Londa Schiebinger, *O Nous den Echei Fylo?* (The mind has no sex?) (Athens: Katoptro, 2006).

Gender Studies in Latvia
Development and Challenges
Irina Novikova

In the early 1990s women's/gender studies as a theoretical standpoint and methodology was a new perspective in the knowledge-production process in Latvia—on the national as well as regional levels. The first women's studies centers were established in the Baltic countries in 1992. These were the Women's Studies Center at the University of Vilnius (director Marija Ausrine Pavilioniene), the Women's Studies Center at the University of Tartu (director Anu Laas), and the Women's Information Center founded by Parsla Eglite in Riga, under the umbrella of the Institute of Demography at the Academy of Sciences of Latvia. The Riga Center collected statistical and scientific survey data, as well as publications, memoirs, photographs, and other information about women of Latvia. In the early 2000s it was dissolved. Apart from this Center various women scholars were involved in research projects on women's issues (Ilze Trapneciere, Dagmara Beitnere, Maiga Kurzmetra, Tana Lace). In 1994 Ella Buceniece and Ausma Cimdinya started a joint research project on feminism and Latvian culture. The first results were articles on the history of Western feminisms, translations from Simone de Beauvoir's *The Second Sex* in the literary monthly *Karogs*,[1] and an annual almanac dedicated to women in Latvian culture, literature, and history.[2] In 2008, Ausma Cimdinya organized an international Simone de Beauvoir conference at the University of Latvia.[3]

The Center for Gender Studies at the University of Latvia (LU DZSC) was established in 1998. The Center—the only educational and research institution in Latvia dealing with various women's and gender issues in society, culture, and politics—was seen by its founders as an interdisciplinary educational, pedagogical, and research project, bringing together contributions from a wide range of disciplines across the social sciences and the humanities. The LU DZSC has regularly organized summer schools funded by the Nordic Research Advisory Body on Nordic Research Policy (NorFA), the International Higher Education Support Program, (HESP), and the Nordic Council of Ministers for different target audiences. The Center is well known in Latvia for its publications[4] and lecture courses that develop gender expertise in the fields of education (primary, secondary, higher) and research. At the same time, individual researchers and lecturers introduced innovative courses in different departmental programs at the University of Latvia (in theology, literary and visual studies, and philosophy). An impetus to the field of women's/gender studies was also given by the development of gender studies in the countries of East Central Europe and Russia. One cannot disregard the importance of the Network of East-West Women (NEWW)[5] with its book donations to the library of the Center for Gender Studies at the initial stage of its existence, the support of the gender studies program of the Central European University in Budapest (and the personal support of one of its first directors, Prof. Joanna Regulska), and the tremendous backing coming from feminist, women's, and gender studies institutions and individuals from the Nordic countries.

There was a failed attempt to launch a gender studies center at Valmiera Collegium, due to a lack of human resources (i.e., at least a group of women interested in feminist/women's/gender studies and motivated in launching such a center). Even at the University of Latvia, the activities of the Center for Gender Studies were initially limited to the library, a number of optional courses for the university students, and publications.

In the process of launching the Center for Gender Studies at the University of Latvia, we (the founding board—Ieva Zake, Sandra Sebre, Maija Rubene, Irina Novikova, Elizabete Picukane, and Skaidrite Lasmane) formulated the following goals and objectives of LU DZSC. First, to develop discourses on the relationship between different feminist theoretical standpoints and the postsocialist/post-Soviet specifics of women's history, experiences, and gender relations in the Latvian context. Second, to develop women's/gender studies discourse in Latvia and to support new theoretical and practical initiatives in the country. Third, to develop cooperation networks among women's/gender studies centers in the Baltic countries in terms of regional cooperation with Nordic countries and with East Central European countries and Russia. Finally, to provide our students with access to contemporary theoretical discourses and methodological approaches in feminist/women's/gender studies.

The task was not only to appropriate and to rethink new knowledge but also to think about using feminist, gender, and women's studies frameworks for the analysis of the postsocialist women's condition and transformations of gender contracts/relations. With the support of different funding agencies (Soros Foundation-Latvia, East-West Women's Network, Mama Cash and NorFA), feminist/gender/women's studies networks and collaborative projects, the Center for Gender Studies has secured regular publications of books in the fields of feminist, gender, women's, and critical men's studies in Latvian and English. In the 1990s, with the support of the Nordic Council of Ministers, the Center published a newspaper *Zhenshchinii Baltii* and an almanac *Gender. Kultura. Obshtestvo* (Gender. Culture. Society), both in Russian, for Baltic Russian-speaking women. The Center supported the introduction of courses based on feminist/women's/gender studies methodologies, in order to launch a gender studies program or module at the university. However, all our initiatives and activities, technical/communications equipment as well as the library and publications have been funded and supported by external agencies, and the University of Latvia provided the Center only with a small room.

The International Projects of LU DZSC

The Center participated in various collaborative research projects: the Leonardo da Vinci project "Gender Sensibility in Vocational Training" (2006–2008); the EU multilateral project "Men and Masculinities" (1999–2002); the EU multilateral project "Network of European Women's Rights," NEWR (2002–2005), initiated and coordinated by the University of Birmingham, UK and led by Prof. Donna Dickenson; the EU multilateral project "Co-ordination Action on Human Rights Violations" (CAHRV, 2003–2006); the EU multilateral project "Media and Power" (2004), a research sub-project on gender stereotypes in Latvian mass media; ATHENA-AOIFFE—Network of Feminist Studies in Europe (coordinator University of Utrecht, with Prof. Rosi Braidotti as the project leader); and a USA-Baltic countries project entitled "Women and Leadership" (1997–1999).

The cooperation with feminist scholars from different regions of Europe—with feminist research networks such as NOISE and ATHENA, with partners from the Nordic region (the Christina Institute for Women's Studies, University of Helsinki), from Ukraine (the Kharkiv Center for Gender Studies), from Lithuania (University of Kaunas) and Estonia (University of Tartu), and via the Central European University—contributed to and stimulated greatly the research and teaching of separate scholars. However, it has not resulted in a sustained process of institutionalization of women's and gender studies in Latvia. I would say that in Latvia in general there has been a lack, if not an absence, of what I call feminist motivation and engagement, which is essential for any successful initiative of launching gender/women's/feminist studies in the country. Another factor was indifference on the part of higher educational authorities. It was expressed, on the one hand, in a meticulous formalization of the gender equality principle in documentation and legislation, but on the other hand, in a politics of reluctance of financing and promoting critical gender-related projects and courses.

The LU DZSC also organized several international summer schools. The first one—a NorFA PhD Summer Research Course on "Gender and Multiculturalism"—was held in Yurmala in 1998, in cooperation with Professor Aili Nenola from the Christina Institute for Women's Studies of the University of Helsinki. In August 2001—the SOROS-OSI Summer University on "Men and Masculinities" was organized in Riga. The NoRFA PhD Summer Research Course on "Gender, Religion, Ethnicity: Comparative Anthropological Perspectives" followed in 2002 in Jūrmala. In June 2003, Jūrmala again hosted a NoRFA Summer Research Course for PhD students, now on "Gender Equality and Men's Practices in European Welfare Regimes: Comparative Perspectives." The participants of these summer schools were young scholars and postgraduate students from the universities of the Nordic countries as well as from the countries of East Central Europe and Russia. These summer schools were useful for individual young scholars in their research projects as well as in promoting general academic awareness of methodological directions, theoretical discussions, and empirical topics in women's, gender, and critical masculinity studies.

Another result of our cooperation with Nordic and Baltic partners was a Baltic-Nordic conference on women's and gender research held in Riga in October 1998. This conference, financed by the Nordic Council of Ministers, hosted around sixty participants from the various universities from the region. The Conference Proceedings were published in 1999.

The LU DZSC offered a number of lecture courses for undergraduates and graduates at the University of Latvia: "Introduction to Gender Studies," "Feminism: Myths and Realities," "Feminism and Psychoanalysis," "Sociology of Gender," "Feminist Literary Theory and Literary History," "Introduction to Human Sexualities," and "Gender and Social Movements," to name but a few. The funding for these (optional) courses was provided by the interfaculty accounting system. However, we started losing these courses two years ago, because the list was downsized to language courses, and professors and associate professors were no longer allowed to teach them (because of a 40 percent cut of the university's annual budget and the nearly bankrupt situation of the Latvian economy).

On the one hand, as our experience has shown, university policy and curriculum designers have generally been pleased to have a center for gender/women's studies as an indication of the "Europeanization" of the Latvian university curriculum and infrastructure. There was no special resistance from the university authorities to launch such centers with the aim of accumulating teaching and research potential of feminist scholars, but there was no further regular and sustained institutional support either. In such a situation, a project either would survive due to individual and group commitments, or would dissolve very quickly, as happened in Valmiera College in the early 2000s. Actually, the collaboration of quite a few researchers in women's and gender studies in Latvia was inter-individual and sometimes collective, by virtue of their participation in international projects (of scholars such as Ausma Cimdina, Elizabete Picukane, Evija Caune, and Irina Novikova).

On the other hand, the sustained marginalization of the gender/women's studies segment in the university policies of academic development was also conditioned by additional factors. The first was a lack of lecturers and researchers with a feminist standpoint in their research and teaching. Gender was very quickly integrated into the scholarly and academic lexicon, and stripped of its political and critical meaning and potential. Feminist and women's studies researchers work primarily in their original academic disciplines and they lack a pool of kindred spirits, so to say, in their professional milieu. Often women's issues are merely included as parts of more general research projects (as in the case of the UNDP Development Reports). Second, even with the goodwill of the university authorities to facilitate the institutionalization of gender/feminist/women's studies in the distant future, the major question for gender and women's studies, particularly in the last years, has been whether a gender studies program would be economically viable for the university. The whole situation was aggravated by the recent substantial cuts in university funding (40 percent for the

University of Latvia and even more for regional universities), because of the severe economic crisis in Latvia, and the downsizing of the teaching staff at the universities.

Another important factor has been the visible absence of an independent women's movement in Latvia. The process of NGO-ization of women's groups and organizations in the 1990s, combined with pressing donor-driven agendas, has not resulted in creating a powerful women's lobby and influence in politics, trade unions, and the business sector. The present-day antipathy of the majority of women toward political lobbying for their interests and rights is rooted in many aspects of both historical and contemporary realities. Among them are the effects of past Soviet "gender equality" politics and the "revival of old values" politics in post-Soviet national state-rebuilding. These factors interacted in the 1990s, which strongly affected women's status in Latvia. The intensive neoliberal restructuring of the national economy of the 1990s–2000s has also proved harmful to the political, social, and reproductive rights of women and to their status as citizens. In order to invigorate gender and women's studies in Latvia an important factor was needed: women's political and social solidarity for women's issues in this country, which has two major language communities and specific citizenship situations—Latvian and Russian speakers, citizens and aliens—and a neoliberalist agenda in its economic and social policies.

At the same time, there was an inadequate commitment to gender equality among policy designers during Latvia's accession process into the European Union and the harmonization of national legislation with the *acquis communautaire* (the accumulated legislation, legal acts, and court decisions that constitute the body of EU law). Gender equality issues are not taught in higher education or in further training institutions preparing civil servants on a regular and nationwide basis; guidelines, handbooks, and manuals on gender mainstreaming and gender equality are not available in the public reading market. Moreover, the very process of gender equality and gender mainstreaming policies is marginalized and dealt with uncritically: there is no assessment and critical evaluation of this process and it is not clear whether it has any positive impact upon the de facto discrimination of women.[6]

For women's and gender studies in Latvia, this context implies more challenges to be addressed. The foremost question is whether gender mainstreaming in Latvia is seen by politicians and women's NGOs as part of the expansion of an equal opportunities agenda, although gender mainstreaming should be dealt with beyond this all-encompassing and politically convenient framework. Other questions are whether political opportunities, mobilizing structures, and strategic framing already exist in Latvia and whether women's organizations can turn themselves into effective and collective social stakeholders in this process, especially if we take into consideration the long-term process of donor-driven agendas and their limitations for women's organizations.

To conclude, in Latvia we still live in the situation of "no-mass based feminist movement as long as feminist ideas are understood only by a well-educated few" (bell hooks).[7] The academic settings are structured so that researchers in gender/women's/feminist studies remain in their peripheral spaces, with no potential for a career in mainstream academia. Thus, women researchers practice a "borderline" strategy by combining research for their hearts and research for their careers. At the same time, the lack of books and other materials in the Latvian language and the absence of teaching-oriented networking initiatives in national women's/gender/feminist studies are challenges that must still be met.

◊ **Notes**

1. Fragments from *The Second Sex* were translated into Latvian by Irēna Auziņa as *"Otrāis dzimums"* (*Le Deuxième Sexe,* 1949), in *Karogs* 12 (1994): 151–163, and *Kentaurs* 21, no 40 (2006): 116–143.

2. Almanac *Feministica Lettica,* vol. 2–3 (Riga: Zinatne, 2001, 2003).

3. See the conference program and information at http://beauvoir100.blogspot.com (accessed 19 April 2010).

4. The LU DZSC has published Irina Novikova, ed., *The Conference Proceedings of the Baltic-Nordic Conference in Women's and Gender Studies* (Riga: ELPA, 1999); idem, *Mūsdienu feministiskās teorijas: antoloģija* (Contemporary feminist theories) (Riga: Jumava, 2001); idem, *Par agru? Par vēlu? Feministiskās idejas, konteksti, pieejas* (Too early? Too late? Contemporary feminist discourses) (Riga: Holda, 2001); Nira Yuval-Davis, *Gender i natsia* (Gender and nation), translated into Latvian by Irina Novikova (Riga: ELPA Press, 2001); Irina Novikova and Dimitar Kamburov, eds., *Men and Masculinities in a Global World: Integrating Postsocialist Perspectives* (Helsinki: Kikimora, 2003); Audrey Guichon, Christien van den Anker, and Irina Novikova, eds., *Women's Social Rights and Entitlements* (Houndmills: Palgrave Macmillan, 2006); Irina Novikova, ed., *Gender Matters in the Baltics* (Riga: University of Latvia Press, 2008).

5. See information on NEWW at: http://www.inch.com/~shebar/neww/neww2.htm (last accessed August 2010).

6. Irna van der Molen and Irina Novikova, "Mainstreaming of Gender Issues in the EU Accession Process with Reference to the Baltic States," *European Journal of Social Policies* 15, no. 2 (2005): 139–156. This publication is based on a joint presentation at the International Conference "Human Development in EU Accession Countries," held at the Stockholm School of Economics in Riga (31 May 2 June 2003); Irina Novikova, "Gender Equality and Gender Mainstreaming: Achievements and Issues," *Sociologija* 14, no. 3 (2004): 6–12.

7. bell hooks, *Feminist Theory from Margin to Center* (Boston: South End Press, 1984), 113.

Gender Studies in Lithuania
Laima Kreivytė

The development of gender studies in Lithuania was a part of the social, political, and cultural emancipation of the 1990s in Eastern Europe. The dominant discourse in the society was built around the idea of "returning to Europe" and the insistence on "European values" with sensitivity to women's rights and gender equality as a part of them. In this respect gender studies in Lithuania were based on theoretical approaches that had been already developed in Western feminist theory and gender studies. Parallel to this process, however, the conservative discourses have been strengthening in contemporary Lithuania. Recently some traditionalist perceptions of family and

sexuality were articulated in the Lithuanian parliament following rigorous public debates. This political and social context is dubious for gender studies development.
On the one hand, the conservative forces refuse to see gender discrimination in the
society and tend to "normalize" their perception of gender relations legally.[1] On the
other hand, this situation reveals an important aspect of Lithuanian gender studies:
"Western" feminist theory cannot be just transported to the local context but must
be rethought and creatively (re)shaped in search of its own words in Lithuania. At
the same time, institutionalized gender studies centers create a necessary ground for
promoting social change.

Gender studies centers in Lithuania are established within the university structures of the major cities: at Vilnius University, Kaunas University of Technology, and
Šiauliai University. The gender studies centers at Kaunas University of Technology
and Šiauliai University act as non-academic units, and are more focused on promoting
gender equality in local communities. The Gender Studies Center at Vilnius University—the first of its kind among the Baltic gender studies institutions—was established
in 1992. Since 2000 the Center has been recognized as an independent division of Vilnius University. It has the status of a department, and it is financed by the university's
budget (wages and administrative expenses). Today the Gender Studies Center functions as an academic unit for interdisciplinary research on women and gender. As a
leading institution in the field of gender studies in Lithuania, the Center also plays an
important role in forming gender policy in the country.[2]

The foundational goal of the Gender Studies Center was to integrate gender studies into the core curriculum at Vilnius University. Every semester the Center offers
undergraduate electives that are open to students from all departments and students
from other institutions of higher education. These courses present different academic
fields and discuss various topics: feminist philosophy, gender sociology, visual culture,
feminist literary critique, feminist theology, personal law, women's history, gender and
information technology, queer theory, masculinity studies, and women's health. The
courses are not only an introduction to gender studies, but give the students an opportunity to incorporate gender perspectives into their main subject of study. Each
year these gender studies electives offered by the Center take in between 300 and 380
students. There are no restrictions placed by the university regarding the fields and
topics of courses offered by the Center. However, there have been obstacles against the
establishment—and recognition—of an independent MA program in women's studies
within Vilnius University. At the same time, the Gender Studies Center managed to
launch an MA program in social work with an emphasis on women's studies. In 2000
we renamed our Center from Women's Studies Center into Gender Studies Center, in
order to broaden the understanding of gender, having in mind the traditionally homophobic Lithuanian community. We found it necessary and important to include in our
curriculum courses on sexual minorities, as well as queer studies. Since 2000, courses
on sexual minorities are offered on a permanent basis. It should be stressed that the
women's perspective remains our primary goal and concern.

The faculty members of the Gender Studies Center have different backgrounds
and academic interests. Professor Dalia Leinartė, the current director of the Center,
does research on the history of the family and various other aspects of women's/gen-

der history in East Central Europe during the nineteenth and twentieth centuries. She has published extensively—both books and articles—in this field.[3] Associate Professor Giedrė Purvaneckienė is an expert on gender, family, and education policy. Since 1994 she has held different positions at high-level political institutions: state counselor on women's issues for the government of Lithuania; parliament member, adviser to the prime minister of the Republic of Lithuania. Associate Professor Solveiga Daugirdaitė is a leading specialist in the field of feminist literary criticism and culture, affiliated with the Institute of Lithuanian Literature and Folklore. This affiliation allows her to promote gender-related topics at the conferences and as part of the research agenda of the institute. Associate Professor Natalija Arlauskaitė teaches courses related to feminist film theory. In 2010 she published a dictionary, *Trumpas feministinės kino teorijos žinynas* (Key concepts of feminist film theory).[4] Associate Professor Artūras Tereškinas received his PhD at Harvard University in 2000. He has written widely about sexuality, gender, masculinity, and popular culture.[5] The main research and teaching interests of Assistant Professor Lijana Stundžė are in the field of gender and communication. Laima Kreivytė is a lecturer at the Center, where she teaches courses on feminist visual culture and queer theory. As a curator, she explores gender issues in the exhibitions she organizes. The international activities of the Vilnius Gender Studies Center include the organization of different courses and public lectures with the participation of gender scholars and activists from Europe and the United States.

The Gender Studies Center has a strong presence in the intellectual life of Lithuania. It organizes conferences, seminars, and film screenings that target not only the academic community but a broader audience as well. These events aim at discussing such gender issues as the relationship between the private and public spheres, gender inequality in politics and the economy, and feminist perspectives on various cultural phenomena: books, films, exhibitions, popular culture. The Center's publishing activities also exert a significant impact on Lithuania's cultural and intellectual life. The publications deal with issues such as violence against women and children, trafficking in women and prostitution, topics that until recently were absent from public discourse and scholarly interests.[6] Dalia Leinartė's book *Adopting and Remembering Soviet Reality: Life Stories of Lithuanian Women, 1945–1970*, represents a long-term oral history study of the experience of women living in Lithuania in post–World War II period.[7]

Between 1999 and 2002 the Gender Studies Center published an annual journal, *Feminizmas, visuomenė, kultūra* (Feminism, society, culture). It was the only interdisciplinary journal in Lithuania and the Baltics that focused on women's and gender issues by bringing together the expertise of leading scholars from the region. In 2002 and 2003 the Center published the electronic journal *Klėja*, which aimed to create a space for young scholars. The national gender studies centers in 2005 together launched a new journal, *Lyčių studijos ir tyrimai* (Gender studies and research), which is included in the international database "Index Copernicus Journals Master List." Since 2002 the Gender Studies Center has been organizing so-called Tuesday seminars. The seminars cover various aspects of women's/gender studies and are open to the faculty of Vilnius University and to the community.[8]

For the further development of gender studies it is crucial to create informational networks in order to collect and exchange resources and data. In 2001 at the Vilnius

Gender Studies Center the Archive of Women's Memories and a Film Archive were established as part of this process. The Archive of Women's Memories documents biographies of Lithuanian women of different age, social status, ethnicity, and location. The collection consists of fifty audiotapes and documents. The Film Archive contains more than sixty films, produced in Soviet Lithuania from 1945 to 1990, and provides examples of various gender representations.[9]

The Gender Studies Center at Vilnius University engages with political life and gender policymaking in Lithuania as well. The most significant recent event, organized in partnership with the Lithuanian parliament, the Ministry of Foreign Affairs of the Republic of Lithuania and the Embassy of the United States, was a public discussion with the participants of the Convening Meeting of the Gender Equality Working Group of the "Community of Democracies," held on 12 April 2010 at Vilnius University. The Community of Democracies is an intergovernmental structure, based on the Charter of the United Nations and the Universal Declaration of Human Rights, in which democracies and developing democracies can take part. It was established in 1999 on the initiative of the United States.[10] On 12 July 2009, Lithuania took over the chairmanship of the Community of Democracies from Portugal. The Community organized a discussion at Vilnius University on "Women and Leadership," focused on women's participation in politics and business, and the impact of their participation in legislative and executive power for the shaping of policy priorities and agenda.[11]

In the course of the last ten years, the Gender Studies Center has been active in promoting gender equality and women's rights nationally and internationally. The Center initiated or took part in several important projects, among them an international training course titled "Increasing Teacher Trainees' Awareness of Sexualized and Gendered Violence: International Study Course on Sexualized and Gendered Violence Aware II," coordinated by the Finnish Oulu University (2003–2004); and a "Preventive Informative Campaign Against Trafficking in Girls and Women in the Baltic States" (Lithuania and Estonia), 2002–2003 (supported by the "Democracy Commission Small Grants Program" of the US Embassy in Lithuania). The Center also organized two international summer schools: "Prevention of Sexual Abuse: Summer Courses for Teachers (Kaliningrad region, Belarus, Russia, Lithuania)," held in 2002 and 2003; and a 2003 international training course: "Violence Prevention Strategy in Schools: Training for Trainers (Kaliningrad Region, Belarus, Russia)."

The faculty of Gender Studies Center includes experts on gender equality issues widely recognized by the government of Lithuania. In 2004 the Lithuanian government asked the chair of the Center Dalia Leinartė to draft the "Third Periodic Report on the Implementation of the Provisions of the United Nations 'On the Elimination of All Forms of Discrimination against Women in Lithuania.'" She was also asked to draft the following reports, "A Review and Appraisal of the Implementation of the Beijing Declaration and Platform for Action Adopted at the Fourth World Conference on Women" (Beijing, 1995), and "The Outcome of the Twenty-Third Special Session of the General Assembly" in Lithuania in 2000. In 2008 Leinartė also drafted the Fourth Periodic Report on the Implementation of the Provisions of the United Nations "On the Elimination of All Forms of Discrimination against Women in Lithuania."

In 2006 the Center started a two-year European Commission-funded project called "Stereotypes of Gender Roles in the Labor Market: Monitoring and Education." The main target groups of the project were social workers, educators, representatives of employment agencies, employers, and nongovernmental workers dealing with social integration and prevention of social disjuncture. More than a hundred lectures were given all over Lithuania during these two years. *"Stereotypes of Gender Roles in the Occupational Field"* was another—quantitative—project in which the Center was involved. It resulted in the publication of *Moterys, darbas, šeima. Lyčių vaidmenys užimtumo sferoje: sociokultūrinis aspektas* (Women, work, family. Gender roles in the labor market: sociocultural perspectives).[12]

Some faculty members of the Gender Studies Center also pursue political careers. Apart from Giedrė Purvaneckienė, it is important to mention the founding director of the Center, Marija Aušrinė Pavilionienė, who held the position from 1994 until 2000. Pavilionienė has recently been elected to her second term in the Lithuanian parliament as a member of the parliamentary committees of European Affairs and Education, Science and Culture, and she promotes a gender sensitive perspective at the level of state policy.

The activities of the Gender Studies Center help to make visible and deconstruct the gender normativity and inequality widespread in Lithuania's social, cultural, political, and economic life. At the same time, the connections between academic research and activism are not settled enough and depend mostly on the individual positions of the faculty. Professors and lecturers engage in public discussions on controversial social issues, analyzing and commenting current initiatives in the spheres of family and sexuality regulations. The most important task that gender studies centers are facing is to establish close connections with activist movements and to strengthen collaboration among themselves. Only then can they effectively oppose the increasing sexism and homophobia in current political discourses.

◊ Notes

1. Here I have in mind the "Law for the Protection of Minors against the Detrimental Effect of Public Information," which came into effect on 1 March 2010, and restricts any public dissemination of information that affect the "mental health, physical, intellectual or moral development" of children under eighteen. One of its paragraphs prohibits information, "which scorns family values and promotes the concept of marriage and family formation, other than stipulated in the Constitution of the Republic of Lithuania and the Civil Code of the Republic of Lithuania." Following the definition of family given in the constitution, this law prohibits any portrayal of marriage as something other than the union of a man and a woman. I call this procedure a "normalization" of gender discrimination because it legally excludes all other forms of partnership but traditional hetero-marriage.

2. In 2005 the Belarusian European Humanities University (EHU), closed in 2004 by the Belarusian authorities, was re-established in Vilnius. The EHU has an active Gender Studies Center, in existence since 1999.

3. Dalia Leinarte, *Adopting and Remembering Soviet Reality: Life Stories of Lithuanian Women, 1945–1970* (Amsterdam and New York: Rodopi, 2010); Dalia Leinartė Marcinkevičienė, *Vedusių visuomenė: santuoka ir skyrybos Lietuvoje XIX a.–XX a. pradžioje* (The society of married people:

Marriage and divorce in Lithuania, nineteenth and early twentieth century) (Vilnius: Vaga, 1999); Dalia Leinartė Marcinkevičienė, *Įžymios Lietuvos moterys: XIX–XX a. pr.* (Famous Lithuanian women: Nineteenth and early twentieth century) *(Vilnius:* Vilnius universiteto leidykla, 1997).

4. Natalija Arlauskaitė, *Trumpas feministinės kino teorijos žinynas* (Key-concepts of feminist film theory) (Vilnius: Vilniaus universiteto leidykla, 2010).

5. Artūras Tereškinas, *Vyrai, vyriškumo formos ir maskulinizmo politika šiuolaikinėje Lietuvoje* (Men and masculinity in contemporary Lithuania) (Vilnius: Vilnius universiteto leidykla, 2004).

6. Marija Pavilionienė and Vida Kanopienė, *Smurtas prieš moteris ir vaikus Lietuvoje* (Violence against women and children in Lithuania) (Vilnius: Vilniaus universiteto leidykla, 1997); Dalia Marcinkevičienė and Rima Praspaliauskienė, *Prekyba moterimis ir prostitucija* (Trafficking of women and prostitution) (Vilnius: Vilniaus universiteto leidykla, 2003).

7. Leinarte, *Adopting and Remembering.*

8. A short description of recent events can be found on the website of the Vilnius University Gender Studies Center, http://www.moterys.lt/index.php?id=521 (accessed 19 August 2010).

9. More information and biographies of women are available in Lithuanian on the website of the Vilnius University Gender Studies Center, *Lyčių studijų centras, Moterų atminties archyvas,* http://www.moterys.lt/index.php?id=309 (accessed 20 August 2010).

10. Council for a Community of Democracies website, http://www.ccd21.org/conferences .htm (accessed 20 August 2010).

11. For more information see "Community of Democracies' working group on gender equality was established in Vilnius, it will promote greater women's involvement into business and politics"; press release 13 April 2010, Ministry of Foreign Affairs of the Republic of Lithuania website, http://www.urm.lt/index.php?-888965767 (accessed 21 August 2010).

12. Ramunė Stukienė, ed., *Moterys, darbas, šeima. Lyčių vaidmenys užimtumo sferoje: sociokultūrinis aspektas* (Women, work, family. Gender roles in the labor market: sociocultural perspectives) *(Vilnius:* Vilniaus universiteto leidykla, 2008); Ingrida Gečienė, *Lyčių vaidmenų stereotipai užimtumo srityje (Gender stereotypes in the labor market)* (Vilnius: Vilniaus universiteto leidykla, 2008).

◆

On the Status of Gender Studies in Macedonia Today

Katerina Kolozova

A Brief History of the Field of Gender Studies in Macedonia

Gender studies emerged as an academic field in its own right in Macedonia in 1999 with the establishment of the Research Center for Gender Studies as part of Euro-Balkan, the Institute of Social Sciences and Humanities Research. Prior to this event, there was a history of feminist critique in the country, in particular in the fields of comparative literature, history of art, sociology, and political sciences, championed by scholars such as Katica Kulafkova, Elizabeta Sheleva, Suzana Milevska, Mirjana Najchevska, and Natasha Gaber. Nonetheless, the promotion of the concept of "gender" and the

idea of "gender studies" occurred with the research projects and conferences organized by the Research Center for Gender Studies at the Euro-Balkan Institute. The Institute of Social Sciences and Humanities Research-Euro-Balkan Institute at that time
was an alternative academic organization, which has evolved into a scientific research
institution recognized by the Ministry of Education and Science, offering accredited
graduate studies in several fields, among them the humanities and social sciences.

The role that the Center, and the Euro-Balkan Institute as a whole, have played in
the promotion of gender studies during the past eleven years has been one of uniting
the large community of Macedonian scholars already active in the fields of feminism
and gender theory. The Center and its projects—during the first four years mostly
funded by the Open Society Institute in Macedonia (the Soros Foundation)—served
as a forum where feminist scholars from different fields and academic institutions
could gather and work together. These activities enabled the emergence of a visible
intellectual scene pursuing feminist and gender-related scholarship, and advocating
its institutionalization within the university education of the country. The strategies
undertaken to provide recognition and visibility for the field, and for the intellectual scene involved in its scholarly activity and vouching for the field's sustainability, were a continuous research production, publishing, and informal education for
academic excellence aimed at scholars (especially by organizing summer schools and
workshops). The strategy employed for the institutionalization of the Center and its
Academic Network was lobbying for gender studies by introducing courses or gender-sensitive approaches in the teaching of the traditional disciplines within existing
university curricula, instead of focusing solely on establishing a department at one of
the state universities. The strategy has proven successful considering that at the moment there are courses in gender studies in almost all departments of social sciences
and the humanities in all of the country's universities. In addition to this, there is a
postgraduate studies department with two programs at the Institute of Social Sciences
and Humanities Research at the Euro-Balkan Institute (established in 2007) as well as
one undergraduate program at the Faculty of Philosophy of the State University of
Skopje (established in 2008, active since 2009–2010).

The Initial and Current Goals of the Field and
Their Relation to the Political Context

By the end of the 1990s and the beginning of the 2000s, the overarching goal was the
so-called institutionalization of gender studies. Today, more than ten years after the
introduction of the field in Macedonia, it seems that this major goal has been achieved:
dozens of courses in gender studies exist in the universities throughout the country,
as well as the possibility to graduate in the field in any of the three cycles of tertiary
education. The courses are numerous and include feminist literary theory courses at
the Faculty of Philology-State University of Skopje taught by Elizabeta Seleva, Maja
Bojadzievska, Kata Kjulafkova, Jasna Koteska; gender studies at the Department of
Philosophy taught by Ferid Muhic; gender and culture at the Institute of Sociological, Political and Juridical Research taught by Katerina Kolozova; gender studies at

the Political Sciences Department of the University American College-Skopje taught by Katerina Kolozova; gender and law taught by Gordana Siljanovska and Dobrinka Taskovska at the Faculty of Law at the State University of Skopje. With the emergence of a new generation of junior scholars in the field, priorities are shifting and changing according to the global debate and theoretical state of the art, focusing on technology and gender, queer studies, cyber theory and gender, and nationalism studies and gender, to name only a few.

Another contextual factor conditions the change in priorities: this is the regional context, articulated among other forms through the Regional Network for Gender Studies in Southeastern Europe (www.gendersee.org.mk) that requires a culturally sensitive approach to the field of gender studies from both the thematic and the epistemological perspectives. The priorities originate from academia but have a political bearing as well: overcoming nationalisms in the region, enhancing regional cooperation, and focusing on topics that go beyond the academic field's overlapping of gender with women's studies, such as queer studies and masculinity studies. The changes in research and teaching priorities have thus stemmed from both trends in international scholarship and global and local activisms. There have been no curricular or research novelties linked to local mainstream politics, that is that have occurred as a direct reaction to mainstream politics. However, gender studies scholars and academics have reacted publicly and vehemently to reactionary political initiatives such as the 2009 debate (albeit not official initiative) concerning limitations on abortion rights, or the failure to include gays and lesbians in the 2010 antidiscrimination law, which may be revised under pressure from the European Union.

The Strengths and Weaknesses of Women's and Gender Studies in Macedonia

Compared to other fields in the humanities and social sciences in Macedonia, the field of gender studies is marked by a strong presence of theoretical discourse. However, an important weakness, shared by all disciplines in the humanities and social sciences, and not something that characterizes the field of gender studies specifically, is that original research and publications (of both periodicals and books) are scarce. This is due to meager research funds in the country. The only journal in the field of gender studies is the bi-lingual—Macedonian and English—*Identiteti: Spisanie za politika, rod I kultura/Identities: Journal for Politics, Gender and Culture,* published by the Euro-Balkan Institute since 2002. European funds are available, but rather competitive. The lack of skills in project proposal writing and the complexity of proposal applications are serious obstacles for research institutes and institutions of higher education to demonstrating greater initiative in applying for EU research funds.

The intellectual isolation, a result of insufficient participation in international conferences and other academic events, contributes to the modest quantity of research but even more so to its lack of originality. The latter problem is caused not only by the economic situation—the lack of funds for participation in academic events that should be provided by the Ministry of Education and Science as well as by the institutions themselves through fundraising activities—but also by the absence of a culture of academic mobility. The reasons for this lack of a culture of mobility are also structural: until recently a high level of international academic activity was not a crucial requirement for maintaining one's position as a university professor. With the introduction in 2009 of science law and rulebook of criteria for election and re-election to academic posts, this situation may improve.

Up-to-date literature—leaving aside course textbooks—is insufficiently available in university libraries. This situation is characteristic for the entire area of the humanities and social sciences. The general situation of the field of gender studies in this respect is somewhat better; academics tend to resort to state-of-the-art literature they obtain on their personal initiative, and their command of English enables them to keep up with the theoretical developments.[1] The other aspects of a structural or systemic nature, which engender what I have called "intellectual isolation," apply to gender scholars as well.

How to Properly Integrate the Notion of Gender into the Scholarship of Southeastern Europe?

It is important for gender studies to affirm its status as a discipline in its own right. This can be achieved by undertaking research in the area of gender studies instead of merely researching gender aspects in mono-disciplinary fields of the traditional positivist kind. Intensive activity in publishing, and initiatives of joint study programs in the areas of gender/women's/feminist studies m be continuously pursued in order to stabilize the position of gender studies as a legitimate academic field. The regional perspective should be as integrated as much as possible in all these initiatives, because it adds an aspect of internationality to gender studies, which helps in increasing the level of recognition and legitimacy in the context of the national academia. Furthermore, integrating the regional perspective is an epistemological priority put forward by the Regional Network for Gender and Women's Studies in Southeastern Europe: to culturally contextualize the field, believing that we can talk of a regional culture rather than a national one. This approach will allow gender studies to become an important tool of critique of nationalisms in the region, and to make a significant contribution to the field of nationalism studies. Besides its political significance, the epistemic stance of contextualizing research regionally rather than nationally is also justified methodologically (in view of the shared cultural heritage of the neighboring and competing empires, the Ottoman and the Habsburg, the shared communist past of most of the countries in the region, are among factors that shape the regional culture as "Balkan").

The Network and the Department of Gender Studies (formerly RCGS) of the Euro-Balkan Institute have undertaken a number of projects to explore, elaborate and further expand this epistemological and methodological position of "regionalizing" gender studies, among which is the Athena Network publication from 2006.[2] In cooperation with the Fribourg University in Switzerland and the Network of Gender and Women's Studies in Southeastern Europe, the Department of Gender Studies at the Euro-Balkan

Institute has initiated a multiyear project focusing on epistemological and thematic issues that will inform a Balkan-specific (or Southeast European) scholarship in the area of gender/feminist studies. Forthcoming are a cluster of epistemological colloquia scheduled for June 2010: "Epistemologies of Identity Studies: Understanding Transition and Gender in the Region of Southeastern Europe" and "Interdisciplinary Methodologies in the Study of Gendered Balkan Identities." Publications of edited books on topics that will issue from these two and other events as part of the long-term project are planned for the years 2011 and 2012.

In this respect, I would also argue that the wider region of Eastern Europe is yet another context with respect to which gender studies should establish a culturally and politically sensitive approach that will inform the field in new and innovative ways, both in terms of epistemology and of contents. Eastern European scholarship, if contextualized, has the unique opportunity to examine the gendered political realities of contemporary neoliberal democracies with reference to the relatively recent communist past. By way of reversing and adopting the perspective of memory, by way of assuming the narrative position of remembering (or re-membering) and of comparatively (to the past) and critically exploring the political realities of today, an entirely new light of interpretation can be shed on both the past and the present. One of the central topics that the entire region of postcommunist Eastern Europe shares is the communist heritage of the pronounced promotion of gender equality in terms of universalism, or rather, universal humanism. The universalism inherited from the communist period—even in the form of strongly promoting gender equality—has been widely discussed by gender scholars in Eastern Europe, who have insisted on gender sensitivity rather than universalist gender-neutral equality. It will be productive to revisit the values of universalism from a perspective that is now informed with the critique of universalist discourse, and to study comparatively the two forms of universalist discourse: the communist one and the Western one in the second half of the twentieth century, as well as in the first decade of the twenty-first century.

The latter issue is my contribution to the debate on the relevance of the great potential for innovative thinking that the effort of building scholarship around this question entails. I am not proposing that the efforts to contextualize be taken as mere sensitivity to local needs. On the contrary, their theoretical innovativeness will inevitably contribute to international gender studies scholarship, which will in turn contribute to gender studies from the region becoming active co-producers of new theoretical concepts instead of being mere "importers," marginal and hence invisible in the global gender studies scene.

◊ Notes

1. Although aware that I am generalizing, I cannot account for every single person who is part of the feminist academic scene.

2. Jelisaveta Blagojevic, Katerina Kolozova, and Svetlana Slapsaks, ed., *Gender and Identity: Theories from and/or on Southeastern Europe* (Belgrade and Skopje: Athena Network Publishing for the Center for Women's Studies and Gender Research-Belgrade and the Research Center for Gender Studies-Skopje, 2006).

Women's and Gender Studies in Turkey
From Developmentalist Modernist to Critical Feminist
Serpil Sancar and Elif Ekin Akşit

The long history of the feminist movement in Turkey has been rediscovered as recently as the 1990s, as a result of the collaboration between academic and political feminisms in Turkey.[1] The legacy of the Ottoman feminists was better understood not only after their strong presence in the nineteenth century had been unearthed, but also when their disappearance after the 1930s was acknowledged. This disappearance was closely related to the alleged solution of the "woman question" in Turkey, similar to what happened in the Soviet Union.[2] Political feminism with its Islamic, socialist, and nationalist tints and with a variety of ethnic roots went underground in Turkey during the single-party years that lasted until 1950.[3] When feminism arose again, in its academic and political forms, it was mostly nourished from a socialist thread. This contribution relates how the multiple forms of feminism that once existed in the Ottoman Empire fully came back in the 1990s with more enriched forms of networks.

The introduction of women's studies in academic life in Turkey can be traced back to the early 1950s within the classical social science disciplines. In this period, the political and academic goals of economic development, national integration, and modernization of the family and society legitimized the feminist entry into the social sciences. Feminists have since been actively writing on topics like family sociology, women's role in history, women's participation in the labor market, women's social role in societal development, reproductive health, and female psychiatry.

Especially during the late 1960s, undergraduate and graduate courses that included women's issues became visible in the curricula of various universities. In this period, significant progress in the field of social sciences was achieved. For the first time, scholars in the social sciences became sensitive to gender-based issues while collecting and interpreting social data, conducting research, and creating statistical files.[4] As a result, knowledge about the education, reproduction, and public activities of the female population was accumulated within certain academic circles and public institutions.[5] The role of women in the transition from traditional extended families to modern nuclear family types and the reduction of birth rates were central topics in this process of knowledge formation.

By the end of the 1970s, the status of women was addressed by feminists (such as Nermin Abadan Unat) within the nationalist paradigm.[6] Although these feminists criticized the gender blindness of the modernization paradigm in the nation-state formation processes, the dominant nation-building policies aiming at integrating remote areas and social groups into the "center" went well together with a sort of modernist "state feminism." These joint efforts produced certain gender equality policies, but the masculine aspects of nationalism, authoritarianism, and militarism remained untouched by this first wave of the institutionalization of women's studies in Turkey.

The second wave emerged at the end of the 1980s and the beginning of the 1990s and came into being with a more integrated relation between academic and political forms of feminism and a critical stance toward masculine domination within modern family- and state systems. Turkish feminism of the 1980s was transformatory and produced new devoted activists from different parts of the society, despite the on-going military regime that crudely shaped the 1980s. As a consequence of this femi-nist movement, new associations and institutions were founded: autonomous groups for political initiatives, women's shelters, libraries, centers, publications, journals and newspapers. Former activists of the feminist movement also founded these new feminist institutions and initiatives. These feminists fulfilled the need to create new epistemological and theoretical paths for knowledge production and policy-oriented studies. Thus, women's studies and feminist aims were integrated. Joint efforts of former feminist activists and feminist scholars led to the establishment of new research centers and produced new curricula, publications, and programs. These were usually called women's studies, "gender" being used only in the English versions of the Turkish establishments or in programs in English, such as the Women's and Gender Studies Program at the Middle East Technical University (METU). Gradually a critical feminist perspective with an interdisciplinary approach became part of the administrative and institutional academic structure. Scholars in the interstices of politics and academia, such as Şirin Tekeli and Aksu Bora, produced collective books that described and empowered this transformation while it was happening.[7]

This inclusion of women's and gender studies in academia in terms of feminist theory, methodology and critiques of male domination, important as it is, was able to survive only in main universities in metropolitan areas such as Istanbul, Ankara and Izmir. Critical feminist studies that go beyond the older developmentalist and modernist perspectives are still hard to find elsewhere. In the last two decades, only four universities have successfully established special women's and gender studies MA programs: two in Ankara (Ankara University, 1993 and METU, 1994), one in Is-tanbul (Istanbul University, 1993), and one in Izmir (Ege University, 1999), which has been less visible in the last years. All these programs were established collectively by feminist academics. The coursework has included theoretical classes as well as meth-odology, yet it is mostly structured in subdisciplinary classes that are designed to rep-resent the instructors' home disciplines. Although the students come from a variety of backgrounds, being more academic in METU and more activist in Ankara and Istan-bul Universities, classes about activism are still hard to find.

The only Turkish PhD program in gender studies was established recently within the KASAUM—Ankara University Center for Women's Studies—and will have its first students in the academic year 2010–2011. This program is also the first one to be named "Gender Studies" in Turkish. The KASAUM has also been the first to publish— since the beginning of 2009—its own periodical, called *Fe Journal* (Journal of feminist critique).[8] This journal aims to achieve interdisciplinarity within the social sciences, as well as to incorporate science and technology studies, and hard-core sciences like medicine, into feminist studies, as its second issue shows.[9] *Fe Journal* is nourished by the Ankara University Women's Studies Center's position in-between academia and politics, and also underlines the transnational connections that both academic and political feminisms have built on.

All the above-mentioned women's studies programs and centers have produced a great number of MA theses in different social science disciplines that transform our understanding of current gender regimes. A number of undergraduate courses were established at different universities within the humanities and social sciences (media studies, cultural studies, economic and social policies, law and jurisdiction, family issues, divinity studies, history, political studies, literature studies). However, under-graduate or graduate courses related to gender issues are still uncommon within med-icine, engineering, architecture, and the natural sciences.

The substantial transformation achieved so far is a result of academic and political activism. Currently, feminist scholars are embedded in social movements for equality, freedom, and human rights as well as respective civil resistance practices against all kinds of domination. Within academia, they have to work extra hours as teaching in women's studies programs is voluntary work for them but they feel rewarded as their contribution transforms the academic context. Although women's and gender studies programs remain fragile and unsteady, feminist scholars are among the most produc-tive in their respective departments. They are well represented in the social sciences division of the Turkish Academy of Sciences with people like the psychologist Çiğdem Kağıtçıbaşı and the historian İsenbike Togan. This success in turn nourishes interest in gender studies and research, and the number of students enrolled in such programs increases.

An example of how feminist research has been pioneering in transforming the es-tablished ways of knowing in Turkey is the field of oral history. Oral history research methods have been primarily used by feminist scholars and have added new layers to the concepts of remembering and forgetting.[10] The rewriting of Turkish history from a gendered perspective has not only resulted in a new kind of scholarship that does more than simply add women's experiences to men's, but it has also produced a unique cri-tique of nationalism and militarism.[11] Turkish feminist scholars have also contributed greatly to studies of nationalism by underlining the role of the household in national-ism and by emphasizing the historical transformations of domestic space.[12]

Moreover, the works that stem from women's studies' research and education lead to direct transformations of women's lives. Despite the fact that women's and gender studies departments have become a part of elite academic power structures, these de-partments have gradually started to challenge the elitist neglect of the lower classes, grassroots, and women. Feminist studies in academia recognize and represent wom-en's experiences as the subject of "scientific" knowledge. Accordingly, some women's studies scholars and centers, such as the Women's Studies Center in Ankara University, have attentively followed the activities of women's organizations and worked in close cooperation with them. Projects for the training and empowerment of women—with the participation of feminist activists, feminist scholars, and women who for example were victims of domestic violence or gender discrimination, or who are part of ex-cluded social groups—have produced a new model for acting together. Most of these projects focus on important problems such as preventing sexual violence, promoting more positive conditions for increasing women's political participation, and supporting women's productive activities. As a result, an increasing number of publications by feminist activists deal with evaluations of women's own perspectives toward the ongoing transformation of gender inequalities.

Feminists have also proven influential in the legal arena. Transformations have been achieved in the Civil and Penal Laws, and a law for the protection of family has been approved as a result of collaboration between feminist scholars and feminist politicians.[13] Men and women are now considered equal in marriage. Furthermore, domestic violence has become an acknowledged crime. Finally, "honor crimes" are no longer downplayed with reference to so-called mitigating circumstances.[14] Gender mainstreaming has also been successful in altering Article 10 of the Constitution in 2004. According to this change, now "the state is obliged to achieve gender equality,"[15] even though the measures that will lead to this achievement still require further efforts on the part of the state.[16]

Thanks to all these academic, project-based, and legal developments, women's knowledge and experience are becoming more visible in Turkey. Women's and gender studies departments, centers, and programs have greatly contributed to this visibility, but much remains to be done. Furthermore, the visibility of women's studies departments and centers themselves is still questionable in Turkey. The fact that women's studies is not limited to certain aspects of women's lives, but can provide insights into all fields of knowledge is being accepted slowly within Turkish academic circles, mostly as a result of the academic achievements of feminist scholars. Yet many prominent social scientists in Turkey still have no interest in or knowledge about the field. At the same time, the very wide scope of women's studies and the interdisciplinary work within the social sciences promise to eliminate gender blindness in academic circles and to increase the receptivity to women's studies. Turkish feminist scholars—like their colleagues elsewhere—have radically questioned the claim that science is a neutral and objective field. The result of these challenges is that academic circles in Turkey are now more open to new types of knowledge.

An example of this development is an event that recently took place at one of the oldest departments in Turkey—the School of Government at Ankara University. Judith Butler gave a speech there about the bonding capacity of being queer in a very crowded conference hall; a result of cooperation among the Human Rights Center of the School of Government, the Women's Studies Center, and KAOS-GL, one of the most active nongovernmental organizations that struggle against the criminalization of gays and lesbians.[17] With a women's studies program at postgraduate levels, a separately functioning women's studies center, and a refereed international journal, the faculty there is the first in Turkey to create a gender studies section within the Department of Political Science and Public Administration.

As a result of these developments, the women's studies division provides an openness to different political and academic formations that include but also go beyond feminist goals and ideas—like LGBT (lesbian, Gay, bisexual, transgender), Kurdish, and religious activism, paving the way to the kind of collective resistance as proposed by Butler. The conceptualization of women's rights issues as identity politics—as most clearly seen in Turkey in the position of the feminist movement within the head-scarf debate—has evolved into an effort to find common ground in order to act in solidarity.

Feminists have sided with Muslim feminists in this debate to redefine the "public sphere" for women in Turkey. The feminist-LGBT coalition and the feminist support for Kurdish demands in Turkey also strengthen such solidarity, presenting solutions to the problems of religious, ethnic, or "race"-based discrimination. A feminist political approach aimed at creating a common agenda—which might be called a "common agenda politics" in contrast to the previous "identity politics"—promises to gain more ground and power in the near future, in particular if the Kurdish, LGBT, and Islamist movements manage to go beyond their own respective identity politics. As a result, the Turkish experience with women's and gender studies promises to present models of coexistence for Europe and the Middle East through the collaboration of academic feminism and the broader feminist movement in Turkey.

◊ Notes

1. Aynur Demirdirek, an active participant in the women's movement published *Osmanlı Kadınlarının Hayat Hakkı Arayışının Hikayesi* (The story of Ottoman women's search for the right to life) (Ankara: İmge Kitabevi, 1993). For an English version, see Aynur Demirdirek, "In pursuit of the Ottoman Women's Movement," in *Deconstructing Images of "The Turkish Woman,"* ed. Zehra F. Arat (New York: Palgrave, 1999). See also Serpil Çakır, *Osmanlı Kadın Hareketi* (The Ottoman women's movement) (Istanbul: Metis Yayınları, 1994).

2. Zafer Toprak, "Halk Fırkasından Önce Kurulan Parti: Kadınlar Halk Fırkası" (The women's republican party), *Tarih ve Toplum* (History and Society) 51 (1988): 30–31; Charlotte Weber,"Between Nationalism and Feminism: The Eastern Women's Congresses of 1930 and 1932," *Journal of Middle East Women's Studies* 4, no. 1 (2008): 83–106; Yaprak Zihnioğlu, *Kadınsız İnkılap* (The revolution without women) (Istanbul: Metis Yayınları, 2003); Elif Ekin Akşit "Latife'nin bir Jesti: Doğu ve Batı Feminizmleri ve Devrimle İlişkileri" (Eastern and Western feminisms and their relation to revolutions), in *Jön Türk Devriminin Yuzuncu Yıl* (The centennial of the young Turk revolution), ed. Sina Akşin, Barış Ünlü and Sarp Balcı (Istanbul: İş Bankası Yayınları, 2010), 235–264.

3. Elif Ekin Akşit, "Osmanlı Feminizmi, Uluslararası Feminizm ve Doğu Kadınları" (Ottoman feminism, transnational feminisms and eastern women), *Doğudan* (From the East) 7 (2008): 84–91, also at http://80.251.40.59/politics.ankara.edu.tr/aksit/dogudan.html (accessed 1 August 2010).

4. For a former discussion of this history, see Serpil Sancar, "Women's Studies in Turkish Academic Life," in *Gender Studies: Trends/Tensions in Greece and other European Universities*, ed. Theodossia-Soula Pavlidou (Thessaloniki: Zitis, 2006); Serpil Sancar, "Üniversitede Feminizm? Bağlam, Gündem ve Olanaklar" (Feminism in the university? The context, the agenda, and the possibilities), *Toplum ve Bilim* (Society and Science) 97 (2003): 164–182.

5. For official data related to women, see www.die.gov.tr and www.kssgm.gov.tr; (accessed 1 August 2010).

6. See, e.g., Nermin Abadan Unat, *Türk Toplumunda Kadın* (Women in Turkish society) (Ankara: Ekin Yayınları ve Türk Sosyal Bilimler Dernegi, 1979).

7. Şirin Tekeli, ed., *Kadın Bakış Açısından 1980'ler Türkiye'sinde Kadın* (Women in Turkey in the 1980s from women's point of view) (Istanbul: İletişim Yayınları, 1990); Aksu Bora and Asena Günal, eds., *90'larda Türkiye'de Feminizm* (Feminism in Turkey in the 1990s) (Istanbul: İletişim Yayınları, 2002); Aynur İlyasoğlu and Necla Akgökçe, eds., *Yerli Bir Feminizme Doğru* (Toward a local feminism) (Istanbul: Sel Yayıncılık, 2001).

8. For more information, see http://cins.ankara.edu.tr/cv.html (accessed 1 August 2010).

9. For more information, see http://cins.ankara.edu.tr/20092.html (accessed 1 August 2010).

10. Eser Köker, Mine Tan, and Özlem Şahin, *Anneanne: Sırlarını Eskitmiş Aynalar* (Grand-mothers: Mirrors with worn-out glazes/secrets) (Istanbul: Çiviyazıları, 2002). The word *sır* means both glaze and secret in Turkish.

11. Nükhet Sirman, "Writing the Usual Love Story: The Fashioning of Conjugal and National Subjects in Turkey," in *Gender, Agency, and Change: Anthropological Perspectives,* ed. V.A. Goddard (London: Routledge, 2000), 202–218; Ayşe Gül Altınay, *The Myth of the Military-Nation: Militarism, Gender, and Education in Turkey* (New York: Palgrave Macmillan, 2004).

12. Gülsüm Baydar, "Tenuous Boundaries: Women, Domesticity and Nationhood in 1930s Turkey," *Journal of Architecture* 7 no. 3 (2002): 229–244; Ferhunde Özbay, "Gendered Space: A New Look at Turkish Modernisation," *Gender & History* 11, no. 3 (1999): 555–568.

13. The Penal Code no. 5237 was accepted on 26 September 2004 and came into force on 1 June 2005. The Civil Code no. 4721 was accepted on 22 November 2001 and came into force 1 January 2002. The Law for the Protection of the Family no. 4320 was accepted on 14 January 1998.

14. Dicle Koğacıoğlu, "The Tradition Effect: Framing Honor Crimes in Turkey," *Differences* 15, no. 2 (2004): 118–151; Ayse Parla, "The 'Honor' of the State: Virginity Examinations in Turkey," *Feminist Studies* 27, no. 1 (2001): 65–88.

15. Türkiye Cumhuriyeti Anayasası, Genel Esaslar, 10. Madde, "Kanun önünde Eşitlik," 18 October 1982 (The Constitution of the Republic of Turkey, general principles, Article 10 "Equality before the law," www.mevzuat.adalet.gov/tr/heml/1113.html.

16. See Serap Yazıcı, "Eşitlik ilkesinin değişen dinamiği ve kadın hakları" (The changing meaning of the principle of equality and women's rights), in *Feminizm ve Hukuk-Hayat, Adalet ve Kadın-Kadın ve Hukuk* (Feminism and law-life, justice, and women-women and law), vol. 3 (Ankara: Ankara Barosu Yayını, 2006); and Ece Göztepe, "Justitia'nın gözbağının arasından görülenler: Anayasal eşitlik ilkesinin devletin aktif yasama faaliyetleriyle hayata geçirilmesinin anlamı" (The meaning of the actualization of the constitutional principle of equality by the law-making actions of the state), in ibid.

17. Usually, most NGOs—including women's organizations—depend mainly on resources from state ministries. Simel Esim and Dilek Cindoglu "Women's Organizations in 1990s Turkey: Predicaments and Prospects," *Middle Eastern Studies* 35, no. 1 (1999): 182.

❖

"The City of Gender Studies" in Central, Eastern, and Southeastern Europe: Concluding Remarks

Krassimira Daskalova

The Forum contributions published in this and the previous issue of *Aspasia* about women's and gender studies in Central, Eastern and Southeastern Europe (CESEE) show that, although women's and gender studies in the region are far from being success stories, the number of individuals, research projects, courses, and university programs dealing with women and gender relations has increased steadily since the fall of state socialism in 1989.[1] The history of official academic recognition of women's and gender studies in the region went hand in hand with the development of civil society and the political actions of various women's and feminist nongovernmental organizations (NGOs).

The contributions to the *Aspasia* 4 and *Aspasia* 5 Forums present the different paths of women's and gender studies in fourteen CESEE countries: Hungary, Poland, Romania, Serbia, and Ukraine (published in *Aspasia* 4) and Bosnia and Herzegovina, Bulgaria, Czech Republic, Estonia, Greece, Latvia, Lithuania, Macedonia, and Turkey (*Aspasia* 5). Some of these contributions are more analytical, while others offer primarily descriptive accounts. All of them provide a good basis for comparing the research, publications, institutionalization, and teaching related to women and gender in CESEE, and together they make an extremely useful addition to other recent publications on this subject, outlining in a convincing way both the main strengths and reasons for ongoing concern.[2]

The Meaning and Power of Naming

Almost everywhere in the region the centers and programs related to the field are called "gender studies." The few exceptions mentioned in the Forum texts are in Serbia, Croatia, and Turkey. The independent Center for the production of feminist knowledge in Belgrade is called Women's Studies Center and the name of the similar institution in Croatia is Center for Women's Studies Zagreb, while in the Turkish case "gender" has been used only in the English translation/version of the programs or when the courses were taught in English. Interestingly enough, the major research centers, networks, and publications in Western Europe use the terms women's studies and gender studies interchangeably. For example, in the cases of the publications of ATHENA (Advanced Thematic Socrates Network), established in 1996 and affiliated with Utrecht University in the Netherlands; and the recently established—on 30 September 2009—professional European Association for Gender Research, Education and Documentation (ATGENDER). Unfortunately, only a few of the articles published in *Aspasia* briefly address the very important issue of naming, in comparison to some recent publications in the West.[3]

Keeping in mind the differences—based on the diverse historical and social contexts—in the developments of all the above-mentioned countries, we should be able to see and outline the similarities in the processes of social restructuring and developments that led to new social inequalities and renewed gender contracts after 1989. Step by step, everywhere in the East European context, both academic women and NGO activists as well as policymakers became aware of the necessity to include the gender dimension in all analyses of contemporary (but also of past) social realities. An important common feature of the gendered transitions in CESEE was that antifeminism was present in the local contexts well before the beginning of any serious gender-related discussions. In addition, all East European societies after 1989 experienced a backlash against feminism before they could develop feminist thinking, and nowhere did an influential women's movement arise.[4] Even today, more than twenty years later, the number of articulate and publicly visible feminists is so small that the few East European feminists still play "Jane of All Trades," as the historian Maria Bucur put it.[5] Many participants in the *Aspasia* Forum argue that the weak feminist conscious-ness, the negativism toward feminists and their actions, and the lack of knowledge about historical feminisms in CESEE, are among the reasons why women's and gender studies here are relative latecomers in the global "city of gender studies."[6] Another reason for the late institutionalization of women's and gender studies in the region (as suggested for example by Theodossia-Soula Pavlidou regarding the Greek case) is that the national governments of the countries in the region were not interested in financing gender research and teaching. No wonder, then, that the first Greek gender studies program was established thanks to an outside push, namely within the EU Third Community Support Framework. Since the beginning of the new millennium, in many East European countries a kind of "room-service feminism" (a concept coined by Mihaela Miroiu) of *acquis communautaire* was imposed by Brussels on its future members. Paradoxically, this (state) feminism, which helped institutionalize women's and gender studies in CESEE (like in the West), also made feminists look like "do-mesticated, 'reasonable' creatures in the new, bearable, innocent *gynaeceum* of gender studies."[7] But no matter how "domesticated" these feminist actions might seem—even going beyond the "normalization" of gender studies[8]—their combination with the ac-tions of local, autonomous feminists led to the introduction throughout the region of various gender sensitive laws aimed at guaranteeing equal opportunities for women and men, stopping domestic violence, the democratization of artificial reproduction, the gender sensitization of East European societies in general, and the establishment of women's and gender studies programs of different kinds. However, as many of the participants in the Forum underline in their contributions, there was an inadequate commitment to gender equality among policy designers in CESEE during the EU ac-cession process. But again: while the enactment of equality policy is useful, with the equalization policy gender studies programs "run the risk of losing [their] potentially subversive power before really entering academia and ending up as a fully integrated component of the academic establishment."[9]

The Strategic Nexus between Academia and Activism

Present within the specific conditions of the former Yugoslav Federation well before 1989, the nexus between academia and activism has produced common research and civic initiatives and helped tone down the negativism toward feminism and gender issues in many CESEE societies; it started to turn feminist knowledge into everyday social practice (see Marina Blagojević's contribution in *Aspasia* 4). Already in the 1980s, but predominantly in the 1990s and the early 2000s, this nexus resulted in the infusion of a gender approach into traditional university education and research. The collabo-ration between academia and women's movements everywhere in CESEE also helped make a greater positive impact by sensitizing policymakers and state legislation con-cerning women and gender relations. This nexus was also the driving force behind the establishment of organizations such as FILIA in Romania and the Belgrade Women's

Studies Center. It produced various common gender sensitive actions and discussions with the participation of both academic women and various women's NGOs (such as those initiated jointly by women academics from Sofia University and the Center for Women's Studies and Politics in Sofia). In addition, the already existing alliance between academic gender studies and queer studies in the region corresponds to a similar alliance between feminist activism and LGBTQ activism.

Autonomy versus Integration

The institutionalization of women's and gender studies in CESEE followed different patterns. In most countries (Bulgaria, Estonia, Lithuania, Poland, Romania, Serbia, Turkey, Ukraine) the institutionalization started with the infusion of traditional disciplines and courses with gender-sensitive issues. After that, in some of the countries, separate gender studies centers were founded within the university structures, while elsewhere such independent/autonomous centers were established outside academia (Serbia and Croatia) as the first step toward a more thorough future integration of the field within local university educational contexts. Today, more than twenty years after the beginning of the transition, it seems that the integrated approach prevails in most of the countries discussed here. No matter how much progress has been made in establishing the discipline in CESEE, there are still countries without any BA, MA, or PhD degree programs in gender studies (like Estonia and Poland) and where gender studies are "homeless," i.e. have not been officially recognized as an academic discipline. But as Agnieszka Graff has shown in her contribution about Poland in *Aspasia* 4, this homelessness can be regarded as both a weakness and strength.

Interdisciplinarity and Internationalization

Interdisciplinarity and internationalization are two common features of all the women's and gender studies programs and centers in the region. Although participants in the Forum are well-aware that the transmission of ideas cannot be automatic and direct and that gender studies in CESEE have to be contextualized both in terms of epistemology and content, many of them emphasize the significance of international networks, regional initiatives and the transfer of (feminist) knowledge. Of particular importance in this respect were the translations of both foundational, classical feminist texts by authors like Virginia Woolf and Simone de Beauvoir, and works by contemporary Western feminist scholars such as Judith Butler, Julia Kristeva, Joan Scott, Elaine Showalter, Lillian Robinson, Andrea Dworkin, and Catherine MacKinnon, to name but a few.

For the promotion of gender studies in the countries of CESEE, the initiatives from and research support by the Central European University (CEU) in Budapest were especially important. Many of the Forum participants have been related in one way or another to the Gender Studies Department of the Central European University: as students, professors, fellows, or participants in the various activities of the Department.

For the Baltic countries, similar support also came from the Christina Institute of the University of Helsinki, Finland, and the Nordic Gender Institute, which organized a number of summer schools and graduate seminars held in different Nordic countries. Given the lack of any institutional (state) support, in some of the CESEE countries the international networks and personal commitment of the pioneers of gender studies were crucial factors in building up the field.

Apart from the positive examples of international exchange and collaboration, however, gender studies specialists in the region are well aware of the unequal position of East European scholars vis-à-vis their Western colleagues in terms of access to specialized literature in the field, international fellowships, research grants, and participation in conferences and other influential and prestigious scholarly events, not to mention the difference in salaries.

Gender Studies Bookshelves: Periodicals and Other Publications[10]

Not surprising, the ordinary women's magazines everywhere in the region appear to be indifferent (or "allergic," as Mihaela Miriou has put it) to feminism and gender issues.[11] Feminists and gender researchers everywhere in CESEE tried to establish their own periodicals to publish texts countering the misogyny and sexism prevalent in popular magazines and popular culture in general.[12] Moreover, with the help of feminist periodicals and publishing houses, various classical and contemporary feminist authors entered academic teaching and influenced scholarly research on women and gender relations in the region.

Some of these periodicals have turned into the most successful and long-lasting East European projects in gender studies. These include the Greek feminist journal *Dini* (Whirlpool), published since 1986; the Serbian feminist journals *Feminističke sveske* (The feminist notebooks), *Ženske studije* (Women's studies, published since 1995), renamed *Genero* in 2002, and *ProFemina;* the Romanian journal *AnALize* (in existence since 1997); the Ukrainian biannual *Gendernye issledovaniia* (Gender studies), published since 1998 by the Kharkiv Center of Gender Studies; the Croatian Women's Studies Journal *TREĆA* (The Third), published since 1998; and the Estonian *Ariadne Lõng* (Ariadne's thread), published since 2000.

In other countries discussed here, however, gender studies periodicals were short-lived. This was the case with the Latvian newspaper *Zhenshchinii Baltii* (Baltic women), published in the 1990s; the Lithuanian journals *Feminizmas, visuomenė, kultūra* (Feminism, society, culture), published from 1999 to 2002; electronic journal *Klėja* (2002–2003); and the two Bulgarian journals *P.S.* (Post Scriptum), which started in 2000 and existed only a couple of years, and *Altera* (2005–2008). Hopefully, the newest Lithuanian journal in the field, *Lyčių studijos ir tyrimai* (Gender studies and research) and the Turkish interdisciplinary *Fe Journal*, published since the beginning of 2009, will last longer.[13] Although in some of the CESEE countries there are no academic periodicals dealing specifically with the study of gender, many other journals there have published both special issues and sections on women's and gender topics, thus proving the growing interest in this subject in CESEE after 1989 (and in some cases even before that).

Several publishing houses appeared since 1989, which firmly support gender studies series and separate publications, among them the Serbian feminist publisher *94*, *Osnovy* (Foundations) in Ukraine, *Polirom* in Romania, and *Polis* in Bulgaria.

Conclusion

Although Centers for Gender Studies or informal groups and initiatives dealing with gender issues and interdisciplinary gender-oriented courses exist in many universities in the countries throughout the region, compared to the situation in Western countries, the weakness of gender studies and feminist knowledge production in the countries of CESEE is beyond doubt. The main reasons for this relative weakness are the lack of state financial support and understanding in the retraditionalized context of post-1989 CESEE, the related lack of enough university professors who are prepared to teach in this highly disadvantageous economic situation, the lack of libraries[14] and lack of sufficient teaching materials in the local languages, and the lack of opportunities for professional realization. Optimism for the future comes from the fact that everywhere in CESEE there is a new generation of scholars, who graduated from various MA gender studies programs, such as the Central European University in Budapest, the European Humanities University (Minsk-Vilnius), and various West European and US and Canadian universities. Their energy and knowledge will enhance the future "city of (East European) gender studies."

One interesting and so far not enough explored direction of research would be to gather oral histories about the paths to academia taken by pioneers of the gender and feminist perspective in the various academic disciplines and countries in Central, Eastern, and Southeastern Europe, stories about the process of creating—as Mihaela Miroiu named it—"A Mind of Our Own."[15] These stories would outline the experiences of various generations of university women and their struggles for professional citizenship, both in becoming professionals on formal equal terms and in their ongoing questioning of male domination in knowledge production, reproduction, and the management of scholarly institutions and life. In the CESEE contexts, where gender expertise has not been considered equally important to other kinds of scholarly expertise, these stories will show that existential and professional loneliness experienced by some of the pioneers was a result of their engagement with feminist perspectives and gendered knowledge.

◊ Notes

1. See the Forum "The Birth of a Field. Women's and Gender Studies in Central, Eastern and Southeastern Europe," ed. Krassimira Daskalova, *Aspasia* 4 (2010): 155–205.

2. Theodossia-Soula Pavlidou, ed., *Gender Studies: Trends/Tensions in Greece and other European Countries* (Thessaloniki: Zitis, 2006). Information about Thessaloniki's Gender Studies Center is available at http://www.phil.auth.gr/gender (accessed 1 August 2010). See also Marlen Bidwell-Steiner and Karin S. Wozonig, ed., *A Canon of Our Own? Kanonkritik und Kanonbildung in den Gender Studies* (A Canon of Our Own? Canon Criticism and Canon Building in Gender Studies) (Innsbruck: Studien Verlag, 2006); Susan Zimmermann, "The Institutionalization of

Women and Gender Studies in Higher Education in Central and Eastern Europe and the Former Soviet Union: Asymmetric Politics and the Regional-Transnational Configuration," *East-Central Europe/L'Europe du Centre-Est: Eine wissenschaftliche Zeitschrift* 34–35, nos. 1–2 (2007/2008): 131–160 (Published within the thematic issue: "Social History in East Central Europe"). See also the contributions dealing specifically with Eastern European developments, such as the one by Eva Fodor and Eszter Varsa, "At the Crossroads of East and West: Gender Studies in Hungary," in *Global Persepctives on Gender Research: Transnational Perspectives,* ed. Christine Bose and Minjeong Kim (New York and London: Routledge, 2009). Although the multivolume edition *The Making of European Women's Studies,* published by ATHENA, Utrecht University, includes some East European cases, their number is hardly enough to justify the all-Europe encompassing title. Actually most available publications about European gender studies are Weastern-centered or Western-based. This lack of sensitivity toward Western biases and Western-centeredness could be pointed out as one of the major characteristics of West European Gender Studies.

3. For an example of a Western publication, see Nina Lykke, "Women's/Gender/Feminist Studies—a Post-disciplinary Discipline?" in *The Making of the European Women's Studies,* ed. Rosi Braidotti, Edyta Just, and Marlise Mensink, vol. 5 (Athena, Utrecht University, 2004), 91–101.

4. In this I differ from for example Jiřina Šmejkalová, in whose view "anti-feminism *arrived* in the local context long before …" (emphasis added). See Jiřina Šmejkalová, "Gender as an Analytical Category of Post-Communist Studies," in *Gender in Transition in Eastern and Central Europe. Proceedings,* ed. Gabriele Jähnert et al. (Berlin: Trafo Verlag, 2001), 49–56, quote on 54. In my view, antifeminism has been part of the modern history of most (including East) European societies during the twentieth century, and early twentieth-century Marxist socialism and the later state-socialist ideology are to blame to a large extent for the negative image of European feminisms and feminists. We can therefore only speak about renewed eruptions of antifeminist gestures and parlance in the 1980s and later. About the role of socialism in the construction of the notion of "bourgeois feminism" and the long history of the latter's stigmatization, see Marilyn J. Boxer, "Rethinking the Socialist Construction and International Career of the Concept 'Bourgeois Feminism'," *American Historical Review* 112, no. 1 (2007): 1–28.

5. Mihaela Miroiu, "A Mind of Our Own. Gender Studies in Romania," *Aspasia* 4 (2010): 157–167, esp. 161.

6. Here I refer to the interview Kornelia Slavova and I conducted with the well-known American feminist scholar Catherine Stimpson during her short visit to Sofia in 2006, published in the Bulgarian cultural weekly *Kultura* (Culture). See Krassimira Daskalova and Kornelia Slavova, "Za 'Gradut na Gender Studies'" (On the "City of Gender Studies"). Interview with Prof. Catherine Stimpson," *Kultura* no. 25 (21 June 2002), 12.

7. Miroiu, "A Mind of Our Own. Gender Studies in Romania," 157–167.

8. Marina Blagojević, "Feminist Knowledge and the Women's Movement in Serbia. A Strategic Alliance," *Aspasia* 4 (2010): 184–197.

9. Theodossia-Soula Pavlidou, "Gender Studies at Greek Universities," this volume, [00-00]

10. I took the idea for the title of this section from Agnieszka Graff's piece "Gender Studies in Poland: A View from Within," *Aspasia* 4 (2010): 167–176, esp. 171.

11. Miroiu, "A Mind of Our Own. Gender Studies in Romania," 163.

12. About the Serbian case of misogyny, see the volumes by Marina Blagojević, ed., *Mapiranje mizoginije u Srbiji: diskursi i prakse.* (Mapping the misogyny in Serbia: discourses and practices), 2 vols. (Belgrade: AŽIN, 2000, 2004). See also the review of these two volumes by Tatyana Kmetova in *Aspasia* vol. 2 (2008): 227–229. About the sexism and misogyny in contemporary Bulgarian popular culture, see Milena Kirova and Kornelia Slavova, eds., *Identichnosti v prehod: rod, medii i populiarna kultura v Bulgaria sled 1989 g.* (Gender identities in transition: Media

and popular culture in Bulgaria after 1989) (Sofia: Polis, 2010). A review of this book by Nade-zhda Alexandrova is included in this volume, see 204–210 and especially 207–210.

13. None of these CESEE women's and gender studies journals is included in the WWW Virtual Library, http://www.iisg.nl/w3vlwomenshistory/journals.html#0402 (accessed 1 August 2010).

14. See the useful remarks of Annamaria Tagliavini, "Behind Every Successful Women's Studies Programme There Is a Specialised Women's Library: On Intra-European Cooperation of Women's Libraries and Documentation Centres," in *The Making of European Women's Studies* 4 (2003).

15. See Miroiu, "A Mind of Our Own. Gender Studies in Romania"; for Western examples, see Karla Erickson, Jennifer Pierce, and Hokulani Aikau, eds., *Feminist Generations: Life Narratives from the Academy* (Minneapolis: University of Minnesota Press, 2007).

About the Forum Authors

Elif Ekin Akşit teaches Ottoman History, Women's History, and Turkish History at the School of Government, Ankara University. She is among the first alumnae of Women's Studies in Turkey—she graduated from the Middle East Technical University—and she has been affiliated with the Women's Studies Center and Program in Ankara University since 2000. Her recent publications include "Politics of Decay and Spatial Resistance," *Social & Cultural Geography* 11, no. 4 (2010): 343–357; and "Fatma Aliye's Stories: Ottoman Marriages beyond the Harem," *Journal of Family History* 35, no. 3 (July 2010): 207–218. Her current research focuses on women's education in the Ottoman Empire and Turkey as well as the place of nationalism in the politics of education in Turkish history and the urban implications of nationalist practices. Email: elifaksit@ yahoo.com

Krassimira Daskalova, PhD, is a professor of cultural history at the Faculty of Philosophy, St. Kliment Ohridski University of Sofia, Bulgaria. She is the author of two books on cultural history and editor or co-editor of thirteen volumes in the fields of women's and gender history and the history of the book and reading. These edited volumes include: with Francisca de Haan and Anna Loutfi, *A Biographical Dictionary of Women's Movements and Feminisms. Central, Eastern, and South Eastern Europe, 19th and 20th Centuries* (Budapest and New York: Central European University Press, 2006); with Sirkku K. Hellsten and Anne Maria Holli, *Women's Citizenship and Political Rights* (London: Palgrave Macmillan, 2006); with Ute Gerhardt, *Ost – West Feminismen. A Special Issue of L'Homme. Europäische Zeitschrift für Feministische Geschichtswissenschaft* 16. Jg. Heft 1, 2005. She has also published—in Bulgarian, English, German, French, Russian, Romanian, and Serbian—more than 40 book chapters and articles in the above mentioned fields. She is editor and book review editor of *ASPASIA: The International Yearbook of Central, Eastern and Southeastern European Women's and Gender History.* Email: krasi@ sclg.uni-sofia.bg

Hana Hašková received her PhD in sociology from Charles University in Prague. She is a researcher at the Department of Gender and Sociology, Institute of Sociology, Academy of Sciences of the Czech Republic, and a lecturer at Charles University. Her research focus is on the gender aspects of reproduction, intimate lives, family policies, and the concept of care. Hašková also studies women's civic organizations in postsocialist Europe. She led several research projects on childbirth practices and childlessness and coordinated the Czech research team on an international project on Enlargement, Gender, and Governance (EU FP 5). Currently she leads a project on changes in partnership and family forms from the life course perspective, and the Czech research team in the international project "Gendered Citizenship in Multicultural Europe: The Impact of Contemporary Women's Movements" (EU FP 6). In 2010, she received the Otto Wichterle prize for young researchers. She is the author of *Fenomén bezdětnosti* (The phenomenon of childlessness) (Prague: Slon, 2009); co-editor of *Práce a péče* (Work and care) (Prague: Slon, 2008), and of *Women and Social Citizenship in Czech Society: Continuity and Change* (Prague: Sociologický ústav Akademie věd České republiky, 2009). Email: hana.haskova@soc.cas.cz

Katerina Kolozova, PhD, is a professor of gender studies and philosophy at the postgraduate study programs of the Institute of Social Sciences and Humanities Research–Euro-Balkan in Skopje. She is also the Institute's director for science and postgraduate studies. Among her recent publications are *The Real and "I": On the Limit and the Self* (Skopje: EuroBalkan Press, 2006); *Gender and Identity: Theories from/on Southeastern Europe,* co-edited with Svetlana Slapsak and Jelisaveta Blagojević (Belgrade/Utrecht: Advanced Thematic Network for Women's Studies in Europe-ATHENA, 2006); *The Invisible Gender: Analysis of the High School Curricula in Macedonia* (Bilingual publication: English and Macedonian, co-authored with Mitko Cheshlarov) (Skopje: Iuventas Publishing, 2006); and *The Lived Revolution: Solidarity with the Body in Pain as the New Political Universal* (Skopje: EuroBalkan Press, 2010). Email: kkolozova@euba.edu.mk

Lamija Kosović is a PhD candidate in philosophy of communication arts at the European Graduate School (New York, US and Caas Fee, Switzerland). She graduated from the New School for Social Research in New York with an MA in media studies. Currently, she works as an academic tutor for the MA program in Gender Studies at the Center for Interdisciplinary Postgraduate Studies of the University of Sarajevo. Her fields of interests are cultural studies and gender, Deleuzian studies, and feminist film criticism. She has published in Bosnian and English in Bosnia and Herzegovina, and internationally. Email: lamija@cps.edu.ba

Laima Kreivytė is an art critic and curator based in Vilnius. In 1999–2000, she studied in the Program of Gender and Culture at the Central European University in Budapest. Currently Kreivytė teaches at the Vilnius Academy of Art, Gender Studies Center of Vilnius University, and at the European Humanities University. Her scholarly interests include feminist visual culture and film theory, queer studies, and popular culture. Email: laima72@yahoo.com

Gorana Mlinarević is a PhD candidate in women's studies at the National University of Ireland, Galway. She graduated from the University of Oxford with a Master's in International Human Rights Law, and from the Universities of Sarajevo and Bologna with a European Regional MA Degree in Human Rights and Democracy in Southeastern Europe. Currently, she works as an academic tutor for the MA program in Gender Studies at the Center for Interdisciplinary Postgraduate Studies of the University of Sarajevo. Her fields of interests are armed conflict, peace building, and gender; gender relations in the Balkans; human rights in general and with a focus on gender issues. She published extensively in Bosnia and Herzegovina, and internationally, in both Bosnian and English. Email: gorana@cps.edu.ba

Irina Novikova is a professor at the Department of Culture and Literature and director of the Center for Gender Studies at the University of Latvia. She is the author of many articles on gender, ethnicity, and citizenship in Latvia, and editor and co-editor of several volumes published both in Latvian and English. These include a book with translations of feminist theoretical texts into Latvian: *Mūsdienu feministiskās teorijas: antoloģija* (Contemporary feminist theories) (Riga: Yumava, 2001) and *Gender Matters in the Baltics* (Riga: University of Latvia Press, 2008). She has also published extensively in the research areas of gender, "race," and genre in literature and films. Her current project is on "race" and gender in Soviet Russian films. Email: iranovi@lanet.lv

Theodosia-Soula Pavlidou is a professor of Linguistics (with specialization in Pragmatics and Sociolinguistics) at the Aristotle University of Thessaloniki. She was a member of the Women's Studies Group, at the same University, since its foundation in 1983. Her interest in language and gender, and women/gender studies dates back to the late 1970s. Her recent publications on these subjects include two edited volumes: *Glossa-Genos-Fylo* (Language–grammatical gender–social gender) (Thessaloniki: Institute for Modern Greek Studies, 2006, second edition) and *Spoudes Fylou: Taseis/Entaseis stin Ellada kai se Alles Europaikes Chores* (Gender Studies: Trends/Tensions in Greece and Other European Countries) (Thessaloniki: Zitis, 2006). She is also the author of "Gender and Interaction," in *The SAGE Handbook of Sociolinguistics*, ed. by Ruth Wodak et al. (London: Sage, 2011), 404–419. Her research also focuses on telephone talk, classroom interaction, politeness, and other topics in cross-cultural pragmatics and conversation analysis. Email: pavlidou@lit.auth.gr

Raili Põldsaar Marling is an associate professor of American studies at the University of Tartu, Estonia. Her interdisciplinary research, which combines discourse analysis, gender studies, and cultural studies, has primarily focused on the comparative analysis of the politics of representation in public discourse, with special focus on gender and power. She has published on gender studies, discourse analysis, American studies, and comparative literature. Raili Põldsaar Marling is an acting editor of *Ariadne Lõng*, Estonian journal of gender studies; an editor of *Aspasia: The International Yearbook of Central, Eastern and Southeastern European Women's and Gender History*; and a board member of the Estonian Women's Studies and Resource Center. Email: Raili.Marling@ut.ee

Serpil Sancar is a professor of political science and gender studies at Ankara University, Faculty of Political Science. She teaches political sociology, social research methodology, feminist theory, masculinity studies, the Turkish gender regime and comparative gender equality policies. She is one of the founders of the Women's Studies Center and has been among the lecturers in the Women's Studies MA Program since their foundation in 1993. Sancar is also the Director of Ankara University's Women's Studies Center. She has recently published one of the first academic works on masculinity based on three years of qualitative research conducted in Turkey, *Erkeklik: İmkansız İktidar? Ailede, Piyasada ve Sokakta Erkekler (Masculinity: Impossible power? Men in the family, the market and on the streets)* (Istanbul: Metis, 2009). Email: Serpil.Sancar@politics.ankara.edu.tr

Kornelia Slavova, PhD, is an associate professor of American culture and literature at the Department of English and American Studies, St. Kliment Ohridski University of Sofia. Her publications are in the field of gender studies, American drama, and cross-cultural studies. She has edited and co-edited several books of gender theory and literary criticism. She is the author of *The Gender Frontier in Sam Shepard's and Marsha Norman's Drama* (Sofia: Polis, 2002) and *The Traumatic Re/Turn of History in Postmodern American Drama* (Sofia: Sofia University Press, 2009). Since 2008 she has been associate editor of the *European Journal of Women's Studies,* published by SAGE. Email: kornelias@abv.bg

review essays

Canon-Building and Popular Culture:
Gender Trouble in Bulgarian Culture Today

Review Essay by **Nadezhda Alexandrova**

Milena Kirova, *Literaturniat kanon. Predizvikatelstva* (The literary canon. Challenges) (Sofia: Sofia University Press, 2009), 287 pp., 15 BGN (hb), ISBN 978-954-07-2811-7.

Milena Kirova, ed., *Neslucheniat kanon. Bulgarski pisatelki ot Vuzrazhdaneto do Vtorata svetovna voina* (The canon that did not happen. Bulgarian women writers from the Bulgarian national revival period to World War II) (Sofia: Altera, 2009), 430 pp., 18 BGN (pb), ISBN 978-954-975-732-3.

Milena Kirova and Kornelia Slavova, eds., *Identichnosti v prehod: rod, medii i populiarna kultura v Bulgaria sled 1989 g.* (Gender identities in transition: Media and popular culture in Bulgaria after 1989) (Sofia: Polis Publishers, 2010), 256 pp., 12 BGN (pb), ISBN 978-954-796-032-9.

These three volumes in the field of feminist literary history and gender studies appeared in Bulgaria within less than a year. Such a wave of academic attention raises the question: Does gender still cause trouble in the present-day Bulgarian culture? I invoke Judith Butler's famous book *Gender Trouble* in order to reflect on the social sensitivity to gender equality in contemporary Bulgarian political and public life. In addition, the books discussed here display the collective efforts of several Bulgarian literary critics to provide visibility to the position of a number of women authors who stand outside or occupy the periphery of the existing Bulgarian literary canon. Finally, the three volumes pay attention to the trouble of translating "gender" into Bulgarian. The existing constellation of non-conflicting translations (gender appears in Bulgarian as *pol, rod, polorod,* or *sociopol*) is evidence that gender is sustainable for academic research only if interdisciplinarity and inclusion are among its key methodological principles. The essays included in these three collections display the willingness shown by researchers from different backgrounds (both men and women) to apply gender methodology in their interpretations of social phenomena and literary texts.

aspasia Volume 5, 2011: 204–210
doi:10.3167/asp.2011.050110

Bulgarian literary and cultural studies still reflect on the outcomes of state social-ism and try to explain the outburst of certain misogynist and masculinist attitudes toward femininity during the twenty years since 1989. Many of the essays included in these volumes discuss the small resonance of public debates on gender equality in Bul-garian society today. Among the reasons for this situation is the humble presence of women authors in the Bulgarian literary canon, which leads to under-representation of women's literary works in the contents of Bulgarian textbooks and school programs. The researchers in these three volumes aim to challenge the theoretical and historic foundations of canon-formation by pointing out the critical genealogies that divert public attention from women's literature.

Milena Kirova's collection of essays called *Literaturniat kanon. Predizvikatelstva* (The literary canon. Challenges) displays her efforts to confront the petrifaction of the Bul-garian literary canon. She is aware of the pessimistic standpoint according to which neither by retrieving the names of unknown (women) authors, nor by promoting rare texts of well-known female writers, can we break the inherently political nature of ca-nonical lineages. However, the essays in the collection present the optimistic strategy of adding "micro-histories" to the grand narratives in order to ensure pluralism of opinion and polyphony of critical voices.

The book has two main sections: "Iz bulgarskiat kanon" (From the Bulgarian lit-erary canon) and "Drugiat kanon: simvolichni istorii v bibleiskjsia text" (The other canon: Symbolic narratives in the biblical text). In the first section the author uses a combination of methods to challenge the mechanisms of canon-formation. In texts such as "Zhiva voda: Repertoarut na placheneto v poeziata na Bulgarskoto vuzrazh-dane" (Living water: The repertoire of crying in the poetry of the Bulgarian national revival) or "Ovcharstvoto kato simvolichna praktika v tvorchestvoto na Ivan Vazov i Zahari Stoianov" (Shepherdness as a symbolic practice of authorship in Ivan Vazov's and Zahari Stoyanov's prose), Kirova explores myths and attitudes of male authors who are included in the canon. She explains the manner in which their literary pres-ence is related to moral imperatives and Bulgarian national doctrines. In other essays, such as "Jordan Yovkov i desniat kulturen proekt" (Jordan Yovkov and the right-wing cultural project) or "Rozi i pepel. Edin zabraven roman (Emanuil Popdimitrov, 'V stranata na rozite')" (Roses and dust: A forgotten novel [Emanuil Popdimitrov, "In the country of roses"]), she explores well-known authors from new and unexpected perspectives.

Another group of essays in the first section discusses the "sins of Bulgarian literary criticism" (178) toward women authors, and includes "Mara Belcheva, ili izborut da budesh Eho" (Mara Belcheva, or the choice of being the nymph Echo), "Mezhdu putia i miastoto: Kakuv e polut na texta-Bagriana?" (Between the road and the topos: What is the gender of Bagryana's text?), and "Romanut na Anna Kamenova 'Haritininiat griah' grehovete na literaturnata kritika" (Anna Kamenova's novel "Haritina's sin" and the sins of literary criticism).

The second section of Kirova's book deals predominantly with texts from the Old Testament. She particularly explores the construction of Jewish identity by a system of values that originated in the moral principles of the ancient Jewish community. Kirova returns to exemplary stories of prohibited or socially acclaimed love, such as the myth of Samson and Dalila. She reflects on the capacity of modern literary studies to include interpretations of Biblical texts, and addresses a contemporary trend in the Western humanities—represented by theorists such as Northrop Frye or Philip Davies—to pose the question of the genre, the narrative, and the figurative resources of a certain Biblical text. This analysis serves as another answer to the initial question about the trouble of canon-formation. By regarding myths and ancient religious texts through panoptical interpretative keys, researchers can reconstruct the foundations of the national literary canon.

The next collection of essays, edited by the same prominent Bulgarian academic—Milena Kirova—can be regarded as a challenge to the field of Bulgarian literary studies. It has the provocative title *Neslucheniat kanon. Bulgarski pisatelki ot Vuzrazhdaneto do Vtorata svetovna voina* (The canon that did not happen. Bulgarian women writers from the Bulgarian national revival period to World War II). The title refers to the existing academic discourse about the lack of Bulgarian feminist criticism.[1] Bulgarian literary history, for example, suffers from amnesia when it comes to discussion about the first Bulgarian poetess. Since the nineteenth-century National Revival, several women have been given this honor. However, even today the names of women authors are rarely included in textbooks. Usually their writing is introduced in a biographical note about the men of letters who happened to be their relatives or lovers and the mentorship on women's poetic style is considered useful and important for the promotion of the woman as an author. In the introduction to the volume the editor explains the "non-happening canon" with the principle of the double mimicry, which Bulgarian women authors have to consider. On the one hand, they follow the canon of the European literary models; on the other hand, they should respect the nation-building imperative of the homemade canon. The volume presents three strategies for opening the canon for women authors. The first is the inscription of women's literature into the existing tradition; the second is the creation of an alternative canon; and the third shares the utopist vision of the canon as a polytopical and polyphonic structure, encompassing micro-topoi and literary presences, which have not received critical attention so far.

The collection includes twenty-five essays that discuss the writing of eighteen women authors. Representatives of women's literature from the period of the National Revival—Stanka Nikolitsa Spasso-Elenina, Elena Muteva, Karamfila Stefanova, and Irina Bachokirova—are discussed in the essays by Katya Staneva and Ivan Radev. The way the public/private dichotomy intersects with the mother/daughter relationship is the focus of the essays—written by Iliana Pavlova, Albena Stambolova, and Milena Tsaneva—on Ekaterina Karavelova and her daughter Lora Karavelova, and on Blennika and her daughter, the literary scholar Milena Tsaneva. Two poetesses who followed the echo of male poetic voices and met males' expectations of femininity are the subjects of Amelia Licheva's (about Ekaterina Nencheva) and Milena Kirova's (about Mara Belcheva) texts. The roles of the muse (Evgenia Mars) and that of the leftist comrade and writer (Vela Blagoeva) are critically addressed by Zhivka Simova and Sofia Angelova. Kalin Mikhailov, Inna Peleva, Kristina Yordanova, Ognyana Georgieva-Teneva, and Tatyana Ichevska discuss the instability of the cultural and political presence of a number of women authors who wrote their works before World War II and even after it, such as Elisaveta Bagryana, Dora Gabe, Fanny Popova-Mutafova, and

Anna Kamenova. The claim of overcoming the reductionism of the literary canon is shared by Albena Vacheva and Ludmila Malinova in their essays on Magda Petkanova and Vesela Vassileva. This long list of contributions does not pretend to cover all aspects of the work of the above-mentioned scholars and literary critics. Instead, it hopes to show the effort of the volume to reach the goal of a polytopical and polyphonic structure. Kirova's edited book makes evident the sustainability of gender as an analytical category by including in the common project researchers with different professional affiliations and academic experiences.

The third book, *Identichnosti v prehod: rod, medii i populiarna kultura v Bulgaria sled 1989 g.* (Gender identities in transition: Media and popular culture in Bulgaria after 1989), which came out at the beginning of 2010, is the result of a research project. The two main goals of the initiative, funded by the Austrian ERSTE Foundation and carried out by the Bulgarian Association of University Women in 2009, were to collect data about the role of the media and popular culture in reconstructing gender identities in postcommunist Bulgaria and to provide analysis of changing gender dynamics through specific case studies. The collection contains twelve texts by male and female scholars who explore the changing conceptions of femininity and masculinity before and after 1989 in several areas such as advertising, consumption, the fashion industry, popular music, sports, TV culture, and popular fiction, as well as media discourses and representations. In their introduction, the two editors—Sofia University professors Milena Kirova and Kornelia Slavova—discuss the dynamic process of transformation of the contemporary gender system in Bulgaria. They claim that if researchers seek to reveal and show the reproduction of intolerance and gender stereotypes in popular culture, the social understanding of the importance of gender equality could increase. However, the last piece in the collection—a brilliant anthropological study of maternity and reproduction by the French scholar Marika Moisseeff—ends with a powerful pessimist stand: "Science fiction provides visibility to an idea, which our 'egalitarian' societies try to conceal: pregnancy as women's prerogative creates the biggest difference between men and women. When it comes to this invariant, which grounds the radical difference between sexes, female exceptionality is associated with demon forces. The uterus is a Pandora's box, the origin of a thousand demons" (254; author's translation).

The metaphor of seeking demons in popular culture can be used for the hidden homophobic and sexist attitudes that are carried on the wings of "the seemingly subversive and democratic potential of popular culture" (129). In a chapter called "Neravenstvoto na polovete v massovata kultura. Sotsiologicheski prochit" (Gender inequalities in popular culture. A sociological survey) Rumiana Stoilova presents the main findings of a sociological survey among more than five hundred students from eight universities in Bulgaria (69.3 percent women and 30.7 percent men). The conclusion of the study is that popular culture icons are gaining greater prominence among young people. This is explained as a reaction to the inaccessibility of Western products before 1989 or as a local influence of the larger processes of globalization and cosmopolitanism. Most students have listed names of famous international performers and pop stars from the areas of music, media, film, and sports as role models. The impact of public figures in terms of influencing their choice of appearance or lifestyle is stronger among women (52.9 percent) than among men (39.4 percent). Interestingly enough, very few of the role models come from the fields of education, science, or politics.

Kirova discusses gender stereotypes in advertising in the chapter titled "Da otgledash neravenstvo, ili gender stereotypi v bulgarskata reklama" (Producing inequality: Gender stereotypes in Bulgarian ads). Popular advertisements serve as a measuring rod for traditional dispositions and social attitudes toward gender roles in the public and private sphere. The most widely shown adverts for alcohol or detergents not only draw on traditional ideas of masculinity and femininity, but also imply sexist perceptions, which might seem to be funny and innocent, but in practice reveal the social inequality between men and women.

An important contribution of this collection is the attempt to follow the developments from the last decades of state socialism until today. In such a manner Slavova discusses fashion as one of the pillars of popular culture in her chapter "Moda i modelirane na identichnostta vuv vreme na prehod" (Fashion and self-fashioning gender identities in times of transition). Her analysis does not approach fashion simply as art or industry but as a system of social encoding. In the early 1950s and 1960s, fashion was a taboo topic in Bulgaria because of its association with the Western bourgeois way of life and its respective values of individualism, self-expression, narcissism, pleasure seeking, and luxury. But since the late 1960s fashion started gaining prominence, exploding in the mid-1990s when more than 150 lifestyle magazines were introduced in Bulgaria.

The dictates of fashion have become possible only through the mediation of television and print media. At the same time, the Bulgarian press is heavily politicized: politics still plays a major role in the everyday life of society and the individual. Media images have a strong impact on young people in terms of their upbringing as well as the formation and maintaining of gender roles, as argued by Totka Monova in her contribution called "Vremeto na zhenite sreshtu vremeto na muzhete: Sotsialni roli i modeli na prehoda vuv fokusa na mediate" (Women's time against men's time: Social roles and models of transition in the media).

Elka Dobreva and Nikolay Atanasov share the same diachronic approach in their studies of the representations of homosexuality in contemporary Bulgarian popular culture. They criticize severely the homophobic nature of this culture. In his text "Mediina vidimost na homoseksualnostta kato paradox" (The visibility of homosexuality in the media as a paradox), Atanasov defines three different periods in approaching the issue and its social treatment. The first period contains the legal measures against homosexuality during communist rule and its practical invisibility in the media until the end of 1989. The second encompasses the "gay-witch" hunting period of the 1990s, during which the stigma of homosexuality for a politician was much more degrading than if he or she was accused of corruption. The third stage began at the turn of the twenty-first century. Atanasov describes the last stage as the time when the homosexual topic lost its ambiguity and was accepted as a market commodity.

As a compensatory move for saving traditional masculinity from the alleged dangers of "sentimental femininity," sport has received huge media coverage in Bulgarian society during the past twenty years. A radical resistance to the lures of femininity and modern unisexual social behavior is the figure of the hypermasculine *mutra* (thug)—a criminal creature involved in shady economic activities, appearing in the 1990s within the circles of former athletes supported by the communist state. Since communist times in Bulgaria there exists a very clear gender division between the images of the iron men (wrestlers) against those of the golden ladies (practicing rhythmic gymnastics). The positive characteristics of sport outlined by Ralitsa Muharska in her text "Mutromuzhestvenost i sport" (Thug masculinity and sport), such as loyalty, bravery, perseverance, and the capacity to concentration, cannot but dissolve in a context where sports is a site of the hegemony of brutal force.

Vladimir Trendafilov discusses the exoticism of the female detective in Bulgarian criminal novels in "Zhenata v bulgarskia kriminalen roman" (The woman in the Bulgarian criminal novel). The dominant method in art and fiction before 1989—socialist realism—argues Trendafilov, did not allow the fabrication of circumstances that did not exist in reality. That is why the few women characters present were not investigators but served mostly as secondary characters against the background of the crime-related action or as victims of the crime under investigation. It was only in the 1990s that an outburst in criminal fiction gave way to female characters that existed not only as seductive spies but also as rational investigators of crimes.

The impact of popular culture on the production of movies and especially on the evolution of the image of the metro-masculine pop star—an exhibition of the miracles of cosmetics and plastic surgery on a man's body—is analyzed in the chapters by Ludmil Dimitrov "Gender stereotipi v dva bulgarski filma i tiahnata retseptsia v Slovenia" (Gender stereotypes in the reception of two Bulgarian movies in Slovenia) and Dimitar Kambourov "Polorodut na pop musikata: edna bulgarska versia" (Sexo-gender of the pop music: a Bulgarian version). There is a contrast between the positive outcome of Dimitrov's experiment to introduce exemplary Bulgarian movies to Slovenian students and the disillusioned tone of Kambourov about the impact of popular music on improvisation and creativity. The latter shared the observation of a dangerous tendency in the realm of popular music. His claim is that the promotion of women's voice and song has intensified the tendency of simplified arrangements, in which the improvisations are limited and the instrumental parts are turned into accompanying elements. What is more, according to Kamburov, together with their voice women singers have inserted their bodies, used as autonomous entities, which can generate and create their own symbolic messages. Popular music has begun to transform into a "space of simple and primitive collective identification" (213).

The linguistic study of Gergana Dacheva also does not sound optimistic about the self-perception of men and women with regard to their social representations through language. Her qualitative analysis of twenty autobiographical narratives shows that women invest greater passion into an effective and powerful way of retelling. They tend to use hyperboles in order to keep the attention of the listener. Unlike women, men often use verbs in the first person singular and do not necessarily keep the channel of communication open through explicit markers of communicativity in conversational situations. This observation reiterates the "gender trouble" that is inherent in language: if our social reproduction as individuals is possible only through linguistic articulation of our thoughts, then every other effort to decrease gender inequalities and hierarchies is predestined to failure.

A way out of the linguistic dilemma is to learn "the morphology" of chemistry. Such an approach is used Kambourov when he chooses to translate gender not just as "sex" (*pol*) or "gender" (*rod*) but as a compound of the two, "sexogender" (*polorod*), just as water is a compound of oxygen and hydrogen. The term "sexogender" sounds like a peculiar neologism. However, it is an example of the efforts to apply interdisciplinarity and include new perspectives in order to sustain the intelligibility of "gender" in the Bulgarian academic context today.

◊ About the Author

Nadezhda Alexandrova, PhD, is an assistant professor at the Faculty of Slavic Studies, St. Kliment Ohridski University of Sofia. Her publications deal with the representations of women's voices in nineteenth-century Bulgarian culture, and also discuss affective relationships and liminal spaces in women's narratives. Email: n.alexandrova@slav.uni-sofia.bg

◊ Note

1. For discussions in English, see for example, Irina Gigova, "The Feminisation of Bulgarian Literature and the Club of Bulgarian Women Writers," and the review essay about Miglena Nikolchina's works by Bilyana Kourtasheva, "Women's Literary History and Its (Im)Possibility," both published in *Aspasia* 2 (2008): 91–119 and 219–223, respectively.

Women and Gender in an Age of Fervent Nation-Building

Case Studies from Southeastern Europe

Review Essay by **Svetla Baloutzova**

Tatyana Stoicheva, *Bulgarski identichnosti i evropeiski horizonti, 1870–1912* (Bulgarian identities and European horizons, 1870–1912) (Sofia: Iztok-Zapad, 2007), 377 pp., 14 BGN (pb), ISBN 954321345-3.

Mari A. Firkatian, *Diplomats and Dreamers: The Stancioff Family in Bulgarian History* (Lanham, MD, UK: University Press of America, 2008), 359 pp., ISBN-13: 978-0-7618-4069-5.

İpek Çalışlar, *Latife Hanim* (Kalem Literary Agency, 2006; Bulgarian translation: Sofia: IK "Uniskorp," 2009), 479 pp., 17 BGN (pb), ISBN 978-954-330-222-2.

The processes of nation-building and modern national identity-formation in Southeastern Europe have generated a growing amount of research work over the last two decades. Yet, despite the strong scholarly interest in Southeastern European political and cultural transformations on the road to modernity, women's experiences in this course are still awaiting their emergence "from the shadow of history."[1] Hence, the three recent monographs by Tatyana Stoicheva, Mari Firkatian, and İpek Çalışlar discussed here present an especially valuable contribution to the still underexplored niche of women's histories in the period of Southeastern Europe's nation-geneses. The professional occupations of the authors—a professor in English literature and culture at St Kliment Ohridski University of Sofia, Bulgaria; a professor in East European history at the University of Hartford Hillyer College, US; and a prominent Turkish writer and journalist—has left its distinct impact on their approaches to the topic of nation-formation. Each author sheds a unique light on national developments in the late nineteenth and early twentieth century, and the role of women in them. Stoicheva embraces the academic merits of postcolonial theory to explore the complexities in

aspasia Volume 5, 2011: 211–215
doi:10.3167/asp.2011.050111

the imagining of Bulgarians, Bulgarianness, and the role assigned to women in the last years of the pre-Liberation and first thirty-four years of the post-Liberation eras (from 1870 to 1912) in modern Bulgarian history. Firkatian opts for a family history approach in which the Stancioffs' relationships with their social environment follow and reflect the prolonged birth pangs of the Bulgarian nation. Likewise, Çalışlar's choice of a modern biography helps her highlight an exceptional woman's life experience, and simultaneously to incorporate indispensable moments from the early stages of the evolving modern Turkish state. All three books are based on original and less known primary sources, while Firkatian's and Çalışlar's monographs also integrate fascinating research results from—both private and public—archival collections, which have not previously been available for scholarly work.

With regard to its historical timespan, *Diplomats and Dreamers* roughly coincides with Stoicheva's choice of a timeframe in *Bulgarski identichnosti i evropeiski horizonti* (Bulgarian identities and European horizons)—the emergence of the Bulgarian nation-state on the European map (in 1878) and its development up to the disastrous series of two Balkan Wars and World War I. From there it takes its readers into the post–World War I period with all its dramatic events and foreseen belligerent outcome, and expands further into a transformed post–World War II reality, thus comparatively paralleling Çalışlar's temporal framework. Yet the focal point in both Firkatian's and Çalışlar's works is rooted in the 1920s. This is the decade when their extraordinary protagonists—Nadejda Stancioff, Bulgaria's first woman-diplomat, and one of the very few on the international agenda, and Latife Hanim, the Western-educated and reform-minded young wife of the founding father of the modern Turkish Republic, Mustafa Kemal Atatürk—channeled their thoughts, efforts, and lives in servitude to their nations' good.

All three books offer precious insights into the peculiarities and ordeals of two nations—the Bulgarian and the Turkish—materializing at the end of the nineteenth century and the early twentieth century from the dismembered Ottoman empire as nation-states in their own rights. Organized in seven chapters, including a detailed, methodologically enlightening introduction and a conclusion, and further enriched with two appendixes and a detailed summary in English, Stoicheva's *Bulgarski identichnosti i evropeiski horizonti* addresses the vital interaction between European culture and local Bulgarian traditions influencing the country's international and native image-making, policy-building, and self-esteem. It illuminates the discourses resulting from the ingrained sense of superiority of Western travelers and responsible for the construction of stratified images of Oriental "barbarity" in their writings. It also exposes the divisions between Western and Eastern Europe enforced on a newly emerging nation by a hierarchical national categorization with a pronounced flavor of Western abrogated arrogance and domination. Stoicheva's study focuses on the diverse constructions of Bulgarian self-perception resulting from this geopolitical rift and elaborates on the choices the natives made—or perhaps, had to make—for themselves, their institutions, and their state. Within this turbulent period of defining and redefining the new "essence" of national identities, new gender models were constructed and new gender regimes were negotiated. Their desperate fluctuation between a hankered-after "European" female perfection and the gendered house-chore requirements of the local, domestic tradition—as explicated in Stoicheva's detailed and amusing analysis of a popular male song—seems hard to be reconciled with and accommodated in the modernizing mind of the Southeast European man.

The pains of nation-building and image-construction, together with the blurring boundaries of desired female identities—the latter staggering between the local East and the paragon West—constitute central points of representation in Firkatian's and Çalışlar's works as well. Here, the complex process of a national identity-gain through women's changing images is subjected to the personal histories of the renowned Bulgarian Stancioff family of diplomats, and the romance, marriage, and subsequent divorce of Latife Uşşaki (née Uşakizâde) to the hero of the Turkish Republic, Kemal Atatürk.

Running through nine chapters, Mari Firkatian takes the story of young Nadejda Stanicoff through the trials of the Bulgarian nation, morally wronged in its cohesion and integrity by the Great Powers and fighting for its reunification in a lengthy string of wars. A staunch supporter and promoter of the Bulgarian national cause on the international arena, Nadejda nevertheless exhibited a complex set of nonconflicting multiple identities. Swaying between a flower-perfumed Orient and a "civilized" Occident due to her mixed family background (with Bulgarian father and French mother) and international education, Miss Stancioff—the future Lady Muir—easily blended in both poles of Europe's geography by effortlessly combining her acute feeling of Bulgarianness with a sense of "we British" (Firkatian, 293). In contrast, Latife Hanim's European education in London and Paris, added to a liberated mode of female conduct and passionate patriotism, turned her into an aspired model of the "new" Turkish woman, but also seemed to have been at the core of her personal mishap, divorce, and final social isolation.

All three studies underscore the essential roles delegated to women in an overtly male-drafted project of nation-construction and symbolic national representation. If visualized in the public field, women come forward as showpieces in a carefully preconceived and arranged show-window. They become instrumental for their nation's beneficial self-perception and are equally influential in promoting their nation's image on the international arena. A special heading in Stoicheva's book elaborates on the female survivors of the blood-drenched April 1876 uprising by linking their hitherto unvoiced experiences as innocent yet discredited victims of rape to the imperative idea of national honor sustained by normative female sexual purity, and if necessary, by social exile.

On the other end of the scale, it was both Nadejda's exceptional personality and the needs of her nation for a young, beautiful, and trustworthy face to obliterate Bulgaria's image of the "the stormy petrels of Europe" (Firkatian, 225), and to epitomize a politically modernized and reliable international actor that propelled her to the position of Bulgarian first secretary to Washington, DC.[2] Similarly, Latife actively participated in the making of the modern, secular Turkish state by symbolizing the image of the "liberated," "European" woman in Turkey's new policy- and image-making. Unveiled, unabashedly accompanying her husband into the hitherto all-male zones of the public space, and actively assisting Atatürk in his state-construction work, Latife was chosen by the national leader to stand for the new and emancipated Turkish nation in his far-reaching political and nation-building project.

Would there be limits on the scope of women's freedom of initiative, freedom of action, and their contribution to the national construction process in the age of Southeastern European ardent nationalism? How permissive would male-designed societies be to women's societal inclusion and social emancipation even when the women's question had been publicly acknowledged as vital for the nation's future, its domestic modernization and international stability? None of the three works engages in an explicit answer, but issues relating to conservatism, traditionalism, and radical reformism in gender relations are touched upon by Firkatian and Çalışlar through Nadejda's successful marriage to a Scottish aristocrat and Latife's marriage failure. Longing after the past and its clearly outlined social boundaries and values, shunning the newly mushrooming results of a rapidly outspreading modernity, Nadejda seemed smoothly to adjust to her new marital duties while continuing her political and social activism. Conversely, the inquisitive reader may wonder whether the coexistence and collision of traditionalism and modern radicalism in the private family sphere might not be held responsible for the breakdown of Latife's marriage. How willing would a reformer's mind be to overcome customarily established gender prescriptions in his/her privacy and sincerely embrace the idea of gender equality in his/her spousal relationship?

Left open to the reader's own decision, it is nevertheless worth mentioning how dangerous Çalışlar's research was perceived to the hitherto untarnished image of Kemal Atatürk in modern Turkish history and historiography. The publication of her groundbreaking work nearly cost the author her civil freedom.

Bulgarski identichnosti i evropeiski horizonti, *Diplomats and Dreamers*, and *Latife Hanim* make significant contributions to the field of women's history in Southeastern Europe as presented against the background of zealous domestic and international affairs on the road to national establishment, independence, and recognition. Written in a sophisticated, yet clear and comprehensible language, the three monographs unobtrusively elaborate on women's diverse yet predictable plights—the latter blending in and being constructed through their nation's own trials and ordeals.

None of the works offers a simplistic or cliché-like message to its audience. Instead, all three books actively engage the readers' curiosity, encouraging them to long for additional, follow-up information on the protagonists' life stories and go on the Internet or to a library for further research. Recommended to a scholarly and a non-academic public alike, the three works make a delightful complementary reading—a truly intellectual pleasure that is worth its experience.

◊ About the Author

Svetla Baloutzova, PhD, is an adjunct lecturer at Sofia University, Faculties of History and Philosophy, and a staff-member of the Centre for Advanced Study Sofia. Her academic interests focus on social history and historical demography, the European family and gender relations over space and time, as well as the evolution of the welfare state. Baloutzova also works in the field of linguistics and British cultural studies. She is the author of *Demography and Nation: Social Legislation and Population Policy in Bulgaria, 1918–1944* (Budapest and New York: Central European University Press, 2010). Email: svetla@cantab.net

◊ **Note**

1. The phrase has been borrowed from Krassimira Daskalova, ed., *Ot siankata na istoriiata: Zhenite v bulgarskoto obshtestvo i kultura. Sbornik s tekstove* (From the shadow of history: Women in the Bulgarian society and culture. A reader) (Sofia: LIK, 1998).

2. Eventually, Nadejda declined this honor as a sign of protest against the right-wing coup d'état in June 1923.

book reviews

Marina Blagojević, *Knowledge Production at the Semiperiphery: A Gender Perspective* (Belgrade: Institut za kriminološka i sociološka istraživanja, 2009), 260 pp., (pb), ISBN 978-86-83287-36-9.

Book Review by **Ana Proykova**
St Kliment Ohridski University of Sofia

Knowledge Production at the Semiperiphery presents the concept of a necessary connection among theory, research, and policy while providing insight into the relationships among knowledge production, gender, and locality. The book contains an introduction, conclusion, and seven chapters, which consider various difficulties that women scientists from the semiperiphery have faced in the process of knowledge production. The author defines semiperiphery, positioned between the center and the periphery, as "essentially shaped by the effort to catch up with the core, on one hand, and to resist the integration into the core, so not to lose its cultural characteristics" (33–34). Blagojević affirms that localization influences the concept of scientific excellence by pushing women out of the margins of a respectful career. From the perspective of the core, the semiperiphery is "different and not similar enough" and should be improved by imitating the already developed models available in the core. Such a neocolonial approach, argues Blagojević, suppresses development of new paradigms coming from the semiperiphery, which has been in a state of permanent reform meaning that "one reform is following the other while the previous one has not been finalized, nor its effects explored" (36). These issues are carefully examined in the first chapter—"Non White Whites, Non-European Europeans and Gendered Non-Citizens: On a Possible Epistemic Strategy from the Semiperiphery of Europe."

The second chapter, "Creators, Transmitters and Users: Women's Scientific Excellence at the Semiperiphery of Europe," analyzes the actual system of exclusions, the mechanisms driving positions, ways of knowing, knowledge production, and possible recognition of excellence of women-scientists from the semiperiphery. The core questions raised by the author are: Who are the producers of the knowledge, who transmit the knowledge, and who is using it? Blagojević is concerned that lack of interest in theoretical discussions about knowledge produced in the semiperiphery makes this knowledge invisible and not that easily accessible to the global context.

The third chapter, "Gender and Knowledge at the Balkan Semiperiphery: Women in Science," covers the problem of women in science seen as unused resources due to the "multiple transitions" of this region. The Balkan area is characterized by the migration, mixing, and dividing of the population. "Ambivalence, paradoxes, discontinuity, detrimental unintended consequences, chaos, and entropy are only a few of the results" (101) of the processes mentioned. The author is concerned that the theories coming from the core (i.e., from the West) suffer from an "epistemic void," meaning the absence of paradigms and canons, and the fragility of theories and concepts when faced with a different lived reality. The author uses another concept, "muteness," to denote communication with the "outsiders" who are unfamiliar with the context the author comes from, while the concept of "numbness" explains the problem of communication toward inside, "the context."

Interesting research on women scientists who returned to Hungary after having been abroad on grants for long periods is the core of the fourth chapter, called "Hungarian Women Scientists Returnees: Becoming a Cultural Minority or Being Integrated into the Elite?" It shows that international professional migration frequently creates additional tensions in home countries and institutions—this alarming finding must be taken into account when "brain circulation" is announced as a panacea. The next chapter—"Women Professionals from the Semiperiphery of Europe: New European Proletariat?"—deconstructs the negative stereotypes about women's position at the semiperiphery. In addition, it shows how mobility can have contradictory effects: financial gains versus social status. Such a complex understanding is currently needed as the EU builds huge research infrastructures where many scientists, including women from the semiperiphery, will work for long periods. The large facilities proposed by the European Strategy Forum on Research Infrastructures (ESFRI) will cover all human activities related to knowledge production. Could society afford the prejudice that the knowledge of women from the semiperiphery is of low value? In fact, the main difference between women scientists in East and West is that women professionals are not fairly compensated in the East and often have to take on additional jobs to make a living. Those women scientists in the Balkans who have families and children share the daily problems of nonprofessional women. This economy of women's sacrifice, Blagojević suggests, should be replaced with an economy of pleasure and rewards.

The chapter "Nomadic Scientists in a Transnational Landscape: Practicing Intersectionality" advocates this new approach in the knowledge production. "The main question is whether scientific nomadism, in social sciences and in gender scholarship, particularly, brings some new quality to knowledge" (199). Blagojević's answer is yes, because of its endless capacity for connection of different pieces. The last chapter, "Shifting the Paradigm: Arguing for the Positive History Approach," is a concrete application of different epistemic lessons developed within feminism. It elaborates on how the gender perspective contributes to reconciliation.

Although quite different topics are covered in one volume, they are connected via the establishment of new approaches, concepts, and paradigms. Blagojević's work will be of interest to social scientists, by providing them with provocative ideas about more research on various contested issues raised by her. It will also be useful for women pursuing high-level positions in transition countries. Finally, it would be instructive for those decision makers who are willing to face and accept nontrivial ideas.

◆ ──

Maria Bucur, *Heroes and Victims: Remembering War in Twentieth-Century Romania* (Bloomington and Indianapolis: Indiana University Press, 2009), 352 pp., $19.60 (pb), ISBN 978-0-253-22134-6.

Book Review by **Malgorzata Fidelis**
University of Illinois at Chicago

── ◆

In *Heroes and Victims: Remembering War in Twentieth-Century Romania,* Maria Bucur examines official and unofficial commemorations of the two World Wars to illuminate the contested and negotiated nature of memory. The book represents an impressive interdisciplinary endeavor, in which the author crosses the conventional boundaries of historical research and uses not only state and regional archival documentation (primarily from Bucharest, Iaşi, Cluj, Sibiu, Braşov, and Satu Mare), but also novels, films, monuments, plaques, street names, oral interviews, and Internet discussions.

Indeed, it is the local dimension of memory that Bucur finds the most significant. She decenters national ideology as the primary lens through which to understand the formation of modern cultural identities in Eastern Europe. "I accept that nationalism was a major filter through which individuals and local communities came to see death in the world wars," Bucur writes. "But I do not consider the role of the state and politics as *the* central force in commemorative discourses, either during the early post-1918 years or even at the height of Communist dictatorship" (5). For Bucur, memory is "always local," and the narratives of war were constantly contested and reformulated by a variety of state and local actors.

Because of its ethnic diversity and shifts in military alliances, Romania presents a fruitful case to study the complexities of memory. Romania's minorities, such as Hungarians, Germans, and Jews, often challenged the state-endorsed narrative centered on ethnic Romanians and the Orthodox religion. The focus on Romania allows Bucur to highlight the significance of the Eastern Front as in the case of World War I, when Romania's staggering human losses (26 percent of the total population) were among the highest on the continent. War energized national feelings, but as Bucur notes, interpreting the events in a uniform "nationalist idiom" was difficult to achieve. Rather, "community-based and individual rituals revolved more directly around dealing with the pain of loss" (59).

One of the major strengths of the book is the author's analysis of the role of gender in shaping memory and understanding death. Bucur demonstrates that notions of heroism and virtue were highly gendered, and that only gradually, under the pressure of mass mobilization, did women become partially incorporated into the broader narrative of war. But the resistance to redefine traditional gender hierarchies proved particularly strong. Bucur found, for example, that people consistently excluded sexual violence from personal and public narratives. Wartime rape remained stigmatized as shameful and marginal to the suffering of Romanian society in general. Bucur also dis-

BOOK REVIEWS   219

cusses the unique role of women in the Romanian Orthodox "cult of the dead," where "funerary rituals were fundamentally gendered, with women playing specific gender roles at every step of the way not only as followers, but also as central gatekeepers of the passage into afterlife" (19). Nevertheless, the commemoration of wars remained heavily focused on male combatants and masculine notions of heroism and sacrifice.

World War II proved to be even more complicated for constructing a uniform memory of the events. Because Romania switched sides in August 1944 from the Axis to the Allied powers, Romanian soldiers had disparagingly different experiences depending on when and where they fought. The Romanian-Hungarian conflict in Transylvania and the extermination of Jews (especially in areas recovered from the Soviets after 1941) on orders of the Romanian military dictator, Ion Antonescu, added to the complexities. The last three chapters of the book offer fascinating insights into how communist leaders and ordinary people grappled with these legacies within the dynamically changing political landscape of communism and postcommunism.

The postwar communist state promoted a vision of war, in which the sacrifices and the victory were associated exclusively with the communists and wartime atrocities and collaboration with "outsiders" (primarily Germans and Hungarians). But it was the Romanian dictator Nicolae Ceauşescu who took public commemorations of war to new levels. He initiated the annual parades on 23 August (the date of the coup against Antonescu in 1944) centered primarily on ethnonationalism and the cult of the leader. In Bucharest, such parades "were meant to look like a river of love and loyalty flowing under the gaze of the loving leader, much as Leni Riefenstahl's images displayed the smiling Aryan marches under Hitler's eyes" (182). Indeed, the ethnonational mythology, as Bucur notes, was a common thread found in personal and official narratives during the communist and post-communist periods. Bucur sees this ubiquitous preoccupation with ethnically based victimization as distinctly Eastern European and different from the Western European narratives where "the memoirs of self-avowed perpetrators from World War II have been acknowledged as important, alongside those of their victims" (222). More recently, pressure from the European Union prompted the Romanian government officially to acknowledge Romanian participation in the Holocaust, but the majority of Romanians still cultivate ethnically based suffering and are "unable to look globally and with empathy toward *all* the victims and perpetrators of violence during that period" (225).

Bucur's work is an important contribution to the growing field of memory studies in Europe. It offers a fresh and much needed perspective on Eastern Europe with particular attention paid to local understandings and responses to social and political upheavals of the twentieth century. By decentering the nation, Bucur offers a more nuanced understanding of Romanian society, in which identities are shaped not only in response to the state, but also in relation to the family and local cultural community. At the same time, Bucur shows how questions of locality can help us understand a broader issue of the post-1945 European identity. Elevating the ethnically based suffering, so omnipresent in Romanian narratives of World War II, points to the difficulty of forging democratic citizenship in postcommunist Eastern Europe. What is then the role of memory in setting common goals for Europe? Is it possible to have a uniform European memory of the twentieth century? Do we need one? Bucur's rich and com-

pelling analysis adds a new component to examining the challenges faced by Europeans today, including questions of citizenship, cultural identity, and self-definition.

◇ ————————————————————————————————————

Irina R. Chikalova, ed., *Zhenshchiny v istorii: vozmozhnost byt uviden-
nymi* (Women in history: The possibility to be visible) (Minsk: BPGU
imeni Maksima Tanka, 2001–2004), vol. 1, 320 pp., (pb), ISBN 985-435-
359-1; vol. 2, 320 pp., (pb), ISBN 985-435-359-2; vol. 3, 308 pp, (pb),
ISBN 985-435-776-7.

Book Review by **Marianna G. Muravyeva**
University of Helsinki, Finland

———————————————————————————————— ◇

Collections of articles in the field of post-Soviet gender studies became a vehicle of
knowledge dissemination during the age of the emergence and transformation of gender studies in post-Soviet scholarship in the 1990s and early 2000s. This genre could
have easily been called "solianka" (salty mix), after the name of the famous Russian
soup. The whole idea of such collections was about bringing together whatever any-
one could say about gender and women and publish it under a puzzling and provoking title. These books did not have a consistent methodology of article representation
and picked themes according to the genuine interest of the participating authors. And
this was a good thing, because today these collections are a unique source to study the
historiography of gender studies development.

Irina R. Chikalova's *Zhenshchiny v istorii* represents a rare example of a gender his-
tory-oriented publication that attempts to create some space for academic discussions
on the most important issues involved. Chikalova, the editor of the series, put together
the three volumes in a period of three years. She is a well-known Belarusian gender
specialist who wrote a number of articles important for the development of gender
studies in the Belarusian context. In her editorial introduction, Chikalova lays out the
methodology of the series, suggesting that before using "gender" as an operative concept one should "extract" women from the darkness of the past and give them "their
own" history (1: 9). So the series is constructed very much along the lines of revealing
women's experiences and their different pasts.

The first volume covers the history of women in Europe, Northern America, and
Belarus in a very wide spatial context, while the second and third volumes single out
certain chronological periods and certain topics. Volume one starts with the theory
and methodology of gender history. Other methodological articles included range
from a classical historiographical approach in volume one, to cognitive philosophical
reflections and postmodern analyses in the other two volumes.

The first volume includes three parts. Part one discusses theory and methodology of gender studies in history (with chapters by Natalia Pushkareva, Olga Shutova,
Irina Yukina, and Natalia Chukhim). Part two deals with the history of women in

Europe and North America (texts by Olga Lentsevitch, Maria Bulanakova, Elena Surta, Oksana Petrovskaya, Tatiana Koroleva, Natalia Novikova, Alexander Ermakov, Olga Chesnokova, Irina Chikalova, Nadezhda Shvedova and Evgenia Israelyan, and Vladislav Froltsov). Finally, part three is devoted to the history of women and family in Belarus (contributions by Natalia Slizh, Svetlana Tolmacheva, Olga Sobolevskaya, Galina Yakovleva, and Angela Bulich).

The second volume has six parts that address sex, culture, and gender (Elena Yanchuk and Igor Kogan); women and Christianity (Viktoria Sukovataya and Dmitri Mazarchuk); the history of women in medieval Europe (Larisa Vonsovitch and Elena Surta); gender aspects of ethnography (Oksana Kis and Vladimir Tugaj); social and political experiences of European women in the Modern times (Maria Bulanakova, Irina Chikalova, Natalia Novikova, Alexandr Ermakov, and Vladislav Froltsov); and the same issues in Russia (Irina Yukina, Irina Chikalova, Yulia Gradskova, Galina Karpova, Elena Yarskaya-Smirnova, and Tatiana Rovenskaya).

The final volume is divided into five parts, dedicated respectively to the gender project in the humanities (Irina Chikvalova, Olga Shutova and Alisa Tolstokorova); matriarchy (Anna Borodina, Dmitry Borodin, and Igor Bogin); motherhood (Natalia Pushkareva, Pavel Romanov, Elena Tsyglakova, and Elena Yarskaya-Smirnova); medieval women (Elena Chernyshova, Valentina Uspenskaya, and Elena Surta); and women in Modern Europe (Anatolij Dulov, Alexandr Ermakov, Irina Nikolaeva, and Irina Chikalova).

This three-volume publication marks the multidisciplinary use of gender in the post-Soviet context and differentiates the notion of history within the context of various social sciences. The majority of articles, however, does not reflect the methodological changes underlined in the theoretical texts, and instead consists of more traditional, descriptive and positivistic women's history, conveying the well-known message that "women existed in the past." The level of scholarship differs very much from text to text, generally mirroring the state of women's and gender history. There is a certain geographical balance between the history of women in Europe and the United States and the history of women in Russia and other post-Soviet countries (mainly Belorussia), which can be explained by the professional background of the authors and the editor herself, who originally was a specialist in contemporary British history. This balance does not help the development of Russian women's history, however, because there is no attempt to connect the regions or to use any comparative approach to reflect on the history of Eastern European women vis-à-vis the Western narrative.

The articles on women in medieval France or in seventeenth-century England (1: 105–114 and 1: 125–131) or on German women during World War II (1: 180–188; 2: 236–254; 3: 130–141) are as informative as they could be in Russian. However, they do not present the best scholarship available and do not suggest any fresh ideas either, as they do not use any new sources or interpretations not to mention archives. Russian (1: 189–192; 2: 150–209; 3: 154–179), Belarus (1: 270–318; 3: 218–235, 255–267), Ukrainian (3: 55–71), and Latvian (3: 72–76) material occupy about one-third of the series and represent much stronger research and scholarship. Unfortunately, the idea of the first volume to have a special part on Belarus women did not continue into the other volumes, which would have been the most interesting material to be uncovered and published (especially, from the point of view of original Belorussian research). Articles on Jewish women in Belarus (1: 290–303), the two short texts on Belarus women in 1930 (1: 304–310) and Belarus women during the Nazi occupation of Belarus in 1941–1944 (2: 255–268) as well as on other aspects of the history of Soviet women reveal the complexity of women's experiences and the efforts of contemporary scholars to provide them with their own voice. Examples of the latter include women's Gulag experiences by O. Chestnokova, the Russian feminist pre-revolutionary ideology by Irina Yukina, gender cultural construction during Stalinism by Yulia Gradskova, Elena Yarskaya-Smirnova and Galina Karpova, and the Russian women's literary movement of the 1980s–1990s by Tatiana Rovenskaya.

The series is an important part of the development of post-Soviet gender history research and represents interesting material for those who would like to understand the complex processes within the field of gender history today. It reflects the current major features of gender studies in the former Soviet Union: the referential practices, the concentration on the West, the selectiveness of editorial policies, the unevenness of the level of scholarship, and the general instability of the field. In addition, this series is a rare discussion forum about gender history coming from Belarus, which makes the series valuable in itself. These three volumes can be useful for all scholars and students interested in post-Soviet gender history and for those who would be interested to read about trends and perspectives of gender studies within the post-Soviet space.

Kristen Ghodsee, *Muslim Lives in Eastern Europe: Gender, Ethnicity, and the Transformation of Islam in Postsocialist Bulgaria* (Princeton and Oxford: Princeton University Press, 2009), 270 pp., $24.95 (pb), ISBN 978-0-691-13955-5, $65.00 (hb), ISBN 978-0-691-13954-8.

Book Review by **Moyuru Matsumae**
University of Morioka, Japan

The activities of Muslims in the Rhodope Mountains in southern Bulgaria have received great attention in Bulgaria in recent years. The media has run stories about the Islamic education in this region funded by foreign organizations, or how young Muslim girls sometimes wear headscarves in the "Arab" style that is different from their traditional dress. The news coverage of the growing presence of Islam has created a sensation in Bulgaria, which has a sizable Muslim minority,[1] but little attention has been paid to the voices of the local people. Why are ordinary people in this small region adopting foreign forms of Islam now?

Muslim Lives in Eastern Europe is an exciting and compelling account showing how people have embraced a new type of Islam instead of returning to its traditional forms. Ghodsee explores in rich detail why this change is occurring only in limited communities of Bulgarian-speaking Muslims in Madan and Rudozem, cities of central Rhodope.

The author conducted fieldwork among Pomaks, people belonging to this minority. Her extensive research revealed that economic changes after the collapse of state socialism in these cities had also strong social consequences, particularly concerning the redefinition of masculinity and femininity. The sudden shift in gender roles is an important but often overlooked factor in the embrace of new forms of Islam.

The book consists of six chapters in addition to the introduction and conclusion. The first two chapters explore the contested history of Bulgarian-speaking Muslims—a distinct ethnic group from the Turks who form the majority of Muslims in Bulgaria— and the everyday life of Pomak men and women in Madan and Rudozem under socialism. The subject of the third chapter is the personal histories of individuals after the fall of socialism. Investigating the wider context as well—for example, the history of the Muslim leadership or, in chapters four and five, the influence of Islamic aid mainly from Arabic countries—Ghodsee examines the arguments of women who follow new forms of dress and behavior (chapter six).

The author focuses especially on how local concepts of masculinity and femininity must be redefined. Under state socialism rural industrialization of Madan and Rudozem forced the population to turn into a modern proletariat and to reconfigure gender identities. The majority of Pomak men in these two cities were employed in the lead-zinc mine enterprise. The miners were highly valued as symbols of the "real" proletariat and well paid. Women also began to enter the formal workforce because such participation was seen as the hallmark of a modern socialist society. Needless to say, this shift required reshaping of existing gender roles. Masculine identities of the Pomak miners, however, were not threatened due to their valorized status and higher wages. What happened after the fall of communism and the mine industry in this region?

It must be noted that it is mostly in the former mining region that people began to embrace the "new" Islam imported from Arabic countries. The sudden economic downturn after the collapse of state socialism is one basic reason for this phenomenon. Ghodsee offers additional convincing reasons for adopting the new forms of Islam. The first is related to the continually contested history of the Pomaks. Religious rivalries over Muslim leadership, the resulting neglect of the needs of ordinary Muslims, and the political division of the Pomaks from Turks during the transitional period led people to look abroad for psychological and financial support. "Arab" Islam has attracted a section of the Pomak population, especially the young, because it is regarded as superior to Turkish Islam and the tradition of the Pomaks.

Another important factor for the growing presence of Islam, according to Ghodsee, is gender instability in this region. After the bankruptcy of the mines, the majority of men lost both their jobs and their highly valued social status as "real men." The local people had to renegotiate forms of masculinity and femininity. The author examines how new mosques in Madan and Rudozem have replaced the mines as centers for making connections among Pomak men. Adopting "Arab" Islam has led a part of the Pomak female population to accept more conservative gender roles, but they actively embrace the new Islam as a way to resist the recent trend of the hyperobjectification of women in Bulgaria, or as a way of dealing with their unemployed and alcoholic husbands. Thus Ghodsee presents Pomak women not as passive victims but as real social actors.

Throughout the book Ghodsee shows how men and women in specific communities accept foreign types of Islam for their locally defined needs and explores closely the issue of agency. It is important to consider these changing practices of the Pomaks in the local context, as they are one part of a global trend that includes growing popularity of Islamic culture. The author also points out that there are interesting continuities between Islam and Marxism, both of which share the idea that gender relations should be organized to promote the common good and recognizes the historical legacies of socialist feminism. At the same time, showing that the older generations tend to support more traditional forms of Islam, Ghodsee calls for recognition of a generational gap among the Pomaks. I think that future studies should try to address further the question of how people renegotiate gender roles across generations.

Muslim Lives in Eastern Europe contains valuable lessons, which can help to understand the dynamism of Islam and postsocialist societies in today's world. It is a useful reading for scholars and students from various fields and all those who work on religious and gender issues.

◊ Note

1. According to the 2001 Bulgarian Census, Muslims make up 12 percent of the population.

Katerina Kolozova, *The Lived Revolution: Solidarity with the Body in Pain as the New Political Universal* (Skopje: Evro-Balkan Press, 2010), 232 pp., €15.00, ISBN 978-9989-136-69-6.

Book Review by **Slavco Dimitrov**
Institute of Humanities and Social Sciences Research "Euro-Balkan,"
Skopje

The Lived Revolution: Solidarity with the Body in Pain as the New Political Universal can be introduced as the result of a radical thinking caused by obsessive attempts of trying to find discourse(s) that can transpose and clone the compulsive, immanent, and even violent taking place of the Real. One can freely "take the risk" of naming this thought radical if the etymological ground of this very term is taken into consideration. Namely, radical would signify the theory or politics of trying to grasp the roots (radix) of a social, political, or philosophical problem, or suffering as the sheer experiential in the contemporary world. Thus, Katerina Kolozova's new book represents a courageous event of undertaking the responsibility of contesting some of the major "dogmatic" axioms of contemporary philosophy (poststructuralism and deconstruction) and hegemonic practices of re-imagining and re-instituting the field of the social cosmos, that is the field of politics (liberal-democracy or the nostalgic yearnings for communism).

Translated into more scholarly parlance, Katerina Kolozova's thought can be aligned in the philosophical corpus of so-called speculative realism, or, as what might also be argued, in the post-Heideggerian leftist postfoundational political philosophy, represented most saliently by authors such as Hannah Arendt, Chantal Mouffe, Ernesto Laclau, Alain Badiou, Slavoj Žižek, Jean Luc Nancy, and Philippe Lacoue-Labarthe.

In order to emphasize the singularity of Katerina Kolozova's work, this book should be introduced as a continuation of the ideas[1] that were comprehensively elaborated in her previous book, *The Real and the "I": On the Limit and the Self*.[2] In that book, comprising six meticulously argued chapters and following the non-philosophical project of Francois Laruelle, Kolozova quakes the axiomatic of the auto-generated, encircled, enclosed and self-censored instance of thinking the concepts of the subject and self, in particular, the way they have been orchestrated and articulated in the post-structuralist and deconstructivist layers of feminist and queer theories on gender and sexuality. While preserving the discursive and epistemic legacy of this line of thought, Kolozova manages to step out of the still prevalent, although veiled logic of dialectics, dualism, and the binarism in these corpora of texts and to introduce a more radical thinking of the major concepts in those theories. Kozolova conserves the stance of the transformative potentialities of postmodernist identity disseminated in the multiplicities of its becomings and different subject positions. At the same time she deconstructs these hegemonic discourses by haunting them with the instances of the One, the Real, the Stable, the "stubborn Sameness," that is the non-totalizable unity of differences as oneness and singularity, as the necessary stance for the transformative complexities and resistant potentials of the subject to be actualized.

Kolozova's new book, *The Lived Revolution: Solidarity with the Body in Pain as the New Political Universal*, can be presented as a Real Event and freshness in the dominant discursive tradition of the "Western" and "Balkan" cultural, feminist, and political philosophy by the very task it attempts realizing. That is the thinking of the political possibilities of resistance and de-/re-institution of the political and cognitive space by the means of the conceptual apparatus of Laruelle's speculative realism or nonphilosophical project.

Kolozova's *The Lived Revolution* is a project that is radical on several levels. Not only because it grasps the roots of some of the major social and theoretical problems and gaps, but also because it offers an exclusively provocative, heuristic, and nonorthodox re-reading of the representations of the contemporary philosophical pantheon, such as Judith Butler, Michel Foucault, Giorgio Agamben, Walter Benjamin, Michael Hardt and Toni Negri, Friedrich Nietzsche, and Karl Marx. Or to put it in a brief parlance, the radicalism comes from Kolozova's reading of these authors with them against them. She manages to sneeze in the fissures of their texts where they have found obstacles and impasses, and thus to derive a thought that can be stigmatized as unthinkable in this same legacy, thought that opens the horizon of thinking the political and disables every attempt of fixing, stabilizing and metaphysically instituting the very ground of politics.

The Laruellian "Real" on which Kolozova epistemologically grounds her project is conceived by the very constancy of its taking place, as an instance, a modality of immanence eluding every attempt of the Transcendental to grasp it, which is the discursive "being-there" of meaning, identity, and signification. The Real here is analyzed in its permanent grounding, degrounding and regrounding of certain political thought and as the heterogeneous origin of politics and sociality. Discourse can thus be thought only in its never-ending essay of transposing, reproducing, and stabilizing the void that the Real is, which is to say, in its continuous transition caused by the Traumatic effects of the Event of the Real's implosion.

The nonphilosophical project steered in this book explores different social-political phenomena by introducing various concepts marked as radical by the means of their, as Kolozova argues, as direct as possible link with the identity-in-the-last-instance, their instance of immanence, that is the Real that they describe with the minimum transcendental material. In the six chapters of her book, Kolozova uses "minimal" language (33) in order to describe and generate a map of different experiences as instances of immanence of the Lived, thus, glossing a syntax of the symptoms of the sheer taking place of the Real—a Symptomatology of the lived revolution. All of these concepts are symptoms of the resistances of the singularities which create the Multitude(s), map of their reactions to the power exercised by the hegemonic and ruling discourses, the unintelligible remainders as the "cry of this body (or, these bodies)" (26) inflicted in various manners by the mechanisms of power.

Hence, ingeniously eluding the dialectical and binary discursive and political constructs of philosophical doctrines and political totalities, Kolozova succeeds in thinking differently allegedly oppositional terms, that is to say thinking them as one, as singular concepts abstracted from their relatedness in the discursive cosmologies. In that direction, Kolozova argues for overcoming the false self-demise versus self-preservation dilemma of Nietzsche, and extracts both of the concepts as singular and radical concepts cloning the immanence of the Real:

> I will argue that the destructive stance is invested in the labor of persistence and survival, of Life; the paths and modes of this investment do not render it reducible to the other term (to Life); the stance of (auto-) destructiveness, seen in its singularity, in its "identity-in-the-last-instance" is integrated in a process which is, in the last instance, one of sustaining, re-creating Life. Executing the task of survival in its paroxysm of pure revolt implies the participation of Death-drive as the inevitable result of the risk always already present in the revolt against the Order-in-Power. Death is unilaterally invested in the tasks of Life without establishing an equation with it (59).

This argument is based on her nonphilosophical heuristic reading of the Butlerian concept of vulnerability, radicalized even more by its very stripping of the necessity of the discursive, transcendental potency of conceptualization. Namely, vulnerability not as the precondition of Humanity that is framed as what needs to be firstly recognized by the grids of the Human's discursive intelligibility for a life, grief, suffering, violence or love to acquire the status of Human and be protected as such. It is rather vulnerability as the sheer experiential, as the prelinguistic root of the "human" struggling against this insurmountable precondition in conformity with the Spinozian "irrevocable call for self-preservation, always already immersed in the struggle for survival" (113). This sheer vulnerability of the bodily exposure is what is always erased and sacrificed for the sign, the Transcendental, the political order to take place, but is at the same time what is always impossible to be annihilated without a remainder and thus prevented from its implosions as the Traumatic in the apparatuses of power.

Grief is what Kolozova also argues to be a stance of revolt and struggle exemplified in the best manner by the solitary work of mourning of the Self. Grief is what enfolds the Self, detaches it from the World and suspends its sociality. Yet this solitude of the mourning Self is nothing but the instance of self-preservation disclosing its dependency and relatedness to the others as the always already there and the yet-to-come. Grief embodies the static stance as the law of resistance, as that pause in the regular everyday flow and suspension of the social order and political cosmos. The stasis, as Kolozova points out, is best illustrated by the body as the stable topos, the site of continuity from where the transformations of subject—imposed by the Normalizing effects of power—are received or enacted. This site of transformations is also the site of the revolutionary stance, the conservatory stance of self-preservation against the disintegration effects of discipline elaborated by Foucault.[3]

Furthermore, Kolozova's thought can be marked as radical because it has the courage to invite us to think within the concepts cloned on the Real, the "radical concepts" which—mapped and connected—produce the potential for utilizing Law's very immanence, that is Benjamin's Divine Violence as the sovereign institution of the political, social, and juridical order. But it would be severe misreading to interpret this as a call on Violence, instead of a hail for depriving it of its very Realness in order to constitute an order that would eradicate its very possibility and instantiate the solidarity with the others inflicted by suffering and pain. Put briefly, Violence as law-making turned against hegemonic politics as violence producing, as the decisive institution of the network of relatedness in the social and political field organized on the principle of the singularity of one's own increasing of the power of acting, self-preservation, and striving for survival.

The six chapters weaving this book, I dare name radical, because they force one to think beyond identity and the Marxist concept of class as the discursive, linguistic, economic, and political commodification of the sheer experience of our bodily existence and labor, of the Real of suffering, pain, vulnerability, pleasure, and joy. Thus it empowers an open-ended alliance of resistances that would overcome the reductionist identitarian logic and install this new but emptied Universal of the Body's vulnerability, its potential of suffering and pain, and its conatus, that is the stance of self-preservation and survival advocated by the acts of solidarity. Finally, *The Lived Revolution* is a book more than useful, not only for scholars coming from various disciplines such as political philosophy, cultural studies, gender studies, and queer theory, but also for social and political activists who gratify undecideability as the principle of permanent reconsideration of the grounds of one's political struggle. It inflames the political passion of activists and political philosophers coming from Eastern Europe to step out the constraints of the disjunction of the leftist melancholia (or capitalist hatred) for the communist political past and the self-mystified teleological claims of the "Western" liberal democracy, and, thus, to rethink politics beyond its present condition and its anticipated possibilities.

◊ **Notes**

1. See also Katerina Kolozova and Zarko Trajanoski, eds., *Conversations with Judith Butler: Crisis of the Subject* (Skopje: "Euro-Balkan" Press, 2001); "Sheer Life Revolting: The Concept of Life and its Political Meaning in Spinoza, Agamben, and Butler," *Identities: International Journal of Politics, Gender and Culture* 7, nos. 1/2 (2009): 71–99; "The Project of Non-Marxism: Monstrously Radical Concepts and Monstrous Representation," *Cultural Logic* (2007), http://clogic .eserver/org2007/Kolozova.prf; "The Grain of the Real within the Identity," *Identities* 3, no. 1 (2004): 6–27; "Non-Dichotomous Possibility of Thinking Unity for the Non-Unitary Subject," *Genero* nos. 2–3 (2003): 49–59.

2. Katerina Kolozova, *The Real and the "I": On the Limit and the Self* (Skopje: "Euro-Balkan" Press, 2006).

3. Michel Foucault, *Discipline and Punish. The Birth of the Prison* (New York: Vintage Books, 1979).

◆ ───

Shana Penn and Jill Massino, eds., *Gender Politics and Everyday Life in State Socialist Eastern and Central Europe* (New York: Palgrave Macmillan, 2009), 292 pp., $85 (hb), ISBN 978-0-230-61300-3.

Book Review by **Krassimira Daskalova**
St. Kliment Ohridski University of Sofia   ◆

This collection, edited by Jill Massino and Shana Penn, consists of an introduction and four sections specifically connected to different realms of women's life under state socialism: "Work, Activism, and Identity," "Sex, Reproduction, Family Relations, and Domestic Space," "Consumption, Leisure, and Culture," and "Gender and Resistance."

Jill Massino's article opens the first part of the book. The text—based on her oral history interviews with women from the Romanian town of Brașov and on various publications from the time of state socialism —shows the importance of women's paid work for the redefinition of women's roles and identities in Ceaușescu's Romania. The author argues that, apart from considering it burdensome, some women were not only highly satisfied with the possibility to earn their own living and be economically independent from their husbands but they used their engagement with paid work in order to renegotiate their relationships within their families (with their husbands) and at work (with their male colleagues). Thus, besides leading to the highly stigmatized "double burden" of women, paid work under state socialism for some women also was rewarding, empowering, and liberating. Massino's work shows that women did have a choice in shaping their own lives within the Romanian totalitarian context, that is that they did have agency.

In an interesting case study, Eszter Tóth presents the changing identities and consumerism of a group of Hungarian women workers, members of the award-winning "Liberation Brigade," most of whom came from the poorest parts of the Hungarian agricultural communities. The oral history interviews the author conducted with these women reveal how both the factory work and the emergence of "goulash communism" stimulated the development of conflicting new socialist identities not only as women-workers and mothers/spouses/caretakers, but also as consumers of the increasingly better quality of life within the Hungarian socialist state.

Basia Nowak's chapter is a further insistence on the necessity to move beyond the Cold War paradigm, which characterized the Soviet bloc in opposition to "the democratic and capitalist West" (46). Building on various publications of everyday state socialism, this text is another example that confirms the non-productivity of simplistic dichotomies (either/or) and rigid conceptual pairs such as "capitalism" versus "communism," "us" versus "them," "state" versus "society," "party authorities" versus "the populace." Using Polish cases of women associated with the *Liga Kobiet* (League of Women), Nowak shows that after World War II, the official character of such mass women's organizations in Eastern Europe notwithstanding, there were still ways for them to serve women's real needs rather than strictly following Party guidelines. She points out that within the Party-states, organizations similar to *Liga Kobiet* "carved out" for themselves "a semiautonomous space, one that functioned for both the official state *and* women, albeit in different ways" (47). Somewhat surprisingly, given the fact that there is not a single line of explanation regarding her understanding of the meaning of feminism, Nowak concludes that the *Liga Kobiet* did not formulate feminist ideas, thus showing a lack of that same nuanced thinking about a complex historical phenomenon as feminism that she was pleading for with her whole text.

Following the work of two state-sponsored women's organizations in Romania and Hungary, Raluca Maria Popa argues that these two National Women's Councils actively shared international commitments to gender equality, development, and peace (the three aims of the UN International Women's Year) via the UN Commission on the Status of Women (CSW), and their membership of and activities within the Women's International Democratic Federation. More than that, women-activists of these organizations emphasized that in socialist countries it was not women's movements or feminisms (as in the West) but the existing and continually changing state gender policy mechanisms that were responsible for the implementation of new gender contracts and shaping equality between women and men. Some of these women, however, knew well the histories of feminisms and even published gender sensitive books on the topic, thanks to their participation in the international feminist movement and meetings with strong female personalities (such as Andrée Michel).

Ulf Brunnbauer's opening article of the second section outlines the changing gender policies of the Bulgarian state. Brunnbauer shows how from the mid-1960s Party rulers responded to the deteriorating demographic picture of the country by imposing another (fraternal) gender contract, one that emphasized women's ("natural") roles as biological reproducers (and mothers and caretakers) and their patriotic duty to the socialist nation, and neglected their free will and autonomy in matters concerning sexuality and reproduction within the Bulgarian Communist Party's intensified pronatalist policies. According to the historical sources used by the author, however, both women members of the Central Committee of the Party and "ordinary women" reacted against the restrictions on abortion which they qualified as "step backward and a dishonor for any 'cultured nation'" (90). Brunnbauer concludes that, although in some respects communist rule was liberating for women in Bulgaria, in still others it was oppressive and due to the "systematic deficiencies of real socialism gender equality was not achieved" (94). This otherwise well-argued presentation of women's situation in state socialist Bulgaria would have been more convincing had the author not insisted on some unsubstantiated claims and had he omitted some factual mistakes.[1]

Donna Harsch's contribution deals with issues of marriage, divorce, and sexuality in the German Democratic Republic in the 1950s and 1960s. The author demonstrates how the policy of the Party-state toward divorce evolved during the period under consideration from favorable to more restrictive and how this change was met by society at large and by young women, in particular, who became more open toward divorce. Interestingly enough, the liberalization of sexuality was equally well received by both the East German state and society or, as Harsch has put it: "society and the socialist state moved in tandem on sexual matters" (113).

Wife beating in comparative perspective, based on practices in state socialist Hungary, Poland, and Romania, is the topic of Isabel Marcus's article. Using extensive interviews (about one hundred in each country) with officials and professionals whose work put them in contact with the issue of wife beating, the author shows that domestic violence was widespread in these countries and presents it as a means of power, control, and subordination of women within family life. Marcus argues that women did use criminal law and family legislation as a way to escape from their abusive relationships.

The next chapter, by Joanna Mishtal, discusses the uneasy relationship between the state and the Catholic Church in Poland before and after 1989 on the issue of women's reproductive rights. During state socialism the Church was among the major anticommunist forces, but starting in the 1980s and continuing during the years of transition, the Polish religious authorities took this symbolic capital and turned it into the most conservative institution, advocating restrictions on women's reproduction and sexual freedom. Actually, the political leaders of Solidarity movement in post-1989 Poland were the ones who delegalized women's reproductive rights and betrayed women. This was a time when identifying with "feminism" was stigmatized as women's emancipation and the notions of women's emancipation and women's rights were linked with the epoch of state socialism, hence rejected. The Polish example confirms the observation that historical turning points do not necessarily affect women's and men's lives in the same way.

Kimberly Zarecor's interesting contribution presents the role of debates about architecture for the establishment of new models of socialist family housing in post-1945 Czechoslovakia. The author concludes that, due to the lack of visible cultural and social changes in European state socialist countries, the new socialist architecture was not able to liberate women from everyday domestic chores.

The third part of the book, dealing with the topic of consumption, leisure, and culture, includes articles by Malgorzata Fidelis and Anna Hofman. In her wonderful analysis of the discourses on the Polish "modern girl," Fidelis shows that under state socialism women's lives consisted not only of hardship, duties, and burdens. She argues instead that, although Eastern European regimes in the early 1960s encouraged—especially young and single—women to indulge in pleasures and entertainments, by the 1968 student revolt the official press started to blame the "modern girl" for extreme consumerism and lack of loyalty toward the socialist state. Fidelis insists that historians should not treat the post-Stalinist period as homogeneous and promoting one single gender model, but rather should try to see the changes and fluctuations in Poland during the 1960s and 1970s. Hofman's text follows the activities of amateur rural women's vocal groups, which performed at state-sponsored festivals in order to show that—thanks to their "special position of musical authority" (190)—they were able to transgress the patriarchal norms of their rural milieu, subvert the hierarchical gender order within the local context, and gain their communities' appreciation as embodiments of local culture. Thus Hofman questions the still existing tendency of producing a fixed, monolithic, rigid, and essentialized image of state socialism.

The last chapter, written by editor Shana Penn, shows how feminism in Poland gradually developed "a discrete life of its own, independent from both communism and the opposition" (p. 216) in the second half of the 1980s with the establishment of the *Polskie Stowarzyszenie Feministyczne* (Polish Feminist Association). After the popular magazine *Kobiety i Zycie* (Women and life) published an article about the organization in August 1988, this feminist group became publicly more visible. Very soon after 1989, however, the Polish media started to promote a back-to-the-home campaign, accusing communism of turning the social order upside down by securing women's access to the public realm and by disempowering men. This was an idea formulated already by Solidarity in the early 1980s and was among the reasons for some women's disillusionment with Solidarity and their subsequent move toward feminism.

Gender Politics and Everyday Life in State Socialist Eastern and Central Europe is a wonderful example of a historical approach that aims at going beyond Cold War terminology and ideas. It shows not only that women did have agency (both institutionally and individually) but also that in various cases their positions were articulated in opposition to official Party discourse and policies. The volume builds on and departs from the existing scholarship dealing with women, gender, and state socialism in Central, Eastern, and Southeastern Europe. It goes beyond social structures and listens to women's own voices and life experiences. It is a valuable contribution to the ongoing international debate regarding the character of state socialism's gender politics and pro-women measures.[2] If taken seriously by mainstream historians in the former Soviet bloc countries, the research and publications of gender scholars could substantially change the way state socialism is being seen and judged. The articles show how complicated and ambiguous everyday socialism was for the various social actors and call for a re-evaluation of existing theories, which generally neglect women's experiences and the gendered character of social reality. This collection will be a useful companion for those professional historians, students, and the general public interested in the recent history of Central, Eastern, and Southeastern Europe.

◊ **Notes**

1. For example, his opinion that "people usually married young, in part to avoid the extra income tax for unmarried adults"; or that "The main reason for early childbirth were avoidance

of the tax in childless married couples" (91, 92). It seems that the author is not well informed about the development of women's/gender history in the country and the outcome of women's activities in the pre-1944 period. Had he used the most relevant publications perhaps he would not have made factual mistakes such as regarding the establishment of the first large women's organization—*Bulgarski Zhenski Suiz* (Bulgarian Women's Union) in 1901.

2. On this, see the Forum "Is 'Communist Feminism' a *Contradictio in Terminis?*" in *Aspasia* 1 (2007): 197–246.

Wendy Rosslyn and Alessandra Tosi, eds., *Women in Russian Culture and Society, 1700–1825* (Basingstoke: Palgrave Macmillan, 2007), 272 pp., £55.00 (hb), ISBN 978-0-23-055323-1.

Book Review by **Polly Mukanova**
St Kliment Ohridski University of Sofia

Women in Russian Culture and Society, 1700–1825 is a valuable collection about one of the most interesting and dynamic periods of Russian cultural and social history: the eighteenth and first quarter of the nineteenth century. The authors examine the role, activities, and influence of women in three spheres: art and literature, social life, and religious life, concentrating on phenomena that have been ignored in the big historical narrative; the narrative in which women have been usually shown as marginal participants in the historical process. Their careful look at the different roles of women in Russian society disproves old theoretical frames.

The first chapter focuses on theatre and writing, in which women participated actively during the period under consideration. The introductory article of Lurana Donnels O'Malley, "Signs from Empresses and Actresses: Women and Theatre in the Eighteenth Century," examines the important role of empresses Anna, Elizabeth, and Catherine the Great in developing the public theatre in Russia. Their patronage was of great importance for women in the theatre, opera, and ballet. Moreover, it was instrumental in encouraging women's participation as audience, playwrights (including Catherine herself), and actresses. Through participation and patronage, women, argues the author, played an active role in the dynamics of Russia's nascent theatrical art. Richard Stites studies another aspect of women's role on the stage in "Female Serfs in the Performing World." Focusing on the mid-eighteenth century to 1825 and including both the capital and the provinces, Stites examines the lives of female actresses, ballet dancers, and singers who were also serfs, and shows how these women could be placed in an unfavorable position by landowners.

In her essay "Women and Literature, Women in Literature: Female Authors of Fiction in the Early Nineteenth Century," Alessandra Tosi introduces lesser known works of the first generation of women novelists. After centuries of being marginalized, during this period Russian noble and elite women were encouraged to write, for which Tosi sees two major reasons: the sentimental ideals of feminine aesthetics, and women's role as hostesses and arbiters of taste within the salon culture. Her text focuses on the restrictive influence of patriarchal values on the development and creative abilities of women. In the next chapter, "Women's Travel and Travel Writing in Russia, 1700–1825," Sara Dickinson discusses the gendered features of travel writing. Traditional culture usually presents women as eternal inhabitants of the domestic sphere, while traveling is considered an activity typical for men. In this patriarchal context Russian women travelers and writers such as Ekaterina Dashkova, Natalia Stroganova and Maria Gladkova were among the first real norm breakers. Finally, Gitta Hammarberg's contribution, "The First Russian Women's Journals and the Construction of the Reader," outlines the main features of the woman reader.

The second thematic cluster of the book sheds light on the social position of women of various ranks and estates. Bigamy is revealed as a phenomenon fairly widespread in Russia from the beginning of the eighteenth century. Olga Kosheleva investigates rare documents, the so-called divorce letters written by husbands to their wives in order to free both from their marriage obligation. Paul Keenan's article focuses on elite women and shows the early effect of Russian Westernization on women's role within court society in the period of Peter the Great and his successors. Galina Ulianova's article, "Merchant Women in Business in the Late Eighteenth and Early Nineteenth Centuries," reconstructs biographies of women entrepreneurs, thus shedding light on female entrepreneurship and the ways women entered the merchantry.

The final part of the collection discusses the interaction of women with the church. Gary Marker's article, "Sacralising Female Rule, 1725–1761," looks at female rule, a distinctive feature of the eighteenth century, and at the justifications advanced for this unprecedented phenomenon in Russian history. William G. Wagner's chapter, "Female Orthodox Monasticism in Eighteenth-Century Imperial Russia: The Experience of Nizhnii Novgorod," draws on previously unused archival sources to investigate the process and impact of contemporary reform on female monasticism in Nizhnii Novgorod. Finally, Wendy's Rosslyn chapter, "Women with a Mission: British Female Evangelicals in the Russian Empire in the Early Nineteenth Century," presents profoundly religious women who—despite a ban by British churches—worked independently as individuals or as missionaries' wives. The British women's work outlined in this chapter was a meaningful part of missionary efforts at a time when barriers to both women's preaching and their formal appointment as missionaries were insuperable. The book finishes with a useful bibliography divided into various themes that covers the scope of the study.

The research presented in this volume is an important contribution about long neglected women's activities, and will be of interest to scholars, students, and other readers of Russian history.

◆ ──

Marius Turda and Paul J. Weindling, eds., *Blood and Homeland: Eugen-*
ics and Racial Nationalism in Central and Southeast Europe, 1900–1940
(Budapest: Central European University Press, 2007), 476 pp., $25.95
(pb), ISBN 978-963-7326-81-3.

Book Review by **Gisela Bock**
Freie Universität Berlin
                                                                        ◆

The title of this volume, taken from the Romanian "Sânge si glie" (blood and soil),
points to its major purpose to analyze the history of eugenics or race hygiene in the
countries of Central and Southeastern Europe (CSEE), the racist or radical or ultra-
nationalist visions of the nation as an ethnically homogeneous state, and the relationship between the two. To do so with respect to CSEE is a task, which has been
neglected until quite recently, even though during the period covered here, the international community of eugenicists, including many outright racists and anti-Semites,
highly appreciated the cooperation with their colleagues from CSEE. (For instance the
International Congress of Population Science, held in 1935 in Berlin, assembled that
community and included among its non-German participants over 30% from CSEE.)
In fact, from its emergence in the 1990s, the international or comparative historiography on eugenics and "race science" had focused mostly on Britain and the United
States, Switzerland and Germany, and the Nordic countries. Moreover, the national
historiographies of the CSEE countries, especially prior to 1989, had completely ignored the subject (with the exception of Austria, where research followed the emergence of such studies on Nazi Germany in the 1980s). The volume opens with a study
of Nazi-German "race psychology" as being at "the core of scientific racism" and as
applied for example to people in the Eastern borderlands of Germany (Egbert Klauke).
Yet primarily the volume is on CSEE: with three essays each on Austria and Romania,
two each on Hungary and Poland, and one each on Estonia, Bulgaria, Czechoslovakia, Yugoslavia, and Greece. Paul Weindling's transnational contribution deals with
three scientists who criticized the scientific pretensions of racial anthropology (though
not eugenics); they were part of a (clearly understudied) "international scientific front
against scientific racism" (276). Two concluding chapters set out to integrate the contributions into broad and long-term visions of "biomedical totalitarianism" (Aristotle
Kallis) and the "conditions of high modernity," which "fostered the synthesis of ultra-
nationalism with eugenics" in interwar Europe (442). Some of the contributions do
confirm this perspective.

The editors' introduction underlines that eugenics as well as racial/racist anthropology in CSEE cannot just be evaluated against the "excellence" of British, German,
and American models and influences; therefore they highlight the "creativity" of au-
tochthonous developments (6). Professionals and other experts, their publications and
ideologies, their organizations and field work is at the center of most articles; some of
them cover the entire period, some only the early years. The result is an impressive

variety of issues and ideas. "Central European eugenics was astonishingly diverse," writes Weindling (263), especially given the fact that eugenics everywhere tended to infiltrate other scientific disciplines and the health sector while absorbing other sectors such as the welfare state (renaming its measures "positive eugenics" and thereby excluding "inferior" groups) or the political handling of prostitution. Maria Bucur presents the hovering between "abolition" and "regulation" for the case of Romania, and Maria Löscher shows the reluctance of (Austrian) Catholicism with respect to eugenics. The issues and languages of ethnic rejuvenation and purification add further to this diversity, for instance in Rory Yeomans' analysis of (modern) eugenically minded Serbian/Yugoslav professionals versus (anti-western) eliminatory fantasies of Croatian race experts.

Ken Kalling reminds us that "the ideology of eugenics is more diverse as a body of knowledge than as a practical application. The approval of eugenic legislation, especially laws relating to sterilization, provides good criteria for testing the eugenic movements in different countries" (253). Poland after World War I, for instance, had left-wing, "progressive" and "non-nationalistic" eugenicists, including Jews and female and male feminists (they do not appear in the other studies) who discussed, as early as 1918, compulsory sterilization. However, by the mid-1930s, such a policy was sternly rejected by the ruling class and the Church (Magdalena Gawin, Kamila Uzarczyk). Sterilization policies were debated everywhere, and eugenicists (though not always racial nationalists) fought for them, unless they preferred "milder" marriage controls. But a law for compulsory sterilization was autonomously introduced only in Estonia (in annexed Austria it was the German law of 1933 that was imposed in 1940). It was rarely applied, almost exclusively on women, and was abolished four years later by the Soviet occupation, when more emphasis was placed on pronatalism than on antinatalism (Ken Kalling). This tendency, by several authors defined (in my view problematically) as a "Latin model," in contrast to a "Nordic model," seems to have dominated in CSEE (as it did in parts of Western Europe), though varying strongly over time. Eugenics in Bulgaria was closely linked to interwar modernization efforts (as in most other countries) and German influence was strong. But the eugenicists lost their fight for compulsory sterilization and the government preferred to follow Germany with a pro-natalist law in favor of families with many children, from which Jews were excluded (Christian Promitzer). Perhaps the pronatalist or "Latin model" was not so much the result of a specifically CSEE eugenics, but of non- or anti-eugenic forces opposing compulsory sterilization policies. Hungary was, according to Marius Turda, among "the vanguard of eugenic thinking in Europe" (186), though he shows it here only for the pre-1918 period. Beyond that and beyond this volume, Turda is studying the further development of sterilization policies (and failed sterilization laws) until the 1940s, for Hungary as well as Romania.

Fortunately, this excellent volume is not the last of its kind but the first of a promising series based on conferences in Budapest, Berlin, Warsaw, Riga, and Vienna, which have altogether assembled over fifty scholars. With this admirable effort, CSEE is now clearly on the "mental map" of this kind of historiography. It would be extremely helpful if future volumes could more strongly highlight two issues. First, a possible transnational dimension among the CSEE countries (beyond nation-based studies). Second, starting perhaps from the many hints at women appearing in this volume—as objects of male eugenic theorists and practitioners, as feminist eugenicists and female race scientists, as participants in or objects of national rejuvenation, mothers or amazons—a more systematic study of gender relations in CSEE eugenics and racial nationalism.

◇ ────────────────────────────────────

Demetra Tzanaki, *Women and Nationalism in the Making of Modern Greece: The Founding of the Kingdom to the Greco-Turkish War* (London: Palgrave Macmillan, St Anthony's Series, 2009), 234 pp., $75 (hb), ISBN 978-0-230-54546.

Book Review by **Haris Exertzoglou**
University of the Aegean, Greece

──────────────────────────────────── ◇

The book by Demetra Tzanaki discusses the relationship of nationalism and womanhood in modern Greece and celebrates female teachers, writers, and publicists who opposed dominant male views by developing alternative concepts of womanhood. The book is divided into three parts and nine chapters. Part one probes the issue of women's education in the Greek kingdom in relation to its institutional framework and the crucial role of private initiatives presented against the indifference of the state. The author also discusses the discursive framework of morality in which the "woman question" in Greece was embedded. In the second part, the author examines the impact of the Greek nationalist agenda on women's education in the Ottoman lands and describes the emergence of a group of women teachers and writers directly involved in girls' education in Smyrna and Constantinople, among them the publisher of the journal *Euridike*, Amalia Ktena. Part three focuses on Callirhoe Parren, an emblematic figure of a woman-journalist, who edited and published *Efimeris ton Kyrion* (The Ladies' journal), one of the leading women's magazines in nineteenth-century Greece. The author discusses how Parren and the small circle of women contributing to the journal argued in favor of claims that would enhance women's position in education, vocational training and the labor market. Callirhoe Parren enlisted the nationalist discourse into her agenda in order to promote a female version of citizenship.

There is much to commend in Tzanaki's effort to discuss the relationship between women and nationalism in Greece, precisely because this link is hardly present, if at all, in almost all mainstream studies of Greek nationalism. Thus Tzanaki follows in the steps of a small but growing group of, mostly female, Greek historians such as Efi Avdela, Eleni Varika, Angelika Psarra, and Eleni Fournaraki, to name but a few, who took upon themselves the task of altering this situation. The author states from the beginning that the object of her study is not Greek women in general but the individuals and groups of literate middle-class women, active in education and journalism, who played a part in the reformulation of concepts of womanhood in nineteenth-century Greece. For this purpose Tzanaki examined extensive textual material—newspapers, journals, magazines, and booklets—and used it to discuss the issues raised by women's activity in Modern Greece. The nub of her argument is that the close correspondence between Greek nationalism and the dominant concepts of womanhood created a space in which female teachers, writers, and publicists raised legitimate claims and opposed the dominant male narrative by developing new knowledge about women and by projecting an alternative concept of female consciousness, solidarity, and citizenship based on the public contribution of women in such fields as education, work, and philanthropy. Certainly this approach is not original as other scholars have already contributed to this line of argumentation. But Tzanaki's book allows the non-specialists to understand that the nature of Greek nationalism was also gendered in the sense that it valued the need of separate spheres between men and women, assigning to the latter particular duties in the "private realm" and rejecting any prospect of women's involvement in public activities, most specifically in politics. However, the definition of the concept of separate spheres in Greek nationalist discourse could not exclude the possibility of reconceptualizing womanhood and this was attested by the ways female teachers and writers appropriated this concept to further their claims on behalf of women in general and to extend their access in the public sphere. Tzanaki suggests that during the second half of the nineteenth century and in direct relation with swinging Greek nationalist aspirations, the meaning of womanhood was transformed and women were conceived to be active rather than passive individuals. These two stages of conceptualizing womanhood, according to the author, are exemplified by the journal *Euridike,* published in Istanbul between 1870 and 1873 and *Efimeris ton Kyrion,* published in Athens between 1887 and 1917.

Although I am sympathetic to the emphasis the book places on the gendered texture of nationalism, I find that it leaves much to be desired in the way nationalism and womanhood are articulated in the specific spatial and chronological framework set by the author.[1] The only reason to discuss this subject within the spatial and chronological parameters proposed by Tzanaki is the underlying belief that the Greek state and the "Greek" Orthodox communities in the Ottoman lands constituted a national whole and could be considered together. Personally I am very skeptical with the way Tzanaki lumps the Greek state and the "Greek" communities in the Ottoman lands under one coherent object of study. Relations between these two parts were indeed close but it does not follow that the diversified and incoherent social and cultural experience of the Ottoman lands can be reduced directly to that of the Greek Kingdom. Such approach implies that the Greek Orthodox communities, under different and conflicting influences as they were, had only one point of reference, that is Athens and not Constantinople or Smyrna and to the extent that the middle classes were involved even London and Paris.

In addition, I find equally questionable Tzanaki's argument about the transformation of an originally religious Greek nationalism into a secular or civic nationalism, and its impact on the concept of Greek womanhood. Greek nationalism was always secular although it often appropriated and used messianic and religious elements particularly within the context of the intense nationalist strife in the Balkans. In addition, concepts of womanhood employed before and after the 1870s—a period that according to the author marked this transformation—did not differ radically. There was always a mix-

ture of moralistic and biological discourses at the basis of the concepts of both womanhood and manhood employed by the Greek nationalist discourse and appropriated by
the women teachers and writers in Smyrna, Constantinople, and Athens. I think that
the author takes at face value these discourses, which caricature women and gender
relations, without considering their metaphoric nature, which in fact allowed female
agents to use them in their rhetoric strategies to access literate women but also to
convince or neutralize a hostile male audience. From this perspective women writers
inhabited a common discursive ground with male intellectuals, who opposed "female
emancipation," and negotiated within this ground the reformulation of womanhood,
motherhood, and the relationship between the "private" and "public" spheres. In
principle, women's education was not questioned in the Ottoman lands: what was the
object of discussion for many years was the kind of secondary education suitable for
women and specifically the parity of language education (i.e., the teaching of ancient
Greek) among boys and girls. Against a more "pragmatic" approach that insisted on
the practical benefits of schooling, women teachers and writers struggled for the kind
of education they believed would enhance the moral and national awareness of girls,
enabling them to withstand the supposedly negative impact of social environment.

Overall, the book is interesting and fits into the growing literature on the subject
but the author should rethink her linear narrative strategy, which places much more
emphasis on the celebration of "female emancipation" and ignores the diversified and
polymorphous field constituting her subject.

◊ **Note**

1. See also the more extensive review of the book by Eleni Fournaraki in *Historien: A Review
of the Past and Other Stories* 9 (2009): 210–217.

news and miscellanea

◆

In Memoriam: Richard Stites (1931–2010)

Richard Stites (2 December 1931–7 March 2010), a pioneer in gender history, took on "unfashionable" themes, researched them diligently, produced imaginative, fascinating monographs, and made his subjects fashionable. He died of cancer in his beloved Helsinki while on research leave, and is buried near the city's Russian Orthodox Cathedral.

Dressed unconventionally, often in a work shirt, no tie, jeans, and a bandana around his long, thinning hair, Stites was a brilliant maverick in the field of Russian history. His doctoral dissertation, completed in 1968, was about the question of the emancipation of women in nineteenth-century Russia. It was not a topic looked upon with great favor at the time. But he got his Harvard degree, and wrote his *The Women's Liberation Movement in Russia: Feminism, Nihilism, and Bolshevism, 1860–1930*, first published by Princeton University Press in 1978, republished in a slightly updated edition in 1991. It was an innovative work, among the first of a wave of books by Western scholars influenced by second wave feminism, which introduced concepts of gender and women's agency into the writing of Russian and Soviet history.

Stites's oeuvre ranged wide within the general area of social history. His extraordinarily long list of influential publications includes scores of articles, at least five edited or co-edited volumes, a major co-authored textbook, *A History of Russia: Peoples, Legends, Events, Forces* (Houghton Mifflin, 2004), and four single-authored books. In addition to *The Women's Liberation Movement in Russia*, these include *Revolutionary Dreams: Utopian Vision and Social Experiment in the Russian Revolution* (Oxford University Press, 1989; 2nd ed. 1991); *Russian Popular Culture: Entertainment and Society since 1900* (Cambridge University Press, 1992); and *Serfdom, Society, and the Arts in Imperial Russia: The Pleasure and the Power* (Yale University Press, 2008). *Revolutionary Dreams* won the 1989 Wayne S. Vucinich Prize of the American Association for the Advancement of Slavic Studies. At his death he was at work on a book that took him into general European history, tentatively titled "The Four Horsemen: Revolution and the Counter-Revolution in Post-Napoleonic Europe," and on another project on "Hitler's International Crusade," about Axis volunteers in the armies invading the Soviet Union in World War II. "His works were one of a kind, outstanding in their writing and in their scholarship," said Richard S. Wortman, an emeritus professor of Russian history at Columbia University. "He dealt with subjects that other people had not yet gone into" (*New York Times* obituary, 12 March 2010).

Stites mastered not only Russian but several Scandinavian languages. In addition to English, he had a reading and speaking knowledge of nine other languages. Recognized by Georgetown University with a "Career Research Achievement Award from the Georgetown University Graduate School" in 2001 and by Helsinki University with an Honorary Doctorate (he spent most of the past thirty summers there, and longer

aspasia Volume 5, 2011: 239–240
doi:10.3167/asp.2011.050113

times when possible), he wore his honors proudly but never boastfully. He was fun loving, adventurous, with a wicked sense of humor.

An exceptional scholar, Stites was a very special human being as well. In my own experience, when I first began my work on the Russian feminist movement, he was unusually generous in sharing his sources and research.

Richard was married and divorced successively to Dorothy, Tatyana, and Elena Stites. He is survived by four children, Tod, Thomas, Andrei, and Alexandra, and by countless colleagues, students, friends, and admirers. He will be sorely missed.

Rochelle Goldberg Ruthchild

Information for Contributors

The Editors welcome contributions. Authors should submit articles as word attachments by e-mail, formatted as Microsoft Word. Electronic submissions are preferred but mailed contributions will be reviewed.

Please note that all correspondence will be transmitted via e-mail. Submissions without complete and properly formatted annotation may be rejected. Manuscripts that have been accepted for publication but do not conform to the *Aspasia* style will be returned to the author for amendment.

> E-mail article submissions to:
> Francisca de Haan: dehaanf@ceu.hu
> Melissa Feinberg: mfeinberg@history.rutgers.edu
>
> E-mail book reviews or review essays submissions to:
> Krassimira Daskalova: krasi@sclg.uni-sofia.bg

Articles should be between 6,000 and 8,000 words, although longer papers may be considered. Book Reviews should not be longer than 800 words. Review Essays should be between 1,500 and 3,500 words, depending on the number of books reviewed.

The document must be set at the US letter or A4 paper size standard. The entire document (including the notes) should be double-spaced with 1-inch (2.5 cm) margins on all sides. A 12-point standard font such as Times or Times New Roman is required and should be used for all text, including headings, notes, and references.

The article must include an abstract of 100 to 150 words and 6 to 8 keywords. The abstract should include the research question or puzzle, identify the data, and give some indication of the findings but should not be a duplication of the introductory paragraph. Each contributing author must also provide a short biographical statement (100 words) and full contact information (including a postal address and e-mail).

STYLE GUIDE

Starting with Volume 5 (2011) the *Aspasia* style guide will be based on the *Chicago Manual of Style*, with full endnotes and without bibliography. For optimal reproduction figures or photos should be submitted as TIFF (resolution at 300 dpi) for photos or EPS (800 dpi), black and white, with all fonts embedded. For further and more detailed instructions please visit the *Aspasia* guidelines online <www.berghahnbooks/journals/asp>.

Upon acceptance, authors are required to submit copyright agreements and all necessary permission letters for reprinting or modifying copyrighted materials. Authors are fully responsible for obtaining all permissions.